Commercial Electrical Wiring

by John E. Traister

Updated to the 1999 *NEC*
by Ronald T. Murray

Craftsman Book Company
6058 Corte del Cedro / P.O. Box 6500 / Carlsbad, CA 92018

Excerpts from the *National Electrical Code* are reprinted with permission from NFPA 70-1999, the *National Electrical Code®*, Copyright 1998, National Fire Protection Association, Quincy, Massachusetts 02269. This reprinted material is not the complete and official position of the National Fire Protection Association on the referenced subject, which is represented only by the standard in its entirety.

National Electrical Code® and *NEC®* are registered trademarks of the National Fire Protection Association, Inc., Quincy, MA 02269.

Library of Congress Cataloging-in-Publication Data

Traister, John E.
  Commercial electrical wiring / by John E. Traister.
      p.   cm.
  Includes index.
  ISBN 1-57218-092-7
  1. Electric wiring, Interior.   2. Commercial buildings--Electric
equipment.   I. Title.
TK3284.T73   2000
621.319'24--dc21
                                      00-047441
©1994 Craftsman Book Company
Third printing 2000

Cover: *Omni Mall* by Johnson Associates Architects
Photo by *Robert Stein Photography*, Plantation, Florida

Contents

Preface

Building construction in the United States alone has reached over $300 billion a year and continues to grow at a phenomenal rate, because the population proliferation demands new living, working, and recreational facilities. A good percentage of this building construction work involves electrical installations.

While many electricians begin their careers wiring residential occupancies (especially those entering the electrical contracting business), there will eventually come a time when these same electricians (or contractors) will want to jump into larger commercial construction projects.

There is another large group of electrical workers that begins apprenticeship training in large industrial establishments, perhaps working for months or years on one comparatively small part of a huge, complex electrical network. In fact, the larger electrical contractors in the United States frequently bid electrical projects valued at $200 million or more. Most industrial electrical workers will feel "out of place" should they be required to undertake a smaller commercial electrical installation. Most will require some initial training to acquaint them with the different wiring techniques, *NEC* requirements, and installation methods.

There are numerous books available on the study of basic electricity and electrical wiring in general. Few, however, are aimed directly at commercial electrical wiring — the type of electrical construction that uses the most electricians, the greatest amount of materials, and for which the greatest amount of working drawings and specifications are prepared. This book, *Commercial Electrical Wiring*, is designed to help this situation.

The intent of *Commercial Electrical Wiring* is to dwell only briefly on introductory material and theories. We quickly jump into practical, on-the-job applications that are used for almost all types of wiring systems for commercial buildings. Furthermore, this book is designed in such a way as to help residential or industrial electrical workers make the transition to commercial wiring methods as smoothly as possible, eliminating much of the field trial-and-error method. Consequently, experienced electrical workers will not have to reinvent the wheel when making the transition.

This book is also designed as a practical study guide for electrical trainees, apprentices, and others who are associated with commercial wiring systems in any capacity — providing guidance in simplified form, and usable at many educational levels.

In summary, this book is a quick reference for those actively engaged in commercial wiring, a learning method for those entering the commercial electrical field from another branch of the electrical construction industry, and a refresher for those electrical workers with wide experience in the field.

I would like to thank my own staff who so willingly and enthusiastically contributed their time and experience to this project — validating data and providing excellent and meaningful illustrations throughout the book.

John E. Traister

Chapter 1
Introduction

Electrical wiring systems for commercial buildings can vary considerably — from a few lighting fixtures and a couple of convenience outlets for a small, one-room vegetable stand (Figure 1-1) to extremely complex installations used in high-rise office or apartment buildings (Figure 1-2). However, in most cases, commercial buildings may be classified into either small, medium, or large sizes, with practically an infinite number of uses for each one; for example, a small building may be used for a real estate office, a roadside produce stand or a laundromat.

SMALL COMMERCIAL OCCUPANCY

In planning the wiring system for a small commercial building — either from a design or installation

Figure 1-2: ... giant, high-rise office buildings

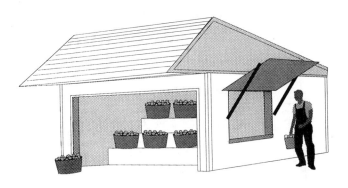

Figure 1-1: Commercial wiring installations range in size from buildings as small a roadside vegetable stand to . . .

standpoint — there are several factors to be considered before material is ordered and the actual installation takes place.

Some of the factors that should be considered for small commercial establishments include the following:

1. Type of general building construction.
2. Is the installation a part of a new building or a modernization of an existing one?
3. Type of ceiling, wall, and floor construction, dimensions, and the like.
4. Wiring methods.
5. Location of service, overhead or underground?
6. Location of service-entrance equipment.
7. Size of service and feeders and sizes and types of service-entrance equipment and panelboards.
8. Wiring of windows and display cases.
9. Type and installation requirements of lighting fixtures. Physical dimensions and construction of recessed lighting fixtures.
10. In the case of a modernization or complete wiring of an existing building, to what extent may the main service be used, and how much will it have to be enlarged?

Taking each of these individually, factor 1 may be determined by the working drawings (Figure 1-3) and specifications, by a job-site investigation, or by consulting with the owners. The same is true for factors 2 and 3.

The wiring method to use (4) may be dictated by the working drawings or specifications. If not, the wiring method to use should comply with the latest edition of the *National Electrical Code (NEC)* and/or local ordinances.

The location of service equipment (5 and 6) may be indicated on the working drawings or the local power company may have to be consulted to determine the best location. Locating the service equipment is often left to the electrician or contractor to decide, but if working drawings are available, they should be followed.

Figure 1-3: Working drawings for the building under construction will furnish answers to most of the workers' questions

Sizing the electric service (7) requires calculations as discussed in Chapter 5 or the designer may have indicated the service size on the working drawings.

Factor 8 — wiring of windows and display cases — can be answered by either the working drawings or by consulting the architect or owner, or it can be worked out between the trades on the job.

Lighting fixtures (9) should be preselected by either the designer, owner, or architect. Installation details may be found in manufacturers' catalogs. Shop drawings (Figure 1-4) are usually provided for commercial installations of any consequence.

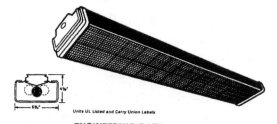

Units UL Listed and Carry Union Labels

ENGINEERING DATA

ROOM SIZE	FIXTURE SPACING, 40-WATT RAPID-START					
	30 FOOTCANDLES			50 FOOTCANDLES		
	1-Lamp Units	1-Lamp Cont. Row	2-Lamp Units	1-Lamp Cont. Row	2-Lamp Units	2-Lamp Cont. Row
SMALL	5' x 5'	6-ft.	6' x 8'	4-ft.	5' x 6'	7-ft.
MEDIUM	6' x 6'	9-ft.	6' x 10'	5-ft.	6' x 6'	9-ft.
LARGE	6' x 8'	12-ft.	7' x 10'	7-ft.	5' x 8'	10-ft.
	70 FOOTCANDLES			100 FOOTCANDLES		
	2-Lamp Cont. Row			2-Lamp Cont. Row		
	5-ft.			3-ft.		
	6-ft.			4-ft.		
	7-ft.			5-ft.		

Figure 1-4: Shop drawings are usually required for all material used in commercial buildings

Factor 10 can be determined either from the working drawings or by a job site investigation.

In general, the designer or electrician performing the work will calculate the total load for the building, determine the number of branch circuits required and service-entrance size, along with feeders, service-entrance equipment, and panelboards. The number of outlets will be determined along with their location. Illumination levels are calculated and then lighting fixtures are selected to provide the required illumination.

Continue by noting connections for any special equipment, such as water heaters or air conditioners. Also determine the requirements for any security/ fire-alarm system, display case connections, and the like.

Determine the lengths of all branch circuits, service, and feeder runs and list the wire size for each. Account for service-entrance equipment and any other major pieces of equipment requiring electrical connections.

The preceding information should provide a good summary of the material needed for the job to be used in estimating costs and the number of men required for the installation and to aid in ordering the required material.

In most cases, small commercial projects utilize rigid steel conduit for the service-entrance regardless of whether it is overhead or underground. Check with the local power company to find out exactly what is required of the contractor or electricians doing the work. Either rigid or EMT conduit is used for all wiring below grade and embedded in concrete slabs. Either EMT or type AC cable is normally used for wiring above grade.

Be extremely careful of any wiring that may be installed in hazardous locations, such as in commercial service stations around the gas pumps and in the garage area.

One main point of concern with this type of project, as well as with most other electrical installations, is to plan the job well so as to perform the work in the shortest possible time, yet keep the quality high and in a workmanlike manner. Other

trades should not be held up in performing their work, and the electrical workers must plan and work accordingly. For example, before the concrete floor may be poured and finished, all conduit, boxes, and fittings must be installed by the electricians. When the ground is graded, wire mesh installed, and the like, the electrical workers usually have only a certain amount of time to complete their portion of the work. Make certain all necessary material is at hand on the job site well before the installation will take place. Have the working drawings or at least a sketch at hand to go by, and work efficiently when the time comes. Also double-check each homerun, circuit, and outlet box location, as once the cement is poured, it would be quite costly to make any changes under it.

At least one electrician should be present during the pouring to ensure that none of the electrical system is damaged; if it is, the damage should be corrected immediately before the concrete sets up.

MEDIUM COMMERCIAL BUILDING

A medium-sized commercial occupancy is planned much like the smaller building just described, except there will be more circuits, a larger service entrance, and so on. In nearly all cases, working drawings will be provided by an architectural-engineering firm to consult during the job. If engineer's drawings are not provided, the electrical contractor should provide some type of layout to be followed. Not only do such drawings aid the workers as the job progresses, but they also give a means of knowing what has been installed at a later date while the building's electrical system is being maintained or repaired.

Depending on the use of the building, the *NEC* or local ordinances may require a different wiring method than would be required in a smaller building.

Most larger commercial buildings utilize a 480/277-volt Y-connected service entrance; all heavy equipment, such as compressors for air conditioning, are designed for use on 480 volts;

electric discharge lighting is all designed for operation on 277-volt, single-phase circuits; dry transformers are required to obtain 120 volts for convenience outlets; and other outlets use 240 and/or 120 volts.

Factors affecting wiring systems in large commercial buildings include the following:

1. Type of building construction, that is, masonry, reinforced concrete, wood frame, and the like.
2. Type of floor, ceiling, and partition construction, height of ceiling, space above ceiling, space under floor, and the like.
3. Wiring methods, type of raceway, sizes of conductors.
4. Type of service-entrance equipment.
5. Type of service and location of service conductors.
6. Connections for equipment not furnished by the electrical contractor but requiring electric service.
7. Type and construction of lighting fixtures, hangers, and supports affecting assembly and installation. Types of lamps.
8. Type and dimensions of floodlighting supporting poles, floodlights and mounting brackets, and so on.
9. Ground conditions affecting trenching for parking lot lighting.

The majority of the factors can be determined by examining the working drawings and specifications, as any commercial building of this size will have a detailed, engineered set of drawings and specifications. If not, the contractor will have to have the system designed and working drawings made to aid the workers on the job. Building inspectors often also require that they be supplied with at least one set of drawings and specifications for use in their office and to check against the actual installation.

In many cases, it is also a good idea for the contractor to examine the job-site conditions prior to bidding or beginning the electrical installation.

A complete take-off of materials will be required for this size of installation for the purpose of estimating the cost of construction, as well as for ordering material and scheduling it for use at the job site.

In many instances, consulting engineers will prepare drawings that leave out much detail, requiring the contractor or his personnel to do extensive research to determine exactly what is taking place. For example, a main distribution panelboard may be indicated on the drawings only by a symbol on the floor plan layout and a catalog number of the equipment. A better drawing, however, will have a complete power riser diagram to supplement the floor plan drawing, showing conduit sizes, wire sizes, number of conductors, and so on. The person installing the system, when only meager symbols are used, usually will have to make a rough layout of the installation and list all details before materials can be ordered or the installation started. Calculations will have to be performed to determine wire size, limit voltage drop, size of conduit, and the like. All these details should be worked out prior to starting the electrical installation.

While commercial electrical installations may vary considerably in detail, in general the majority of them will follow a definite pattern. For example, each will have a service entrance, a distribution panelboard, lighting, and convenience outlets. Furthermore, nearly all will have emergency lighting and signal systems. All will have branch circuits, feeders, and the like.

Therefore, when the electrical technician is called on to design or install a commercial electrical installation, there should really be no "strange" jobs once he or she has worked on a few commercial installations. Then, by following sound basic planning techniques and giving careful attention to details, the trained technician should never be completely stumped, even on jobs of a type that have not been previously handled.

A certain amount of research will have to be done on all electrical jobs. Even seasoned professionals constantly refer to reference material for practically every new project. For example, while the professional engineer may remember the required footcandle level of, say, an office area, he or she will have to refer to manufacturers' catalogs to obtain the illuminating characteristics of certain lighting fixtures. Tables will be consulted to determine voltage drop on various sizes of wire over a given distance and carrying a certain load. Short-circuit calculations will be made to specify the required overcurrent protection — just to name a few. However, the pattern or sequence in which these unknowns are determined is practically the same on every commercial job.

Workers on the job have a further responsibility. While the better engineered drawings and specifications are coordinated to a certain extent with the architectural drawings and the work of other trades, none can be absolutely complete all the time. It is up to the workers on the job to be sure that conduit runs will not interfere with the equipment of other trades. Furthermore, they must make certain that the electrical equipment will not weaken the structural members of the building. The electricians are also required to lay out the circuit runs to use the least amount of material, yet see that the finished job is done in a professional manner.

Sometimes it becomes necessary to vary from the working drawings considerably during the installation, but before doing so, the consulting engineer or architect should be consulted for approval.

In summary, regardless of the technician's position — designer, electrician, supervisor, or whatever — a certain amount of planning is required on all commercial electrical installations. This planning is begun, before any work is started, and then continues on a day-to-day basis until the project is completed. Then a certain amount of planning is required to perform the final tests of the entire system.

Even on projects with detailed engineered drawings and specifications, planning and coordination during the construction phase are still necessary. Certain phases of the electrical installation will have to be carefully planned so as not to hold up any other trades from doing their work. Local inspectors will have to be notified at certain times so that they can inspect the work before it is covered up. Material and tools will have to be ordered so that they will be on the job site when needed. The design group will have to make periodic checks to ensure that the equipment specified is in fact being used, while the workers or the foreman on the job will have to make certain that installed equipment is not damaged by other trades while performing their respective work.

While other factors beyond your control may adversely affect the final electrical installation, job planning and carrying out this plan are largely the basis on which the work will be performed. Be certain that the planning is done on a sound basis.

Chapter 2
The National Electrical Code®

Almost every part of any commercial electrical installation is in some way governed by the *National Electrical Code (NEC)*. You will use the *NEC* in your everyday work, not only on commercial projects, but on all electrical wiring systems used in any type of building construction. Consequently, a thorough knowledge of the *NEC* is one of the first requirements in becoming a trained electrical worker or contractor.

In general, the *NEC* is a set of rules and guidelines specifying the safe installation of electrical wiring and equipment. The *NEC* is probably the most widely used and generally accepted code in the world. It is used as an electrical installation, safety, and reference guide in the United States, and in many other parts of the world.

Purpose And History Of The *NEC*

Owing to the potential fire and explosion hazards caused by the improper handling and installation of electrical wiring, certain rules in the selection of materials, quality of workmanship, and precautions for safety must be followed. To standardize and simplify these rules and provide a reliable guide for electrical construction, the *National Electrical Code (NEC)* was developed. The *NEC* (Figure 2-1), originally prepared in 1897, is frequently revised to meet changing conditions, im-

Figure 2-1: The *NEC* has become the Bible of the electrical construction industry

proved equipment and materials, and new fire hazards. It is a result of the best efforts of electrical engineers, manufacturers of electrical equipment, insurance underwriters, fire fighters, and other concerned experts throughout the country.

The *NEC* is now published by the National Fire Protection Association (NFPA), Batterymarch Park, Quincy, Massachusetts 02269. It contains specific rules and regulations intended to help in the practical safeguarding of persons and property from hazards arising from the use of electricity.

Although the *NEC* itself states, "This Code is not intended as a design specification nor an instruction manual for untrained persons," it does provide a sound basis for the study of electrical installation procedures — under the proper guidance. The probable reason for the *NEC*'s self-analysis is that the code also states, "This Code contains provisions considered necessary for safety. Compliance therewith and proper maintenance will result in an installation essentially free from hazard, but not necessarily efficient, convenient, or adequate for good service or future expansion of electrical use."

The *NEC*, however, has become the Bible of the electrical construction industry, and anyone involved in electrical work, in any capacity, should obtain an up-to-date copy, keep it handy at all times, and refer to it frequently.

Whether you are installing a new electrical system or repairing an existing one, all electrical work must comply with the current *National Electrical Code* (*NEC*) and all local ordinances. Like most laws, the *NEC* is easier to work with once you understand the language and know where to look for the information you need.

In this chapter, you will learn the key terms and basic layout of the *NEC*. A brief review of the individual *NEC* sections that apply to electrical systems will be covered. Sample installations will be discussed in chapters to follow throughout this book.

This chapter, however, is not a substitute for the *NEC*. You need a copy of the most recent edition and it should be kept handy at all times. The more you know about the code, the more you are likely to refer to it. The more your refer to it, the chances of violating any of the *NEC*'s installation regulations become less and less. You are then well on your way to becoming an expert in your field.

NEC TERMINOLOGY

There are two basic types of rules in the *NEC*: mandatory rules and advisory rules. Here is how to recognize the two types of rules and how they relate to all types of electrical systems.

- Mandatory rules—All mandatory rules have the word *shall* in them. The word *"shall"* means *must*. If a rule is mandatory, you must comply with it.

- Advisory rules—All advisory rules have the word *should* in them. The word *"should"* in this case means *recommended but not necessarily required*. If a rule is advisory, compliance is discretionary. If you want to comply with it, do so. But you don't have to if you don't want to.

Be alert to local amendments to the *NEC*. Local ordinances may amend the language of the *NEC*, changing it from *should* to *shall*. This means that you must do in that county or city what may only be recommended in some other area. The office that issues building permits will either sell you a copy of the code that's enforced in that county or tell you where the code is sold. In rare instances, the electrical inspector having jurisdiction over the area may issue these regulations verbally.

There are a few other "landmarks" that you will encounter while looking through the *NEC*. These are summarized in Figure 2-2, and a brief explanation of each follows:

Explanatory material: Explanatory material in the form of Fine Print Notes is designated (FPN). Where these appear, the FPNs normally apply to the *NEC* Section or paragraph immediately preceding the FPN.

Change bar: A change bar in the margins indicates that a change in the *NEC* has been made since the last edition. When becoming familiar with each new edition of the *NEC*, always review these changes. There are also several illustrated publications on the market that point out changes in the *NEC* with detailed explanations of each. Such publications make excellent reference material.

Mandatory rules are characterized by the use of the word:

SHALL

A recommendation or that which is advised but not required is characterized by the use of the word:

SHOULD

Explanatory material in the form of Fine Print Notes is designated:

(FPN)

| A change bar in the margins indicates that a change in the *NEC* has been made since the last edition.

● A bullet indicates that something has been deleted from the last edition of the *NEC.*

Figure 2-2: *NEC* terminology

Bullets: A filled-in circle called a "bullet" indicates that something has been deleted from the last edition of the *NEC*. Although not absolutely nec-essary, many electricians like to compare the previous *NEC* edition to the most recent one when these bullets are encountered, just to see what has been omitted from the latest edition. The most probable reasons for the deletions are errors in the previous edition, obsolete items, or else the intended item was not approved at the time of publication.

Extracted text: Material identified by the superscript letter "x" includes text extracted from other NFPA documents as identified in Appendix A of the *NEC*.

As you open the *NEC* book, you will notice several different types of text used. Here is an explanation of each.

1. Black Letters: Basic definitions and explanations of the *NEC*.
2. Bold Black Letters: Headings for each *NEC* application.
3. Exceptions: These explain the situations when a specific rule does not apply. Exceptions are written in *italics* under the Section or paragraph to which they apply.
4. Tables: Tables are often included when there is more than one possible application of a requirement. See Figure 2-3.
5. Diagrams: A few diagrams are scattered throughout the *NEC* to illustrate certain *NEC* applications. See Figure 2-4.

Size of Largest Service-Entrance Conductor or Equivalent Area for Parallel Conductors		Size of Grounding Electrode Conductor	
Copper	Aluminum or Copper-Clad Aluminum	Copper	Aluminum or Copper-Clad Aluminum
2 or smaller	1/0 or smaller	8	6
1 or 2	2/0 or 3/0	6	4
2/0 or 3/0	4/0 or 250 kcmil	4	2
Over 3/0 through 350 kcmil	Over 250 kcmil through 500 kcmil	2	1/0
Over 350 kcmil through 600 kcmil	Over 500 kcmil through 900 kcmil	1/0	3/0
Over 600 kcmil through 1100 kcmil	Over 900 kcmil through 1750 kcmil	2/0	4/0
Over 1100 kcmil	Over 1750 kcmil	3/0	250 kcmil

Figure 2-3: Typical *NEC* table

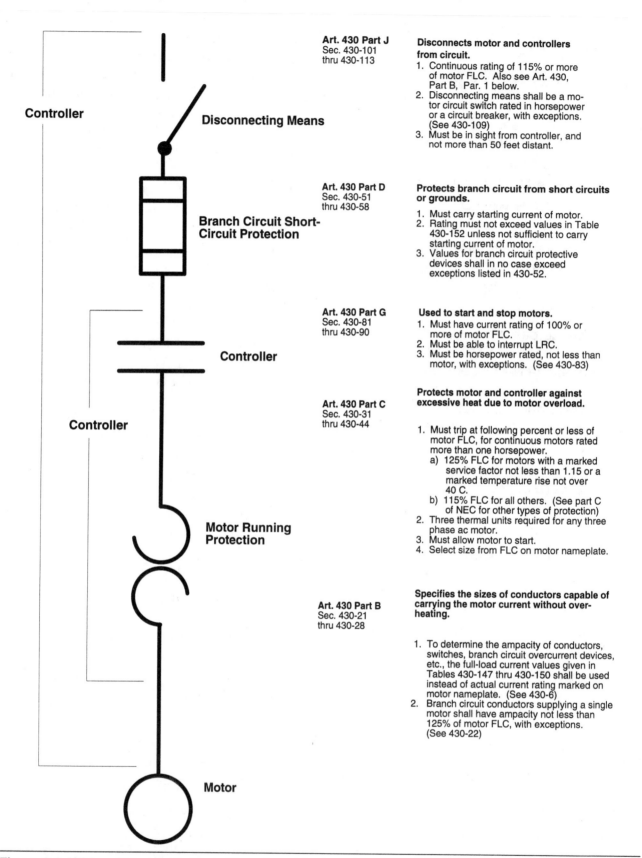

Controller

Disconnecting Means

Art. 430 Part J
Sec. 430-101
thru 430-113

Disconnects motor and controllers from circuit.
1. Continuous rating of 115% or more of motor FLC. Also see Art. 430, Part B, Par. 1 below.
2. Disconnecting means shall be a motor circuit switch rated in horsepower or a circuit breaker, with exceptions. (See 430-109)
3. Must be in sight from controller, and not more than 50 feet distant.

Branch Circuit Short-Circuit Protection

Art. 430 Part D
Sec. 430-51
thru 430-58

Protects branch circuit from short circuits or grounds.
1. Must carry starting current of motor.
2. Rating must not exceed values in Table 430-152 unless not sufficient to carry starting current of motor.
3. Values for branch circuit protective devices shall in no case exceed exceptions listed in 430-52.

Controller

Art. 430 Part G
Sec. 430-81
thru 430-90

Used to start and stop motors.
1. Must have current rating of 100% or more of motor FLC.
2. Must be able to interrupt LRC.
3. Must be horsepower rated, not less than motor, with exceptions. (See 430-83)

Controller

Art. 430 Part C
Sec. 430-31
thru 430-44

Protects motor and controller against excessive heat due to motor overload.

1. Must trip at following percent or less of motor FLC, for continuous motors rated more than one horsepower.
   a) 125% FLC for motors with a marked service factor not less than 1.15 or a marked temperature rise not over 40 C.
   b) 115% FLC for all others. (See part C of NEC for other types of protection)
2. Three thermal units required for any three phase ac motor.
3. Must allow motor to start.
4. Select size from FLC on motor nameplate.

Motor Running Protection

Art. 430 Part B
Sec. 430-21
thru 430-28

Specifies the sizes of conductors capable of carrying the motor current without overheating.

1. To determine the ampacity of conductors, switches, branch circuit overcurrent devices, etc., the full-load current values given in Tables 430-147 thru 430-150 shall be used instead of actual current rating marked on motor nameplate. (See 430-6)
2. Branch circuit conductors supplying a single motor shall have ampacity not less than 125% of motor FLC, with exceptions. (See 430-22)

Motor

Figure 2-4: Wiring diagram similar to those found in the *NEC*

LEARNING THE LAYOUT OF THE *NEC*

The *NEC* is divided into the Introduction (Article 90) and nine chapters. Chapters 1, 2, 3, and 4 apply generally; Chapters 5, 6, and 7 apply to special occupancies, special equipment, or other special conditions. These latter chapters supplement or modify the general rules. Chapters 1 through 4 apply except as amended by Chapters 5, 6, and 7 for the particular conditions.

Chapter 8 covers communications systems and is independent of the other chapters except where they are specifically referenced therein.

Chapter 9 consists of tables and examples.

There is also the *NEC* Contents at the beginning of the book and a comprehensive Index at the back of the book. You will find frequent use for both of these helpful "tools" when searching for various installation requirements.

Each chapter is divided into one or more *Articles*. For example Chapter 1 contains Articles 100 and 110. These Articles are subdivided into Sections. For example, Article 110 of Chapter 1 begins with Section 110-2. Approval. Some sections may contain only one sentence or a paragraph, while others may be further subdivided into lettered or numbered paragraphs such as (a), (1), (2), and so on.

Begin your study of the *NEC* with Articles 90, 100 and 110. These three articles have the basic information that will make the rest of the *NEC* easier to understand. Article 100 defines terms you will need to understand the code. Article 110 gives the general requirements for electrical installations. Read these three articles over several times until you are thoroughly familiar with all the information they contain. It's time well spent. For example, Article 90 contains the following sections:

- Purpose (90-1)
- Scope (90-2)
- Code Arrangement (90-3)
- Enforcement (90-4)
- Mandatory Rules and Explanatory Material (90-5)
- Formal Interpretations (90-6)
- Examination of Equipment for Safety (90-7)
- Wiring Planning (90-8)

Once you are familiar with Articles 90, 100, and 110, you can move on to the rest of the *NEC*. There are several key sections you will use often in servicing electrical systems. Let's discuss each of these important sections.

Wiring Design and Protection: Chapter 2 of the *NEC* discusses wiring design and protection, the information electrical technicians need most often. It covers the use and identification of grounded conductors, branch circuits, feeders, calculations, services, overcurrent protection and grounding. This is essential for *any* type of electrical system, regardless of the type.

Chapter 2 is also a "how-to" chapter. It explains how to size the proper grounding conductor or electrode. If you run into a problem related to the design/installation of a conventional electrical system, you can probably find a solution for it in this chapter.

Wiring Methods and Materials: Chapter 3 has the rules on wiring methods and materials. The materials and procedures to use on a particular system depend on the type of building construction, the type of occupancy, the location of the wiring in the building, the type of atmosphere in the building or in the area surrounding the building, mechanical factors and the relative costs of different wiring methods.

The provisions of this article apply to all wiring installations except remote control switching (Article 725), low energy power circuits (Article 725), signal systems (Article 725), communication systems and conductors (Article 800) when these items form an integral part of equipment such as motors, motor controllers, phase converters and similar types of equipment.

There are four basic wiring methods used in most modern electrical systems. Nearly all wiring methods are a variation of one or more of these four basic methods:

- Sheathed cables of two or more conductors, such as NM cable and BX armored cable (Articles 330 through 339)

- Raceway wiring systems, such as rigid and EMT conduit (Articles 342 to 358)

- Busways (Article 364)

- Cable tray (Article 318)

Article 310 in Chapter 3 gives a complete description of all types of electrical conductors. Electrical conductors come in a wide range of sizes and forms. Be sure to check the working drawings and specifications to see what sizes and types of conductors are required for a specific job. If conductor type and size are not specified, choose the most appropriate type and size meeting standard *NEC* requirements.

Articles 318 through 384 give rules for raceways, boxes, cabinets and raceway fittings. Outlet boxes vary in size and shape, depending on their use, the size of the raceway, the number of conductors entering the box, the type of building construction and atmospheric conditions of the areas. Chapter 3 should answer most questions on the selection and use of these items.

The *NEC* does not describe in detail all types and sizes of outlet boxes. But manufacturers of outlet boxes have excellent catalogs showing all of their products. Collect these catalogs. They are essential to your work.

Article 380 covers the switches, pushbuttons, pilot lamps, receptacles and convenience outlets you will use to control electrical circuits or to connect portable equipment to electric circuits. Again, get the manufacturers' catalogs on these items. They will provide you with detailed descriptions of each of the wiring devices.

Article 384 covers switchboards and panelboards, including their location, installation methods, clearances, grounding and overcurrent protection.

Equipment For General Use

Chapter 4 of the *NEC* begins with the use and installation of flexible cords and cables, including the trade name, type letter, wire size, number of conductors, conductor insulation, outer covering and use of each. The chapter also includes fixture wires, again giving the trade name, type letter and other important details.

Article 410 on lighting fixtures is especially important. It gives installation procedures for fixtures in specific locations. For example, it covers fixtures near combustible material and fixtures in closets. The *NEC* does not describe how many fixtures will be needed in a given area to provide a certain amount of illumination.

Article 430 covers electric motors, including mounting the motor and making electrical connections to it. Motor controls and overload protection are also covered.

Articles 440 through 460 cover air conditioning and heating equipment, transformers and capacitors.

Article 480 gives most requirements related to battery-operated electrical systems. Storage batteries are seldom thought of as part of a conventional electrical system, but they often provide standby emergency lighting service. They may also supply power to security systems that are separate from the main ac electrical system.

Special Occupancies

Chapter 5 of the *NEC* covers *special occupancy* areas. These are areas where the sparks generated by electrical equipment may cause an explosion or fire. The hazard may be due to the atmosphere of the area or just the presence of a volatile material in the area. Commercial garages, aircraft hangers and service stations are typical special occupancy locations.

Articles 500 through 501 cover the different types of special occupancy atmospheres where an explosion is possible. The atmospheric groups were established to make it easy to test and approve equipment for various types of uses.

Articles 501-4, 502-4 and 503-3 cover the installation of explosion-proof wiring. An explosion-proof system is designed to prevent the ignition of a surrounding explosive atmosphere when arcing occurs within the electrical system.

There are three main classes of special occupancy locations:

- Class I (Article 501): Areas containing flammable gases or vapors in the air. Class I areas include paint spray booths, dyeing plants where hazardous liquids are used and gas generator rooms.

- Class II (Article 502): Areas where combustible dust is present, such as grain-handling and storage plants, dust and stock collector areas and sugar-pulverizing plants. These are areas where, under normal operating conditions, there may be enough combustible dust in the air to produce explosive or ignitable mixtures.

- Class III (Article 503): Areas that are hazardous because of the presence of easily ignitable fibers or flyings in the air, although not in large enough quantity to produce ignitable mixtures. Class III locations include cotton mills, rayon mills and clothing manufacturing plants.

Articles 511 and 514 regulate garages and similar locations where volatile or flammable liquids are used. While these areas are not always considered critically hazardous locations, there may be enough danger to require special precautions in the electrical installation. In these areas, the *NEC* requires that volatile gases be confined to an area not more than 4 feet above the floor. So in most cases, conventional raceway systems are permitted above this level. If the area is judged critically

hazardous, explosion-proof wiring (including seal-offs) may be required.

Article 520 regulates theaters and similar occupancies where fire and panic can cause hazards to life and property. Drive-in theaters do not present the same hazards as enclosed auditoriums. But the projection rooms and adjacent areas must be properly ventilated and wired for the protection of operating personnel and others using the area.

Chapter 5 also covers residential storage garages, aircraft hangars, service stations, bulk storage plants, health care facilities, mobile homes and parks, and recreation vehicles and parks.

Special Equipment

Electrical installation requirements in Chapter 6 are frequently encountered by commercial and industrial electrical workers.

Article 600 covers electric signs and outline lighting. Article 610 applies to cranes and hoists. Article 620 covers the majority of the electrical work involved in the installation and operation of elevators, dumbwaiters, escalators and moving walks. The manufacturer is responsible for most of this work. The electrician usually just furnishes a feeder terminating in a disconnect means in the bottom of the elevator shaft. The electrician may also be responsible for a lighting circuit to a junction box midway in the elevator shaft for connecting the elevator cage lighting cable and exhaust fans. Articles in Chapter 6 of the *NEC* give most of the requirements for these installations.

Article 630 regulates electric welding equipment. It is normally treated as a piece of industrial power equipment requiring a special power outlet. But there are special conditions that apply to the circuits supplying welding equipment. These are outlined in detail in Chapter 6 of the *NEC*.

Article 640 covers wiring for sound-recording and similar equipment. This type of equipment normally requires low-voltage wiring. Special outlet boxes or cabinets are usually provided with the equipment. But some items may be mounted in or on standard outlet boxes. Some sound-recording electrical systems require direct current, supplied from rectifying equipment, batteries or motor generators. Low-voltage alternating current comes from relatively small transformers connected on the primary side to a 120-volt circuit within the building.

Other items covered in Chapter 6 of the *NEC* include: X-ray equipment (Article 660), induction and dielectric heat-generating equipment (Article 665) and machine tools (Article 670).

If you ever have work that involves Chapter 6, study the chapter *before work begins*. That can save a lot of installation time. Here is another way to cut down on labor hours and prevent installation errors. Get a set of rough-in drawings of the equipment being installed. It is easy to install the wrong outlet box or to install the right box in the wrong place. Having a set of rough-in drawings can prevent those simple but costly errors.

Special Conditions

In most commercial buildings, the *NEC* and local ordinances require a means of lighting public rooms, halls, stairways and entrances. There must be enough light to allow the occupants to exit from the building if the general building lighting is interrupted. Exit doors must be clearly indicated by illuminated exit signs.

Chapter 7 of the *NEC* covers the installation of emergency lighting systems. These circuits should be arranged so that they can automatically transfer to an alternate source of current, usually storage batteries or gasoline-driven generators. As an alternative in some types of occupancies, you can connect them to the supply side of the main service so disconnecting the main service switch would not disconnect the emergency circuits. See Article 700. *NEC* Chapter 7 also covers a variety of other equipment, systems and conditions that are not easily categorized elsewhere in the *NEC*.

Chapter 8 is a special category for wiring associated with electronic communications systems including telephone and telegraph, radio and TV, fire and burglar alarms, and community antenna systems.

USING THE *NEC*

Once you become familiar with the *NEC* through repeated usage, you will generally know where to look for a particular topic. While this chapter provides you with an initial familiarization of the *NEC* layout, much additional usage experience will be needed for you to feel comfortable with the *NEC*'s content. Until you have gained this practical experience, you can still use the *NEC* to answer questions pertaining to your work. Here's how to locate information on a specific subject.

Step 1. Look through the Contents. You may spot the topic in a heading or subheading. If not, look for a broader, more general subject heading under which the specific topic may appear. Also look for related or similar topics. The Contents will refer you to a specific page number.

Step 2. If you do not find what you're looking for in the Contents, go to the Index at the back of the book. This alphabetic listing is finely divided into different topics. You should locate the subject here. The Index, however, will refer to you either an Article or Section number (not a page number) where the topic is listed.

Step 3. If you cannot find the required subject in the Index, try to think of alternate names. For example, instead of *wire*, look under *conductors;* instead of *outlet box*, look under *boxes, outlet*, and so on.

The *NEC* is not an easy book to read and understand at first. In fact, seasoned electricians sometimes find it confusing. Basically, it is a reference book written in a legal, contract-type language and its content does assume prior knowledge of most subjects listed. Consequently, you will sometimes find the *NEC* frustrating to use because terms aren't always defined, or some unknown prerequisite knowledge is required. To minimize this problem, it is recommended that you obtain one of the several *NEC* supplemental guides that are designed to explain and supplement the *NEC*. One of the best is The National Electrical Code Handbook, available from the NFPA, Batterymarch Park, Quincy, MA 02269 or from your local book store.

Practical Application

Let's assume that you are installing track lighting in a commercial office. The owner wants the track located behind the curtain of their sliding glass balcony doors. To determine if this is a *NEC* violation or not, follow these steps:

Step 1. Turn to the Contents of the *NEC* book, which begins on page 70-2.

Step 2. Find the chapter that would contain information about the general application you are working on. For this example, *Chapter 4—Equipment for General Use* should cover track lighting.

Step 3. Now look for the article that fits the specific category you are working on. In this case, Article 410 covers lighting fixtures, lampholders, lamps, and receptacles.

Step 4. Next locate the *NEC* Section within the *NEC* Article 410 that deals with the specific application. For this example, refer to Part R—Lighting Track.

Step 5. Turn to the page listed. The 1999 *NEC* gives page 235.

Step 6. Read *NEC* Section *410-100, Definition* to become familiar with track lighting. Continue down the page with *NEC* Section 410-101 and read the information contained therein. Note that paragraph (c) under *NEC* Section 410-101 states the following:

(c) Locations Not Permitted. Lighting track shall not be installed (1) where subject to physical damage; (2) in wet or damp locations; (3) where subject to corrosive vapors; (4) in storage battery rooms; (5) in hazardous (classified) locations; (6) where concealed; (7) where ex-

tended through walls or partitions; (8) less than 5 feet above the finished floor except where protected from physical damage or track operating at less than 30 volts RMS open-circuit voltage.

Step 7. Read *NEC* Section 410-101, paragraph (c) carefully. Do you see any conditions that would violate any *NEC* requirements if the track lighting is installed in the area specified? In checking these items, you will probably note condition (6), "where concealed." Since the track lighting is to be installed behind a curtain, this sounds like an *NEC* violation. But let's check further.

Step 8. Let's get an interpretation of the *NEC*'s definition of "concealed." Therefore, turn to Article 100—Definitions and find the main term "concealed." It reads as follows:

Concealed: Rendered inaccessible by the structure or finish of the building . . . .

Step 9. After reading the *NEC*'s definition of "concealed," although the track lighting may be out of sight (if the curtain is drawn), it will still be readily accessible for maintenance. Consequently, the track lighting is really not concealed according to the *NEC* definition.

When using the *NEC* to determine correct electrical-installation requirements, please keep in mind that you will nearly always have to refer to more than one Section. Sometimes the *NEC* itself refers the reader to other Articles and Sections. In some cases, the user will have to be familiar enough with the *NEC* to know what other *NEC* Sections pertain to the installation at hand. It's a confusing situation to say the least, but time and experience in using the *NEC* frequently will make using it much easier. Knowing where to look for what will eventually become second nature.

Now let's take another example to further acquaint you with navigating the *NEC*.

Suppose you are installing Type SE (service-entrance) cable on the side of a small commercial building. You know that this cable must be secured, but you aren't sure of the spacing between cable clamps. To find out this information, use the following procedure:

Step 1: Look in the *NEC* Table of Contents and follow down the list until you find an appropriate category.

Step 2: Article 230 under Chapter 2 will probably catch your eye first, so turn to the page where Article 230 begins in the *NEC*.

Step 3: Glance down the section numbers, 230-1, Scope, 230-2, Number of Services, etc. until you come to Section 230-51, Mounting Supports. Upon reading this section, you will find in paragraph (a) — Service -Entrance Cables — that "Service-entrance cable shall be supported by straps or other approved means within 12 inches (305 mm) of every service head, gooseneck, or connection to a raceway or enclosure and at intervals not exceeding 30 inches (762 mm)."

After reading this section, you will know that a cable strap is required within 12 inches of the service head and within 12 inches of the meter base. Furthermore, the cable must be secured in between these two termination points at intervals not exceeding 30 inches.

DEFINITIONS

Many definitions of terms dealing with the *NEC* may be found in *NEC* Article 100. However, other definitions are scattered throughout the *NEC* under their appropriate category. For example the term *lighting track*, as discussed previously, is not listed in Article 100. The term is listed under *NEC* Section 410-100 and reads as follows:

"Lighting track is a manufactured assembly designed to support and energize lighting fixtures that are capable of being readily repositioned on the track. Its length may be altered by the addition or subtraction of sections of track."

Regardless of where the definition may be located — in Article 100 or under the appropriate *NEC* Section elsewhere in the book — the best way to learn and remember these definitions is to form a mental picture of each item or device as you read the definition. For example, turn to page 70-19 of the 1999 *NEC* and under Article 100 — Definitions, scan down the page until you come to the term "Appliance." Read the definition and then try to form a metal picture of what appliances look like. Some of the more common appliances appear in Figure 2-5. They should be familiar to all readers.

Continue scanning the page until you come to the term "Attachment Plug (Plug Cap) (Cap)." After reading the definition, you will probably have already formed a mental picture of attachment plugs. See Figure 2-6 on page 24 for some more of the more common attachment plugs.

Each and every term listed in the *NEC* should be understood. Know what the item looks like and how it is used on the job. If a term is unfamiliar, try other reference books such as manufacturers' catalogs for an illustration of the item. Then research the item further to determine its purpose in electrical systems. Once you are familiar with all the common terms and definitions found in the *NEC*, navigating through the *NEC* (and understanding what you read) will be much easier.

TESTING LABORATORIES

There are many definitions included in Article 100. You should become familiar with the definitions. Since a copy of the latest *NEC* is compulsory for any type of electrical wiring, there is no need to duplicate them here. However, the following two on page 24 are definitions that you should become especially familiar with.

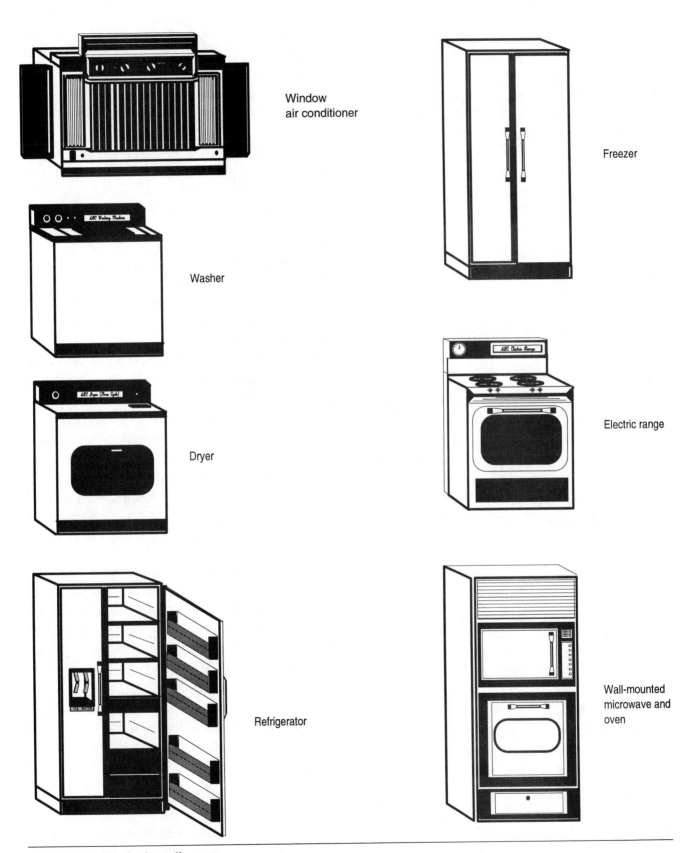

Window
air conditioner

Freezer

Washer

Dryer

Electric range

Refrigerator

Wall-mounted
microwave and
oven

Figure 2-5: Typical appliances

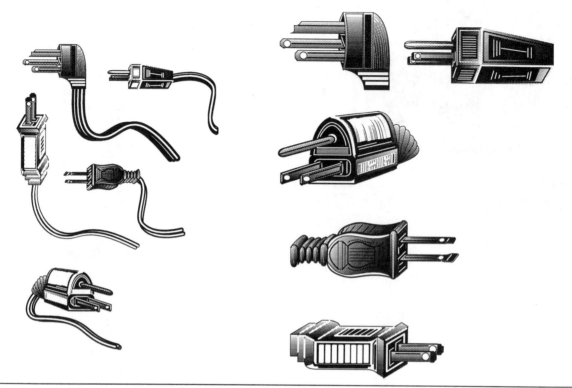

Figure 2-6: Attachment plugs in common use

- Labeled - Equipment or materials to which has been attached a label, symbol or other identifying mark of an organization acceptable to the authority having jurisdiction and concerned with product evaluation, that maintains periodic inspection of production of labeled equipment or materials, and by whose labeling the manufacturer indicates compliance with appropriate standards or performance in a specified manner.

- Listed - Equipment or materials included in a list published by an organization acceptable to the authority having jurisdiction and concerned with product evaluation, that maintains periodic inspection of production of listed equipment or materials, and whose listing states either that the equipment or material meets appropriate designated standards or has been tested and found suitable for use in a specified manner.

Besides installation rules, you will also have to be concerned with the type and quality of materials that are used in electrical wiring systems. Nationally recognized testing laboratories (Underwriters' Laboratories, Inc. is one) are product safety certification laboratories. They establish and operate product safety certification programs to make sure that items produced under the service are safeguarded against reasonable foreseeable risks. Some of these organizations maintain a worldwide network of field representatives who make unannounced visits to manufacturing facilities to countercheck products bearing their "seal of approval." See Figure 2-7.

However, proper selection, overall functional performance and reliability of a product are factors that are not within the basic scope of UL activities.

Figure 2-7: Underwriters' Laboratories label

To fully understand the *NEC*, it is important to understand the organizations which govern it.

NRTL (Nationally Recognized Testing Laboratory)

Nationally Recognized Testing Laboratories are product safety certification laboratories. They establish and operate product safety certification programs to make sure that items produced under the service are safeguarded against reasonable foreseeable risks. NRTL maintains a worldwide network of field representatives who make unannounced visits to factories to countercheck products bearing the safety mark.

NEMA (National Electrical Manufacturers Association)

The National Electrical Manufacturers Association was founded in 1926. It is made up of companies that manufacture equipment used for generation, transmission, distribution, control, and utilization of electric power. The objectives of NEMA are to maintain and improve the quality and reliability of products; to ensure safety standards in the manufacture and use of products; to develop product standards covering such matters as nam-

ing, ratings, performance, testing, and dimensions. NEMA participates in developing the *NEC* and the National Electrical Safety Code and advocates their acceptance by state and local authorities.

NFPA (National Fire Protection Association)

The NFPA was founded in 1896. Its membership is drawn from the fire service, business and industry, health care, educational and other institutions, and individuals in the fields of insurance, government, architecture, and engineering. The duties of the NFPA include:

- Developing, publishing, and distributing standards prepared by approximately 175 technical committees. These standards are intended to minimize the possibility and effects of fire and explosion.

- Conducting fire safety education programs for the general public.

- Providing information on fire protection, prevention, and suppression.

- Compiling annual statistics on causes and occupancies of fires, large-loss fires (over 1 million dollars), fire deaths, and firefighter casualties.

- Providing field service by specialists on electricity, flammable liquids and gases, and marine fire problems.

- Conducting research projects that apply statistical methods and operations research to develop computer modes and data management systems.

The Role Of Testing Laboratories

Testing laboratories are an integral part of the development of the *NEC*. The NFPA, NEMA, and NRTL all provide testing laboratories to conduct research into electrical equipment and its safety. These laboratories perform extensive testing of new products to make sure they are built to code standards for electrical and fire safety. These organizations receive statistics and reports from agencies all over the United States concerning electrical shocks and fires and their causes. Upon seeing trends developing concerning association of certain equipment and dangerous situations or circumstances, this equipment will be specifically targeted for research.

Summary

The *National Electrical Code (NEC)* specifies the minimum provisions necessary for protecting people and property from hazards arising from the use of electricity and electrical equipment. Anyone involved in any phase of the electrical industry must be aware of how to use and apply the code on the job. Using the *NEC* will help you to safely install and maintain the electrical equipment and systems that you come into contact with.

The *NEC* is composed of the following components:

Appendix: Appendix A includes material extracted from other NFPA documents. Appendix B is not part of the requirements of the *NEC* and contains additional material for informational purposes only. Appendix A and Appendix B are located at the end of the book.

Article: Beginning with Article 90 — Introduction, and ending with Article 820 — Community Antenna Television and Radio Distribution Systems, the *NEC* Articles are the main topics.

Chapter: The *NEC* includes nine chapters. Chapter 1 — General, Chapter 2 — Wiring and Protection, Chapter 3 — Wiring Methods and Materials, Chapter 4 — Equipment for General Use, Chapter 5 — Special Occupancies, Chapter 6 — Special Equipment, Chapter 7 — Special Conditions, Chapter 8 — Communications Systems and Chapter 9 — Tables and Examples. The Chapters form the broad structure of the *NEC*.

Contents: Located among the first pages of the book, the contents section provides a complete outline of the Chapters, Articles, Parts, Tables, and Examples. The contents section, used with the index, provides excellent direction for locating answers to electrical problems and questions.

Diagrams and Figures: Diagrams and Figures appear in the *NEC* to illustrate the relationship of Articles and Parts of the *NEC*. For example, Diagram 230-1, Services, shows the relationship of Articles and Parts relating to installation of services.

Examples: Examples in Chapter 9 provide methods to perform load calculations for various types of buildings, feeders, and branch circuits.

Exceptions: Exceptions follow code sections and allow alternative methods, to be used under specific conditions, to the rule stated in the section.

FPN: Fine Print Note: A fine print note is defined in Section 90-5(c). Explanatory material is in the form of Fine Print Notes (FPN).

Index: Located at the end of the book, the index provides a detailed alphabetical listing of the items covered in the *NEC*. The index is used to locate specific information needed to answer electrical problems and questions.

Notes: Notes typically follow tables and are used to provide additional information to the tables or clarification of tables.

Part: Certain Articles in the *NEC* are divided into Parts. Article 220 — Branch Circuit and Feeder is divided into Part A, B, C, and D.

Section: Parts and Articles are divided into Sections. A reference to a section will look like 300-19, *Supporting Conductors in Vertical Raceways,* or 300-19(a) *Spacing Intervals — Maximum.* Sections provide more detailed information within Articles.

Table: Tables are located within Chapters to provide more detailed information explaining code content. For example, Table 310-16 lists ampacities for insulated conductors for copper, aluminum, and copper-clad aluminum conductors with insulation types, sizes, temperature ratings, and ampacity correction factors. Such tables will prove invaluable on all commercial installations.

Chapter 3
Commercial Construction Documents

In all large construction projects and in many of the smaller ones, an architect is commissioned to prepare complete working drawings and specifications for the project. These drawings usually include:

- A plot plan indicating the location of the building on the property.

- Floor plans showing the walls and partitions for each floor or level.

- Elevations of all exterior faces of the building.

- Several vertical cross sections to indicate clearly the various floor levels and details of the footings, foundation, walls, floors, ceilings, and roof construction.

- Large-scale detail drawings showing such construction details as may be required.

For projects of any consequence, the architect usually hires consulting engineers to prepare structural, electrical, and mechanical drawings — the latter encompassing pipe-fitting, instrumentation, plumbing, heating, ventilating, and air conditioning drawings.

Plot Plan

This type of plan of the building site is as if the site is viewed from an airplane and shows the property boundaries, the existing contour lines, the new contour lines (after grading), the location of the building on the property, new and existing roadways, all utility lines, and other pertinent details. Descriptive notes may also be found on the plot plan listing names of adjacent property owners, the land surveyor, and the date of the survey. A legend or symbol list is also included so that anyone who must work with site plans can readily read the information. See Figure 3-1.

Floor Plans

The plan view of any object is a drawing showing the outline and all details as seen when looking directly down on the object. It shows only two dimensions — length and width. The floor plan of a building is drawn as if a slice was taken through the building — about window height — and then the top portion removed to reveal the bottom part where the slice was taken. See Figure 3-2.

Let's say that we first wanted a plan view of a commercial laundry. The part of the building

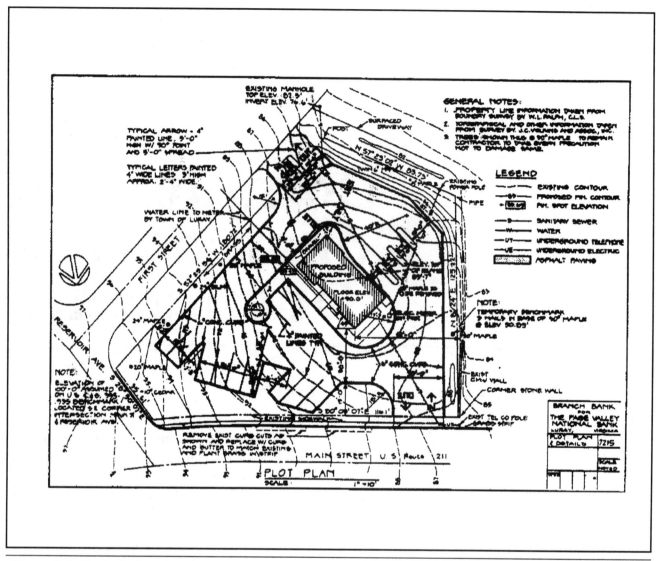

Figure 3-1: Typical plot plan

about the middle of the first-floor windows is imagined to be cut away. By looking down on the uncovered portion, every detail, partition, window and door opening, and the like can be seen. This would be called the first floor plan. A cut through the second floor windows (if applicable) would be the second floor plan, etc. A single-floor building, as shown in Figure 3-3, will only have one basic floor plan, while a high-rise office building may contain a dozen or more floor plans. However, this floor plan will normally be duplicated for each separate trade; that is, electrical floor plan, plumbing floor plan, and so on.

Elevations

A plan view may represent a flat surface, a curved surface, or a slanting one, but for clarifications it is usually necessary to refer to elevations and sections of the building. The *elevation* is an outline of an object that shows heights and may show the length or width of a particular side, but not depth. Figure 3-4 shows elevation drawings for a building. Note that these elevation drawings show the heights of windows, doors, porches, the pitch of roofs, etc. — all of which cannot be shown conveniently on floor plans.

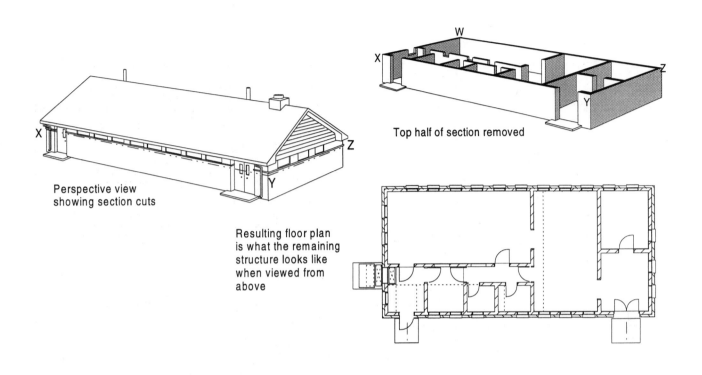

Perspective view
showing section cuts

Top half of section removed

Resulting floor plan
is what the remaining
structure looks like
when viewed from
above

Figure 3-2: Principles of floor-plan layout

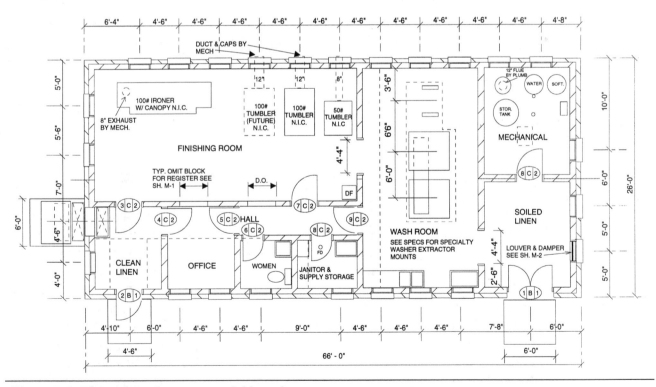

Figure 3-3: Floor plan of a commercial laundry

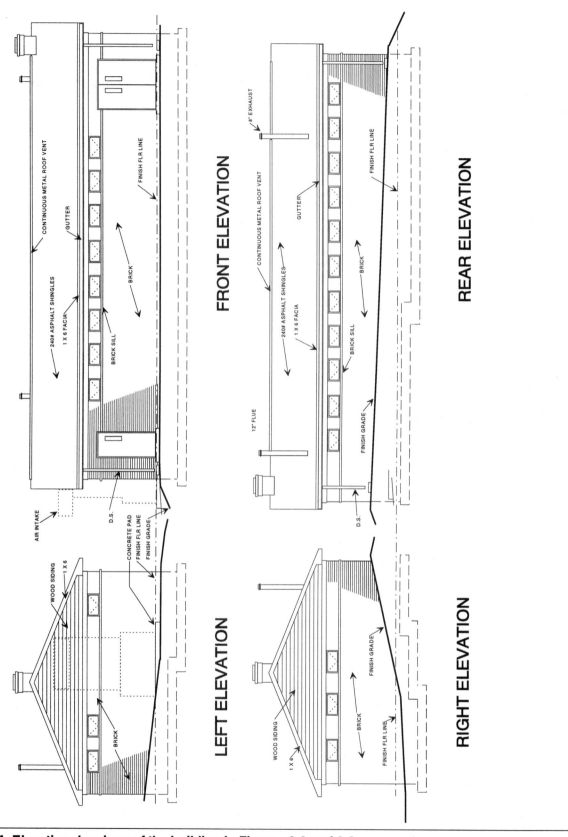

Figure 3-4: Elevation drawings of the building in Figures 3-2 and 3-3

Sections

A *section* or sectional view of an object is a view facing a point where a part of an object is supposed to be cut away, allowing the viewer to see the object's inside. The point on the plan or elevation showing where the imaginary cut has been made is indicated by the section line, which is usually a very heavy double dot-and-dash line. The section line shows the location of the section on the plan or elevation. It is, therefore, necessary to know which of the cutaway parts is represented in the sectional drawing when an object is represented as if it was cut in two. Arrow points are thus placed at the ends of the sectional lines.

In architectural drawings it is often necessary to show more than one section on the same drawing. The different section lines must be distinguished by letters, numbers, or other designations placed at the ends of the lines as shown in Figure 3-5, in which one section is lettered A-A; detail section B, etc. These section letters are generally heavy and large so as to stand out on the drawings. To further avoid confusion, the same letter is usually placed at each end of the section line. The section is named according to these letters—that is, Section A-A, Detail Section B, and so forth.

A longitudinal section is taken length-wise while a cross section is usually taken straight across the width of an object. Sometimes, however, a section is not taken along one straight line. It is often taken along a zigzag line to show important parts of the object.

A sectional view, as applied to architectural drawings, is a drawing showing the building, or portion of a building, as though cut through, as if by a saw, on some imaginary line. This line may be either vertical (straight up and down) or horizontal. Wall sections are nearly always made vertically so that the cut edge is exposed from top to bottom. In some ways the wall section is one of the most important of all the drawings to construction workers, because it answers the questions on how a structure is built. The floor plans of a building show how each floor is arranged, but the wall

sections tell how each part is constructed and usually indicate the material to be used. The electrician needs to know this information when determining wiring methods that comply with the *National Electrical Code (NEC)*.

ELECTRICAL DRAWINGS

The ideal electrical drawing should show in a clear, concise manner exactly what is required of the workers. The amount of data shown on such drawings should be sufficient, but not overdone. This means that a complete set of electrical drawings could consist of only one $8\frac{1}{2}'' \times 11''$ sheet, or it could consist of several dozen $24'' \times 36''$ (or larger) sheets, depending on the size and complexity of the given project. A shop drawing, for example, may contain details of only one piece of equipment, while a set of working drawings for an industrial installation may contain dozens of drawing sheets detailing the electrical system for lighting and power, along with equipment, motor controls, wiring diagrams, schematic diagrams, equipment schedules and a host of other pertinent data.

In general, electrical working drawings for a given project serve three distinct functions:

- To give electrical contractors an exact description of the project so that materials and labor may be estimated in order to form a total cost of the project for bidding purposes.

- To give workers on the project instructions as to how the electrical system is to be installed.

- To provide a "map" of the electrical system once the job is completed to aid in maintenance and troubleshooting for years to come.

Electrical drawings from consulting engineering firms will vary in quality from sketchy, incomplete drawings to neat, very complete drawings that are easy to understand. Few, however, will cover every exact detail of the electrical system. Therefore, a good knowledge of installation practices

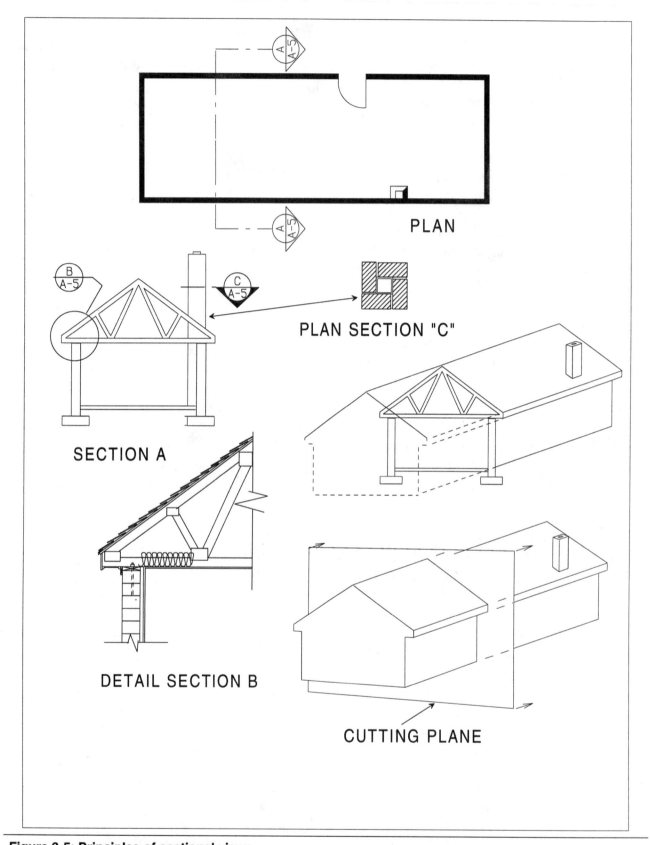

PLAN

PLAN SECTION "C"

SECTION A

DETAIL SECTION B

CUTTING PLANE

Figure 3-5: Principles of sectional views

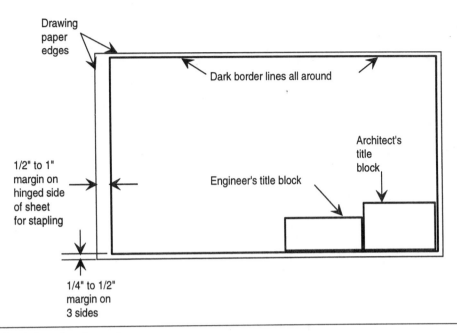

Figure 3-6: Typical drawing layout

must go hand-in-hand with interpreting electrical working drawings.

Sometimes electrical contractors will have electrical drafters prepare special supplemental drawings for use by the contractors' employees. On certain projects, these supplemental drawings can save supervision time in the field once the project has begun.

PRINT LAYOUT

Although a strong effort has been made to standardize drawing practices in the building construction industry, seldom will blueprints — prepared by different architectural or engineering firms — be identical. Similarities, however, will exist between most sets of blueprints, and with a little experience, you should have little trouble interpreting any set of drawings that might be encountered.

Most drawings used for building construction projects will be drawn on drawing paper from $11'' \times 17''$ to $24'' \times 36''$ in size. Each drawing sheet will have border lines framing the overall drawing and a title block as shown in Figure 3-6. Note that the type and size of title blocks vary with each firm preparing the drawings. In addition, some drawing sheets will also contain a revision block near the title block, and perhaps an approval block. This information is normally found on each drawing sheet, regardless of the type of project or the information contained on the sheet.

Title Block

The title block for a blueprint is usually boxed in the lower right-hand corner of the drawing sheet; the size of the block varies with the size of the drawing and also with the information required.

In general, the title block of an electrical drawing should contain the following:

- Name of the project
- Address of the project
- Name of the owner or client
- Name of the architectural and/or engineering firm
- Date of completion
- Scale(s)

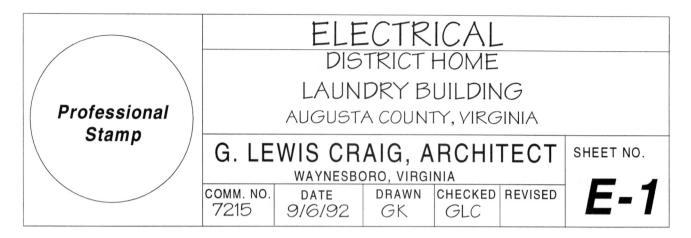

Figure 3-7: Typical architect's title block, drawn to actual size

- Initials of the drafter, checker, and designer, with dates under each if different from the drawing date
- Job number
- Sheet number
- General description of the drawing

Every architectural/engineering firm has its own standard for drawing titles, and they are often preprinted directly on the tracing paper or else printed on "stick-on" paper which is placed on the drawing. See Figure 3-7.

Often the consulting engineering firm will also be listed, which means that an additional title block will be applied to the drawing — usually next to the architect's title block. Figure 3-8 shows completed architectural and engineering title blocks as they appear on an actual drawing for a commercial bank building. Although not shown, the architect's professional stamp appears above the title block.

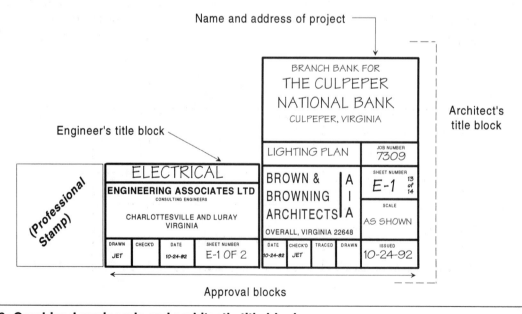

Figure 3-8: Combined engineer's and architect's title blocks

COMM. NO.	DATE	DRAWN	CHECKED	REVISED
7215	9/6/92	GK	GLC	

Figure 3-9: One type of approval block used on electrical working drawings

Approval Block

The "approval block," in most cases, will appear on the drawing sheet as shown in Figure 3-9. The various types of approval blocks — *drawn, checked, etc.* — will be initialed by the appropriate personnel. This type of approval block is usually part of the title block and appears on each drawing sheet.

On some projects, authorized signatures are required before certain systems may be installed, or even before the project begins. An approval block such as the one shown in Figure 3-10 indicates that all required personnel has checked the drawings for accuracy, and that the set meets with everyone's approval. If a signature is missing, the project should not be started. Such an approval block usually appears on the front sheet of the blueprint set and may include:

- Professional stamp — registered seal of approval by the architect or consulting engineer.
- Design supervisor's signature — the person who is overseeing the design.
- Drawn (by) — signature or initials of the person who drafted the drawing and the date it was completed.
- Checked (by) — signature or initials of the person(s) who reviewed the drawing and the date of approval.
- Approved — signature of initials of the architect/engineer and the date of the approval.
- Owner's approval — signature of the project owner or the owner's representative along with the date signed.

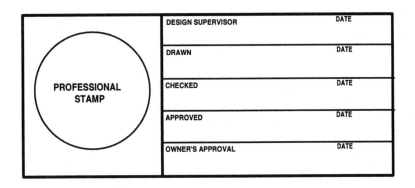

Figure 3-10: Alternate approval box

Figure 3-11: One method of showing revisions on electrical working drawings

- Any other pertinent data required to facilitate the reading and accuracy of the drawing.

Revision Block

Sometimes electrical drawings will have to be partially redrawn or modified during construction of a project. It is extremely important that such modifications are noted and dated on the drawings to ensure that the workers have an up-to-date set of drawings to work from. In some situations, sufficient space is left near the title block for dates and description of revisions as shown in Figure 3-11. In other cases, a revision block is provided (again, near the title block) as shown in Figure 3-12. But these two samples are by no means the only types or styles of revision blocks that will be seen on electrical working drawings — not by any means. Each architect/engineer/designer/drafter

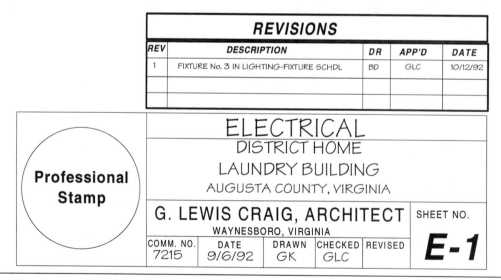

Figure 3-12: Alternate method of showing revisions on working drawings

has his or her own method of showing revisions, so expect to find deviations from those shown.

Caution: When a set of electrical working drawings has been revised, always make certain that the most up-to-date set is used for all future layout work. Either destroy the old, obsolete set of drawings or else clearly mark on the affected sheets, "Obsolete Drawing — Do Not Use." Also, when working with a set of working drawings and written specifications for the first time, thoroughly check each page to see if any revisions or modifications have been made to the originals. Doing so can save much time and expense to all concerned with the project.

DRAFTING LINES

All drafting lines have one thing in common — they are all the same color. However, good easy-to-read contrasting lines can be made by varying the width of the lines or else "breaking" the lines in some uniform way.

Figure 3-13 shows common lines used on architectural drawings. However, these lines can vary. Architects and engineers have strived for a common "standard" for the past century, but unfortunately, their goal has yet to be reached. Therefore, you will find variations in lines and symbols from drawing to drawing, so always consult the legend or symbol list when referring to an architectural or electrical drawing. Also carefully inspect each drawing to ensure that line types are used consistently.

A brief description of the drafting lines shown in Figure 3-13 follows:

Light Full Line - This line is used for section lines, building background (outlines), and similar uses where the object to be drawn is secondary to the electrical system. Care should be taken in choosing the width of such lines, as they sometimes fade out on some printers.

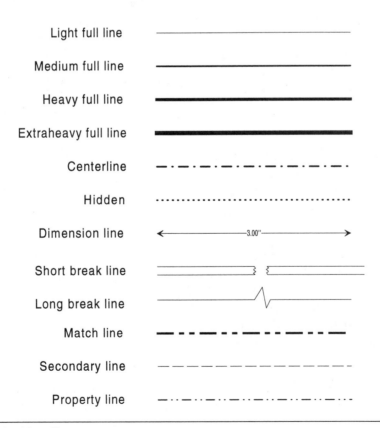

Figure 3-13: Typical drafting lines

Medium Full Line - This type of line is frequently used for hand lettering on drawing. It is further used for some drawing symbols, circuit lines, and the like.

Heavy Full Line - This line is used for borders around title blocks, schedules and for hand lettering drawing titles. Some types of symbols are frequently drawn with the heavy full line.

Extraheavy Full Line - This line is used for border lines on architectural/engineering drawings.

Centerline - A centerline is a broken line made up of long and short dashes alternately spaced. It indicates the centers of objects such as holes, pillars, or fixtures. Sometimes, the centerline indicates the dimensions of a finished floor.

Hidden Line - A hidden line consists of a series of short dashes closely and evenly spaced. It shows the edges of objects that are not visible in a particular view. The object outlined by hidden lines in one drawing is often fully pictured in another drawing.

Dimension Lines - These are thinly drawn lines used to show the extent and direction of dimensions. The dimension is usually placed in a break inside of the dimension lines. Normal practice is to place the dimension lines outside the object's outline. However, sometimes it may be necessary to draw the dimensions inside the outline.

Short Break Line - This line is usually drawn freehand and is used for short breaks.

Long Break Line - This line which is drawn partly with a straightedge and partly with freehand zigzags, is used for long breaks.

Match Line - This line is used to show the position of the cutting plane. Therefore, it is also called cutting plane line. A match or cutting plane line is an extra heavy line with long dashes alternating with two short dashes. It is used on drawings of large structures to show where one drawing stops and the next drawing starts.

Secondary Line - This line is frequently used to outline pieces of equipment or to indicate reference points of a drawing that is secondary to the drawing's purpose.

Property Line - This is a line made up of one long and two short dashes alternately spaced. It indicates land boundaries on the site plan.

Other uses of the lines just mentioned include the following:

Extension Lines - Extension lines are lightweight lines that start about $1/16$ inch away from an object's edge and extend out. A common use of extension lines is to create a boundary for dimension lines. Dimension lines meet extension lines with arrowheads, slashes, or dots. Extension lines that point from a note or other reference to a particular feature on a drawing are called leaders. They usually end in either an arrowhead or a dot and may include an explanatory note at the end.

Section Lines - These are often referred to as cross-hatch lines. Drawn at a 45-degree angle, these lines show where an object has been cut away to reveal the inside.

Phantom Lines - Phantom lines are solid, light lines that show where an object will be installed. A future door opening or a future piece of equipment can be shown with phantom lines.

Electrical Drafting Lines

Besides the architectural lines shown in Figure 3-13, consulting electrical engineers, designers, and drafters use additional lines to represent circuits and their related components. Again, these lines may vary from drawing to drawing, so check the symbol list or legend for the exact meaning of lines on the drawing with which you are working. Figure 3-14 shows lines used on some electrical drawings. Again, these lines may vary from drawing to drawing, so always verify the meaning of lines in the symbol list or legend.

ELECTRICAL SYMBOLS

The electrician must be able to correctly read and understand electrical working drawings which requires a thorough knowledge of electrical symbols and their application.

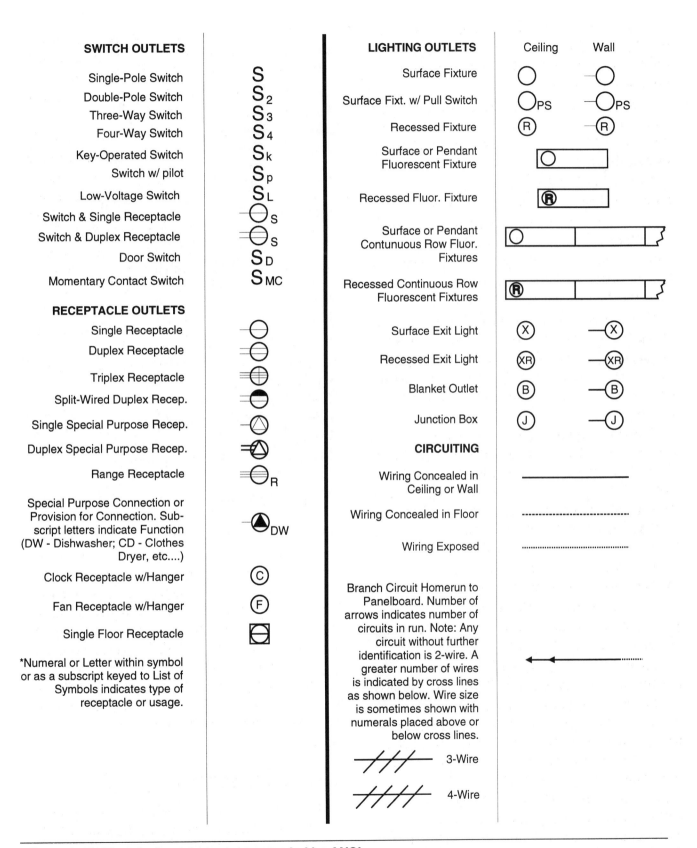

SWITCH OUTLETS

Single-Pole Switch — S
Double-Pole Switch — S_2
Three-Way Switch — S_3
Four-Way Switch — S_4
Key-Operated Switch — S_k
Switch w/ pilot — S_p
Low-Voltage Switch — S_L
Switch & Single Receptacle
Switch & Duplex Receptacle
Door Switch — S_D
Momentary Contact Switch — S_{MC}

RECEPTACLE OUTLETS

Single Receptacle
Duplex Receptacle
Triplex Receptacle
Split-Wired Duplex Recep.
Single Special Purpose Recep.
Duplex Special Purpose Recep.
Range Receptacle

Special Purpose Connection or Provision for Connection. Subscript letters indicate Function (DW - Dishwasher; CD - Clothes Dryer, etc....)

Clock Receptacle w/Hanger
Fan Receptacle w/Hanger
Single Floor Receptacle

*Numeral or Letter within symbol or as a subscript keyed to List of Symbols indicates type of receptacle or usage.

LIGHTING OUTLETS

	Ceiling	Wall
Surface Fixture		
Surface Fixt. w/ Pull Switch		
Recessed Fixture		
Surface or Pendant Fluorescent Fixture		
Recessed Fluor. Fixture		
Surface or Pendant Contunuous Row Fluor. Fixtures		
Recessed Continuous Row Fluorescent Fixtures		
Surface Exit Light		
Recessed Exit Light		
Blanket Outlet		
Junction Box		

CIRCUITING

Wiring Concealed in Ceiling or Wall

Wiring Concealed in Floor

Wiring Exposed

Branch Circuit Homerun to Panelboard. Number of arrows indicates number of circuits in run. Note: Any circuit without further identification is 2-wire. A greater number of wires is indicated by cross lines as shown below. Wire size is sometimes shown with numerals placed above or below cross lines.

3-Wire

4-Wire

Figure 3-14: Electrical symbols recommended by ANSI

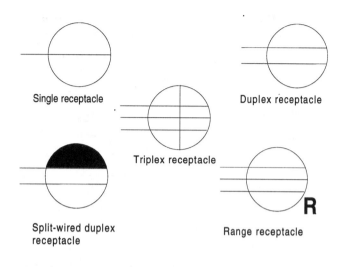

Figure 3-15: Various types of receptacle symbols used on electrical drawings

An electrical symbol is a figure or mark that stands for a component used in the electrical system. For example, Figure 3-14 shows a list of electrical symbols that are currently recommended by the American National Standards Institute (ANSI). It is evident from this list of symbols that many have the same basic form, but, because of some slight difference, their meaning changes. For example, the outlet symbols in Figure 3-15 each have the same basic form — a circle — but the addition of a line or an abbreviation gives each an individual meaning. A good procedure to follow in learning symbols is to first learn the basic form and then apply the variations for obtaining different meanings.

It would be much simpler if all architects, engineers, electrical designers, and drafters used the same symbols. However, this is not the case. Although standardization is getting closer to a reality, existing symbols are still modified and new symbols are created for almost every new project.

The electrical symbols described in the following paragraphs represent those found on actual electrical working drawings throughout the United States and Canada. Many are similar to those recommended by ANSI and the Consulting Engineers Council/US; others are not. Understanding how these symbols were devised will help you to interpret unknown electrical symbols in the future.

Some of the symbols used on electrical drawings are abbreviations, such as WP for weatherproof and AFF for above finished floor. Others are simplified pictographs, such as "A" in Figure 3-16 for a double floodlight fixture or like "B" in Figure 3-16 for an infrared electric heater with two quartz lamps.

In some cases, the symbols are combinations of abbreviation and pictograph, such as "C" in Figure 3-16 for fusible safety switch, "D" for a double-throw safety switch, and "E" for a nonfusible safety switch. In each example, a pictograph of a switch enclosure has been combined with an ab-

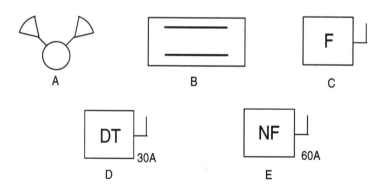

Figure 3-16: Pictographs and notations used to form electrical symbols

breviation, F (fusible), DT (double throw), and NF (nonfusible), respectively. The numerals indicate the busbar capacity in amperes.

Lighting-outlet symbols have been devised that represent incandescent, fluorescent, and high-intensity discharge lighting; a circle usually represents an incandescent fixture; and a rectangle is used to represent a fluorescent fixture. All these symbols are designed to indicate the physical shape of a particular fixture and while the circles representing incandescent lamps are frequently enlarged somewhat, symbols for fluorescent fixtures are usually drawn as close to scale as possible.

The type of mounting used for all lighting fixtures is usually indicated in a lighting-fixture schedule, which is shown on the drawings or in the written specifications.

The type of lighting fixture is identified by a numeral placed inside a triangle or other symbol, and placed near the fixture to be identified. A complete description of the fixture identified by the symbols must be given in the lighting-fixture schedule and should include the manufacturer, catalog number, number and type of lamps, voltage, finish, mounting, and any other information needed for a proper installation of the fixtures.

Switches used to control lighting fixtures are also indicated by symbols — usually the letter "S" followed by numerals or letters to define the exact type of switch. For example, S_3 indicates a three-way switch; S_4 identifies a four-way switch; S_P indicates a single-pole switch with a pilot light, etc.

Main distribution centers, panelboards, transformers, safety switches, and other similar electrical components are indicated by electrical symbols on floor plans and by a combination of symbols and semipictorial drawings in riser diagrams.

A detailed description of the service equipment is usually given in the panelboard schedule or in the written specifications. However, on small projects the service equipment is sometimes indicated only by notes on the drawings.

Circuit and feeder wiring symbols are getting closer to being standardized. Most circuits concealed in the ceiling or wall are indicated by a solid line; a broken line is used for circuits concealed in the floor or ceiling below; and exposed raceways are indicated by short dashes or else the letter "E" placed in the same plane with the circuit line at various intervals.

The number of conductors in a conduit or raceway system may be indicated in the panelboard schedule under the appropriate column, or the information may be shown on the floor plan.

Symbols for communication and signal-systems, as well as symbols for light and power, are drawn to an appropriate scale and accurately located with respect to the building; this reduces the number of references made to the architectural drawings. Where extreme accuracy is required in locating outlets and equipment, exact dimensions are given on larger-scale drawings and shown on the plans.

Each different category in an electrical system is usually represented by a distinguishing basic symbol. To further identify items of equipment or outlets in the category, a numeral or other identifying mark is placed within the open basic symbol. In addition, all such individual symbols used on the drawings should be included in the symbol list or legend. The symbols shown in Figure 3-17 (begnning on page 42) are those recommended by the Consulting Engineers Council/US. You should become familiar with all of these symbols and refer to them often when the need arises.

SCALE DRAWINGS

In most electrical drawings, the components are so large that it would be impossible to draw them actual size. Consequently, drawings are made to some reduced scale — that is, all the distances are drawn smaller than the actual dimension of the object itself, all dimensions being reduced in the same proportion. For example, if a floor plan of a building is to be drawn to a scale of $\frac{1}{4}'' = 1'-0''$, each $\frac{1}{4}''$ on the drawing would equal 1 foot on the building itself; if the scale is $\frac{1}{8}'' = 1'-0''$, each

SWITCH OUTLETS

Single Pole Switch	S
Double Pole Switch	S_2
Three-Way Switch	S_3
Four-Way Switch	S_4
Key-Operated Switch	S_K
Switch and Fusestat Holder	$S_F H$
Switch and Pilot Lamp	S_P
Fan Switch	S_F
Switch for Low-Voltage Switching System	S_L
Master Switch for Low-Voltage Switching System	S_{LM}
Switch and Single Receptacle	⊖ S
Switch and Duplex Receptacle	⊜ S
Door Switch	S_D
Time Switch	S_T
Momentary Contact Switch	S_{MC}
Ceiling Pull Switch	Ⓢ
"Hand-Off-Auto" Control Switch	HOA
Multi-Speed Control Switch	M
Pushbutton	•

RECEPTACLE OUTLETS

Where weatherproof, explosionproof, or other specific types of devices are to be required, use the upper-case subscript letters to specify. For example, weatherproof single or duplex receptales would have the upper case WP subscript letters noted alongside of the symbol. All outlets must be grounded.

Single Receptacle Outlet	
Duplex Receptacle Outlet	
Triplex Receptacle Outlet	
Quadruplex Receptacle Outlet	
Duplex Receptacle Outlet Split Wired	
Triplex Receptacle Outlet Split Wired	
250 Volt Receptacle Single Phase Use Subscript Letter to Indicate Function (DW - Dishwasher, RA - Range) or Numerals (with explanation in symbols schedule)	
250 Volt Receptacle Three Phase	
Clock Receptacle	Ⓒ
Fan Receptacle	Ⓕ
Floor Single Receptacle Outlet	
Floor Duplex Receptacle Outlet	
Floor Special-Purpose Outlet	*
Floor Telephone Outlet - Public	
Floor Telephone Outlet - Private	

* Use numeral keyed explanation of symbol usage

Figure 3-17: Recommended electrical symbols

Example of the use of several floor outlet symbols to identify a 2, 3, or more gang outlet:

Underfloor Duct and Junction Box for Triple, Double or Single Duct System as indicated by the number of parallel lines

Example of use of various symbols to identify location of different types of outlets or connections for underfloor duct or cellular floor systems:

Cellular Floor Header Duct

CIRCUITING

Wiring Exposed (not in conduit)	——— E ———
Wiring Concealed in Ceiling or Wall	———————
Wiring Concealed in Floor	– – – – – – –
Wiring Existing*	- - - - - - - - -
Wiring Turned Up	———————○
Wiring Turned Down	———————●
Branch Circuit Home Run to Panel Board	2 1 ———▶

Number of arrows indicates number of circuits. (A number at each arrow may be used to identify circuit number.)**

BUS DUCTS AND WIREWAYS

Trolley Duct***	T	T
Busway (Service, Feeder or Plug-in)***	B	B
Cable Trough Ladder or Channel***	C	C
Wireway***	W	W

PANELBOARDS, SWITCHBOARDS AND RELATED EQUIPMENT

Flush Mounted Panelboard and Cabinet***

Surface Mounted Panelboard and Cabinet***

Switchboard, Power Control Center, Unit Substation (Should be drawn to scale)***

Flush Mounted Terminal Cabinet (In small scale drawings the TC may be indicated alongside the symbol)*** — TC

Surface Mounted Terminal Cabinet (In small scale drawings the TC may be indicated alongside the symbol)*** — TC

Pull Box (Identify in relation to Wiring System Section and Size)

Motor or Other Power Controller (May be a starter or contactor)***

Externally Operated Disconnection Switch***

Combination Controller and Disconnection Means***

Figure 3-17: Recommended electrical symbols (Cont.)

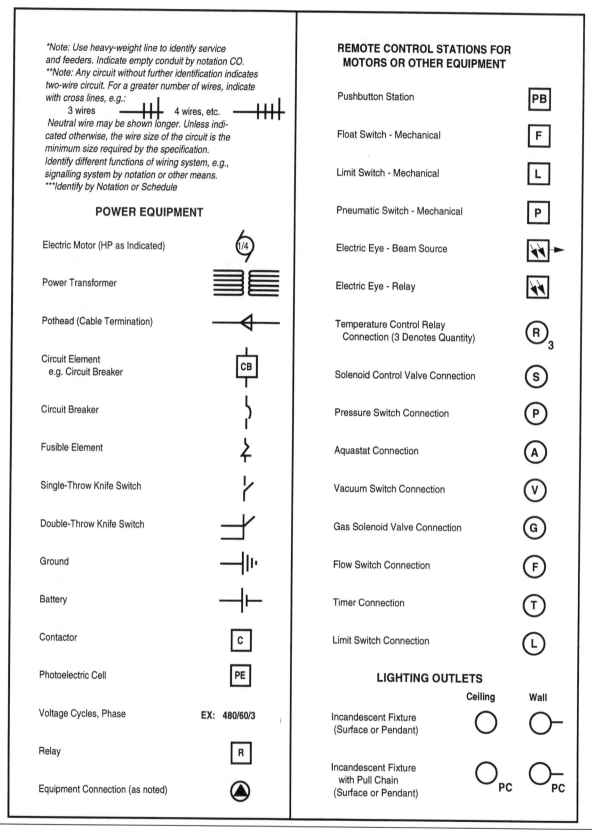

Figure 3-17: Recommended electrical symbols (Cont.)

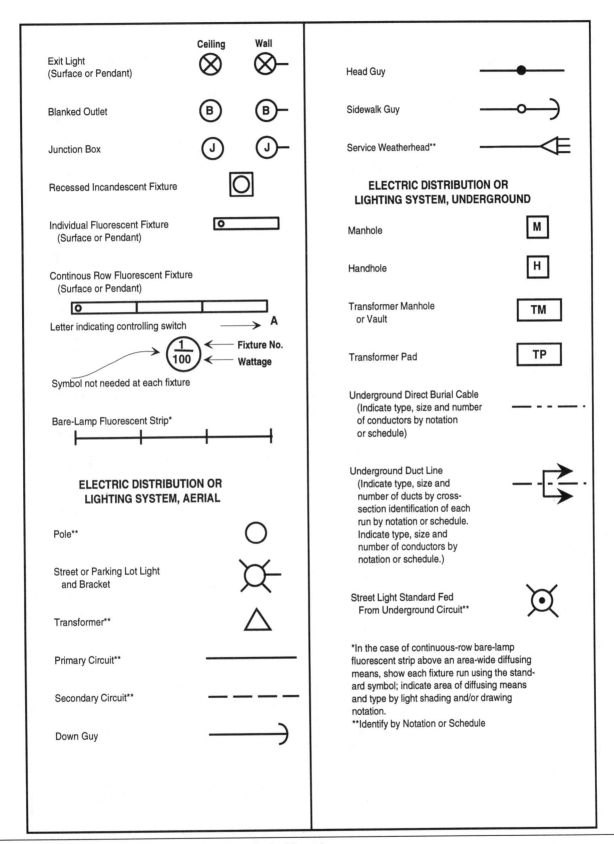

Figure 3-17: Recommended electrical symbols (Cont.)

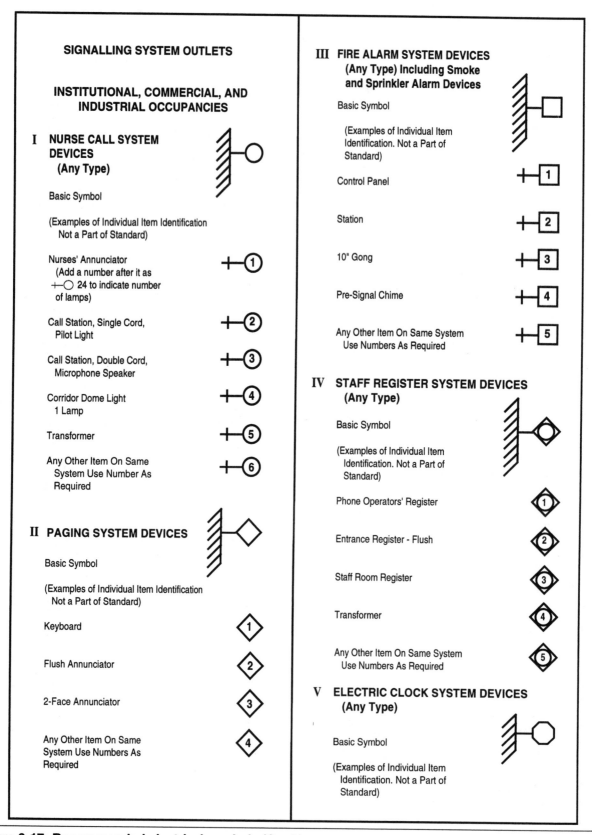

SIGNALLING SYSTEM OUTLETS

INSTITUTIONAL, COMMERCIAL, AND INDUSTRIAL OCCUPANCIES

I   NURSE CALL SYSTEM DEVICES
  (Any Type)

Basic Symbol

(Examples of Individual Item Identification Not a Part of Standard)

Nurses' Annunciator
  (Add a number after it as
  ⊢◯ 24 to indicate number
  of lamps)

Call Station, Single Cord,
  Pilot Light

Call Station, Double Cord,
  Microphone Speaker

Corridor Dome Light
  1 Lamp

Transformer

Any Other Item On Same
  System Use Number As
  Required

II   PAGING SYSTEM DEVICES

Basic Symbol

(Examples of Individual Item Identification
  Not a Part of Standard)

Keyboard

Flush Annunciator

2-Face Annunciator

Any Other Item On Same
  System Use Numbers As
  Required

III   FIRE ALARM SYSTEM DEVICES
  (Any Type) Including Smoke
  and Sprinkler Alarm Devices

Basic Symbol

(Examples of Individual Item
  Identification. Not a Part of
  Standard)

Control Panel

Station

10" Gong

Pre-Signal Chime

Any Other Item On Same System
  Use Numbers As Required

IV   STAFF REGISTER SYSTEM DEVICES
  (Any Type)

Basic Symbol

(Examples of Individual Item
  Identification. Not a Part of
  Standard)

Phone Operators' Register

Entrance Register - Flush

Staff Room Register

Transformer

Any Other Item On Same System
  Use Numbers As Required

V   ELECTRIC CLOCK SYSTEM DEVICES
  (Any Type)

Basic Symbol

(Examples of Individual Item
  Identification. Not a Part of
  Standard)

Figure 3-17: Recommended electrical symbols (Cont.)

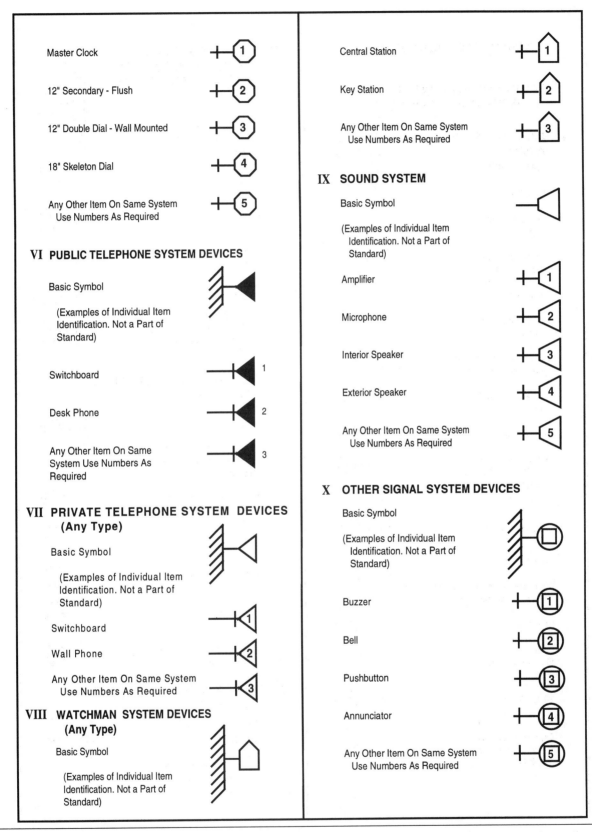

Master Clock

12" Secondary - Flush

12" Double Dial - Wall Mounted

18" Skeleton Dial

Any Other Item On Same System
Use Numbers As Required

VI PUBLIC TELEPHONE SYSTEM DEVICES

Basic Symbol

(Examples of Individual Item
Identification. Not a Part of
Standard)

Switchboard

Desk Phone

Any Other Item On Same
System Use Numbers As
Required

VII PRIVATE TELEPHONE SYSTEM DEVICES
(Any Type)

Basic Symbol

(Examples of Individual Item
Identification. Not a Part of
Standard)

Switchboard

Wall Phone

Any Other Item On Same System
Use Numbers As Required

VIII WATCHMAN SYSTEM DEVICES
(Any Type)

Basic Symbol

(Examples of Individual Item
Identification. Not a Part of
Standard)

Central Station

Key Station

Any Other Item On Same System
Use Numbers As Required

IX SOUND SYSTEM

Basic Symbol

(Examples of Individual Item
Identification. Not a Part of
Standard)

Amplifier

Microphone

Interior Speaker

Exterior Speaker

Any Other Item On Same System
Use Numbers As Required

X OTHER SIGNAL SYSTEM DEVICES

Basic Symbol

(Examples of Individual Item
Identification. Not a Part of
Standard)

Buzzer

Bell

Pushbutton

Annunciator

Any Other Item On Same System
Use Numbers As Required

Figure 3-17: Recommended electrical symbols (Cont.)

RESIDENTIAL OCCUPANCIES

Signalling system symbols for use in identifying standardized residential-type signal system items on residential drawings where a descriptive symbol list is not included on the drawing. When other signal system items are to be identified, use the above basic symbols for such items together with a descriptive symbol list.

Pushbutton	▣
Buzzer	◺
Bell	◖□
Combination Bell - Buzzer	◖□◿
Chime	CH
Annunciator	◇—
Electric Door Opener	D
Maid's Signal Plug	M
Interconnection Box	□
Bell-Ringing Transformer	BT
Outside Telephone	▶
Interconnecting Telephone	▷
Television Outlet	TV

Use this space to sketch new symbols that are encountered in your work

Figure 3-17: Recommended electrical symbols (Cont.)

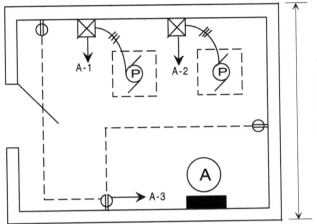

The distance between the arrowheads to the left measures 3-1/2" on the drawing, but since the drawing is made to a scale of 1/2" = 1' - 0", this measurement actually represents 7' - 0"

PUMP HOUSE FLOOR PLAN
1/2" = 1' - 0"

Figure 3-18: Typical floor plan showing drawing title and scale

$\frac{1}{8}$″ on the drawing equals 1 foot on the building, and so forth.

When architectural and engineering drawings are produced, the scale decided upon is very important. Where dimensions must be held to extreme accuracy, the scale drawings should be made as large as practical with dimension lines added. Where dimensions require only reasonable accuracy, the object may be drawn to a smaller scale (with dimension lines possibly omitted) since the object can be scaled with the appropriate scale.

In dimensioning drawings, the dimension written on the drawing is the actual dimension of the building, not the distance that is measured on the drawing. To further illustrate this point, look at the floor plan in Figure 3-18; it is drawn to a scale of $\frac{1}{2}$″ = 1′ - 0″. One of the walls is drawn to an actual length of $3\frac{1}{2}$″ on the drawing paper, but since the scale is $\frac{1}{2}$″ = 1′ - 0″ and since $3\frac{1}{2}$″ contains 7 halves of an inch ($7 \times 0.5 = 3\frac{1}{2}$), the dimension shown on the drawing will therefore be 7′- 0″ on the actual building.

From the previous example, we may say that the most common method of reducing all the dimensions (in feet and inches) in the same proportion is to choose a certain distance and let that distance represent 1 foot. This distance can then be divided into twelve parts, each of which represents an inch. If half inches are required, these twelfths are further subdivided into halves, etc. We now have a scale that represents the common foot rule with its subdivisions into inches and fractions, except that the scaled foot is smaller than the distance known as a foot and, likewise its subdivisions are proportionately smaller.

When a measurement is made on the drawing, it is made with the *reduced foot rule* or *scale;* when a measurement is made on the building, it is made with the *standard foot rule*. The most common reduced foot rules or scales used in electrical drawings are the *architect's scale* and the *engineer's scale*. Sometimes drawings may be encountered that use a *metric scale*, but the principle of using this scale is similar to the architect's or engineer's scales. All types are covered in the next sections.

Architect's Scale

Figure 3-19 on page 50 shows two configurations of architect's scales — the one on the left is designed so that 1″= 1′ - 0″; the one on the right

49

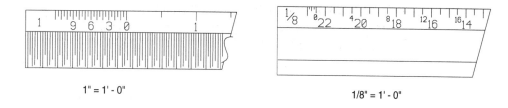

Figure 3-19: Two different configurations of architect's scales

has graduations spaced to represent ⅛″= 1′ - 0″. Now let's zoom in for a closer look.

Note on the 1″ scale in Figure 3-20 that the longer marks to the right of the zero (with a numeral beneath) represent feet. Therefore, the distance between the zero and the numeral 1 equals 1 foot. The shorter mark between the zero and 1 represents ½ of a foot, or 6″.

Referring again to Figure 3-19, look at the marks to the left of the zero. There are four different lengths of marks in this group. The longest marks are spaced three scaled inches apart and have the numerals 0, 3, 6, and 9 for use as reference points. The next longest group of lines each represent scaled inches, but are not marked with numerals. In use, you can count the number of marks to

the left of the zero to find the number of inches, but after some practice, you will be able to tell the exact measurement at a glance. For example, the measurement "A" on Figure 3-20 represents 5″ because it is the fifth "inch" mark to the left of the zero; it is also one "inch" mark short of the 6-inch line on the scale.

The next size line that is shorter than the "inch" line is the half-inch line, and the shortest lines in the group represent ¼″. On smaller scales, however, the basic unit is not divided into as many divisions. For example, the smallest subdivision on the ⅛″ = 1′ - 0″ scale represents 2″. When larger scales are utilized, such as ¾″ = 1′ − 0″, the inch section of the scale is divided in half-inch marks.

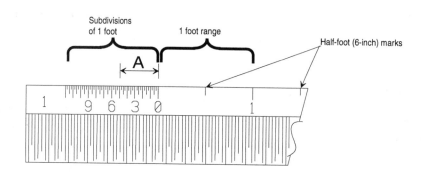

Figure 3-20: Close-up view of the 1-inch architect's scale

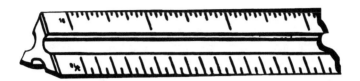

Figure 3-21: Typical triangular architect's scale

Types of Architect's Scales

Architect's scales are available in several types, but the most common include the triangular scale and the "flat" scale. The quality of architect's scales also vary from cheap plastic scales (costing a dollar or two) to high-quality wooden-laminated tools such as produced by K&E, Dietzgen, and others.

The triangular scale (Figure 3-21) is frequently found in drafting and estimating departments or engineering and electrical contracting firms, while the flat scales are more convenient to carry on the job site by workers.

Triangular-shaped architect's scales have 12 different scales — two on each edge — as follows:

- Common foot rule (12 inches)
- $1/16'' = 1' - 0''$
- $3/32'' = 1' - 0''$
- $3/16'' = 1' - 0''$
- $1/8'' = 1' - 0''$
- $1/4'' = 1' - 0''$
- $3/8'' = 1' - 0''$
- $3/4'' = 1' - 0''$
- $1'' = 1' - 0''$
- $1/2'' - 1' - 0''$
- $1\frac{1}{2}'' = 1' - 0''$
- $3'' = 1' - 0''$

Two separate scales on one face may seem confusing at first, but after some experience, reading these scales becomes "second nature."

In all but one of the scales on the triangular architect's scale, each face has one of the scales spaced exactly one-half of the other. For example, on the 1-inch face, the 1-inch scale is read from left to right, starting from the zero mark. The half-inch scale is read from right to left — again starting from the zero mark.

On the remaining foot-rule scale ($1/16'' = 1' - 0''$), each $1/16''$ mark on the scale represents 1 foot.

Figure 3-22 on page 52 shows all the scales found on the triangular architect's scale.

The "flat" architect's scale is ideal for electrical workers on most projects. It is easily and conveniently carried in the shirt pocket, and the conventional four scales ($1/8$-, $1/4$-, $1/2$-, and 1-inch) are adequate for the majority of projects that will be encountered.

Every drawing should have the scale to which it is drawn, plainly marked on it as part of the drawing title, as illustrated in Figure 3-18 on page 49. However, it is not uncommon to have several different drawings on one drawing sheet — all with different scales. Therefore, always check the scale of each different view found on a drawing sheet.

Engineer's Scale

The civil engineer's scale is used fundamentally in the same manner as the architect's scale, the

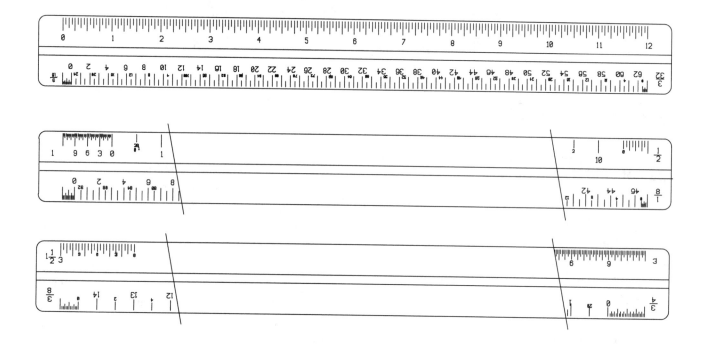

Figure 3-22: The various scales on a triangular architect's scale

principal difference being that the graduations on the engineer's scale are decimal units rather than feet, as on the architect's scale.

The engineer's scale is used by placing it on the drawing with the working edge away from the user. The scale is then aligned in the direction of the required measurement. Then, by looking down over the scale, the dimension is read.

Civil engineer's scales are common in the following graduations:

- $1'' = 10$ units
- $1'' = 20$ units
- $1'' = 30$ units
- $1'' = 40$ units
- $1'' = 60$ units
- $1'' = 80$ units
- $1'' = 100$ units

The purpose of this scale is to transfer the relative dimensions of an object to the drawing or vice versa. It is used mainly on site plans (sometimes called *plot plans*) to determine distances between property lines, manholes, duct runs, direct-burial cable runs, and the like.

Site plans are drawn to scale using the engineer's scale rather than the architect's scale. On small lots, a scale of, say, 1 inch = 10 feet or 1 inch = 20 feet is used. This means that 1 inch (actual measurement on the drawing) is equal to 10 feet, 20 feet, and so on, on the land itself.

On larger drawings, where a large area must be covered, the scale could be 1 inch = 100 feet or 1 inch = 1000 feet, or any other integral power of 10. On drawings with the scale in multiples of 10, the engineering scaled marked *10* is used. If the scale is 1 inch = 200 feet, the engineer's scale marked *20* is used, and so on.

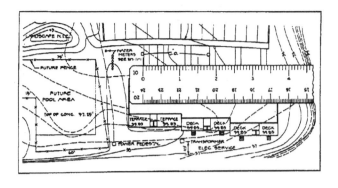

Figure 3-23: Practical use of the engineer's scale

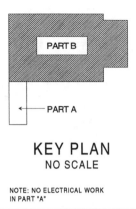

KEY PLAN
NO SCALE

NOTE: NO ELECTRICAL WORK
IN PART "A"

Figure 3-25: Typical key plan

Although site plans appear reduced in scale, depending on the size of the object and the size of the drawing sheet to be used, the actual true-length dimensions must be shown on the drawings at all times. When you are reading the drawing plans to scale, think of each dimension in its full size and not in the reduced scale it happens to be on the drawing. See Figure 3-23.

The Metric Scale

Metric scales (Figure 3-24), are divided into centimeters (cm), with the centimeters divided into 10-divisioned millimeters (mm), or into 20-divisioned half millimeters. Scales are available with metric divisions on one edge while inch divisions are inscribed on the opposite edge. Many contracting firms that deal in international trade have adopted a dual-dimensioning system expressed in both metric and English symbols. Furthermore,

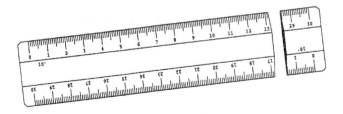

Figure 3-24: Typical metric scale

drawings prepared for government projects frequently require metric dimensions.

Key Plan

A "key plan" usually appears on the drawing sheet immediately above the architect's or engineer's title block (Figure 3-25). The purpose of this key plan is to identify that part of the project to which the drawing sheet applies. In this case, the project involves two buildings: Building "A" and Building "B." Since the outline of Building "B" is cross-hatched in the key plan, this is the building to which the drawing applies. Note that this key plan is not drawn to scale — only its approximate shape.

Although Building "A" is also shown on this key plan, a note below the key-plan title states, "Note: No electrical work in Part A."

On some larger installations, the overall project may involve several buildings requiring appropriate key plans on each drawing to help the workers orient the drawings to the appropriate building. In some cases, separate drawing sheets may be used for each room or department in a large commercial project — again requiring key plans on each drawing sheet to identify applicable drawings for each room or area.

Again, the exact way of signifying a key plan will vary with those preparing the drawings. If in

LIGHTING FIXTURE SCHEDULE

SYMBOL	TYPE	MANUFACTURER AND CATALOG NUMBER	MOUNTING	LAMPS
⊤⎯	A	LIGHTOLIER 10234	WALL	2-40W T-12WWX
▭	B	LIGHTOLIER 10420	SURFACE	2-40W T-12 WWX
⊗	C	ALKCO RPC-210-6E	SURFACE	2-8W T-5
⊢○	D	P 7 S AL 2936	WALL	1-100W A
○	E	P 7 S 110	SURFACE	1-100W A

Figure 3-26: Lighting-fixture schedule for an office/warehouse building

doubt, also check with a responsible person to insure that your interpretation is correct.

DRAWING SCHEDULES

A schedule is a systematic method of presenting notes or lists of equipment on a drawing in tabular form. When properly organized and thoroughly understood, schedules are not only powerful time-saving devices for those preparing the drawings, but they can also save the workers on the job much valuable time.

For example, the lighting-fixture schedule shown in Figure 3-26 lists the fixture type and identifies each fixture type on the drawing by number. The manufacturer and catalog number of each type are given along with the number, size, and type of lamp for each.

Sometimes all of the same information found in schedules will be duplicated in the written specifications, but combing through page after page of written specifications can be time consuming and workers do not always have access to the specifications while working, whereas they usually do have access to the working drawings. Therefore, the schedule is an excellent means of providing

essential information in a clear and accurate manner, allowing the workers to carry out their assignments in the least amount of time.

Other schedules that are frequently found on electrical working drawings include:

- Connected load schedule
- Panelboard schedule
- Electric-heat schedule
- Kitchen-equipment schedule
- Schedule of receptacle types

There are also other schedules found on electrical drawings, depending upon the type of project. Most, however, will deal with lists of equipment; that is, such items as motors, motor controllers, and the like.

ELECTRICAL DETAILS AND DIAGRAMS

Electrical diagrams are drawings that are intended to show, in diagrammatic form, electrical components and their related connections. They are seldom, if ever, drawn to scale, and show only the electrical association of the different components; that is, the various devices, and their connection to each other.

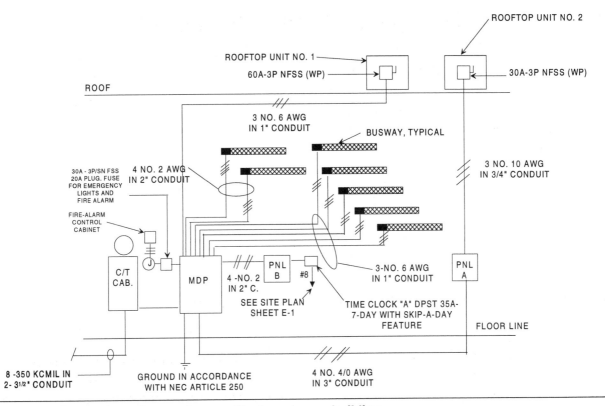

Figure 3-27: Power-riser diagram for an office/warehouse building

Power-Riser Diagrams

Single-line block diagrams are used extensively to show the arrangement of electric service equipment. The power-riser diagram in Figure 3-27, for example, was used on an office/warehouse building and is typical of such drawings. The drawing shows all pieces of electrical equipment as well as the connecting lines used to indicate service-

PANELBOARD SCHEDULE									
PANEL No.	CABINET TYPE	PANEL MAINS			BRANCHES				ITEMS FED OR REMARKS
		AMPS	VOLTS	PHASE	1P	2P	3P	PROT.	FRAME	

PANEL No.	CABINET TYPE	AMPS	VOLTS	PHASE	1P	2P	3P	PROT.	FRAME	ITEMS FED OR REMARKS
MDP	SURFACE	600A	120/208	3φ,4-W	-	-	1	225A	25,000	PANEL "A"
					-	-	1	100A	18,000	PANEL "B"
					-	-	1	100A		POWER BUSWAY
					-	-	1	60A		LIGHTING BUSWAY
					-	-	1	70A		ROOFTOP UNIT #1
					-	-	1	70A		SPARE
					-	-	1	600A	42,000	MAIN CIRCUIT BRKR

Figure 3-28: Panelboard schedule used with the riser diagram in Figure 3-27

entrance conductors and feeders. Notes are used to identify the equipment, indicate the size of conduit necessary for each feeder, and the number, size, and type of conductors in each conduit.

A panelboard schedule (Figure 3-28) is included with the power-riser diagram to indicate the exact components contained in each panelboard. This panelboard schedule is for the main distribution panel. Schedules are also shown on actual drawings (not here) for the other two panels (PNL A and PNL B).

In general, panelboard schedules usually indicate the panel number, the type of cabinet (either flush- or surface-mounted), the panel mains (ampere and voltage rating), the phase (single- or 3-phase), and the number of wires. A 4-wire panel, for example, indicates that a solid neutral exists in the panel. Branches indicate the type of overcurrent protection; that is, the number of "poles," the trip rating, and the frame size. The items that each overcurrent device feeds is also indicated in one of the columns.

SCHEMATIC DIAGRAMS

Complete schematic wiring diagrams are normally used only in highly unique and complicated electrical systems, such as control circuits. Components are represented by symbols, and every wire is either shown by itself or included in an

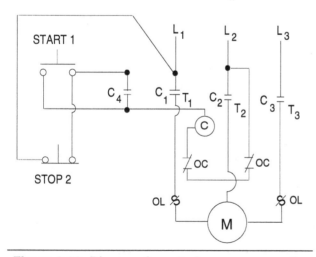

Figure 3-29: Diagram for a 3-phase, ac magnetic motor starter

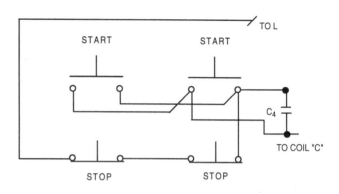

Figure 3-30: Circuit in Figure 3-29 being controlled by two sets of start-stop pushbuttons

assembly of several wires which appear as one line on the drawing. Each wire should be numbered when it enters an assembly and should keep the same number when it comes out again to be connected to some electrical component in the system. Figure 3-29 shows a complete schematic wiring diagram for a three-phase, ac magnetic non-reversing motor starter.

Note that this diagram shows the various devices in symbol form and indicates the actual connections of all wires between the devices. The three-wire supply lines are indicated by L_1, L_2, and L_3; the motor terminals of motor M are indicated by T_1, T_2, and T_3. Each line has a thermal overload-protection device (OL) connected in series with normally open line contactors C_1, C_2, and C_3, which are controlled by the magnetic starter coil, C. Each contactor has a pair of contacts that close or open during operation. The control station, consisting of start pushbutton 1 and stop pushbutton 2, is connected across lines L_1 and L_2. An auxiliary contactor (C_4) is connected in series with the stop pushbutton and in parallel with the start pushbutton. The control circuit also has normally closed overload contactors (OC) connected in series with the magnetic starter coil (C).

Any number of additional pushbutton stations may be added to this control circuit similarly to the way three- and four-way switches are added to control a lighting circuit. In adding pushbutton stations, the stop buttons are always connected in series and the start buttons are always connected in parallel. Figure 3-30 shows the same motor starter circuit in Figure 3-29, but this time it is controlled by two sets of start-stop buttons.

Schematic wiring diagrams have only been touched upon in this chapter; there are many other details that the electrician needs to know to perform his or her work in a proficient manner. Later chapters cover wiring diagrams in more detail — especially in Chapter 14 — Relays and Motor Controllers.

Drawing Details

A detail drawing is a drawing of a separate item or portion of an electrical system, giving a complete and exact description of its use and all the details needed to show the workman exactly what is required for its installation. The power floor plan for an office/warehouse (Figure 3-31 on page 58) has a sectional cut (Section A-A) through the busduct. This is a good example of where an extra, detailed drawing is desirable. This section is shown in Figure 3-32 on page 59, and provides additional details for installing the busduct.

A set of electrical drawings will sometimes require large-scale drawings of certain areas that are not indicated with sufficient clarity on the small-scale drawings. For example, a site plan may show exterior pole-mounted lighting fixtures that are to be installed by the contractor. The site plan itself will probably show only small circles to indicate the approximate location of these fixtures. In most cases, more detail is required for a correct installation. The detail in Figure 3-33 on page 59 shows how the concrete base is constructed on one electrical project.

WRITTEN SPECIFICATIONS

The written specifications for a building or project are the written descriptions of work and duties required of the owner, the architect, and the con-

sulting engineer. Together with the working drawings, these specifications form the basis of the contract requirements for the construction of the building or project. Those who use the construction drawings and specifications must always be alert to discrepancies between the working drawings and the written specifications. Such discrepancies occur particularly when:

- Architects or engineers use standard or prototype specifications and attempt to apply them without any modification to specific working drawings.

- Previously prepared standard drawings are changed or amended by reference in the specifications only and the drawings themselves are not changed.

- Items are duplicated in both the drawings and specifications, but an item is subsequently amended in one and overlooked on the other contract document.

In such instances, the person in charge of the project has the responsibility to ascertain whether the drawings or the specifications take precedence. Such questions must be resolved, preferably before the work is installed, to avoid added cost to either the owner, the architect/engineer, or the contractor.

How Specifications Are Written

Writing accurate and complete specifications for building construction is a serious responsibility for those who design the buildings because the specifications, combined with the working drawings, govern practically all important decisions made during the construction span of every project. Compiling and writing these specifications is not a simple task, even for those who have had considerable experience in preparing such documents. A set of written specifications for a single project usually will contain thousands of products, parts and components, and methods of installing them, all of which must be covered in either the drawings and/or specifications. No one can memorize all of the

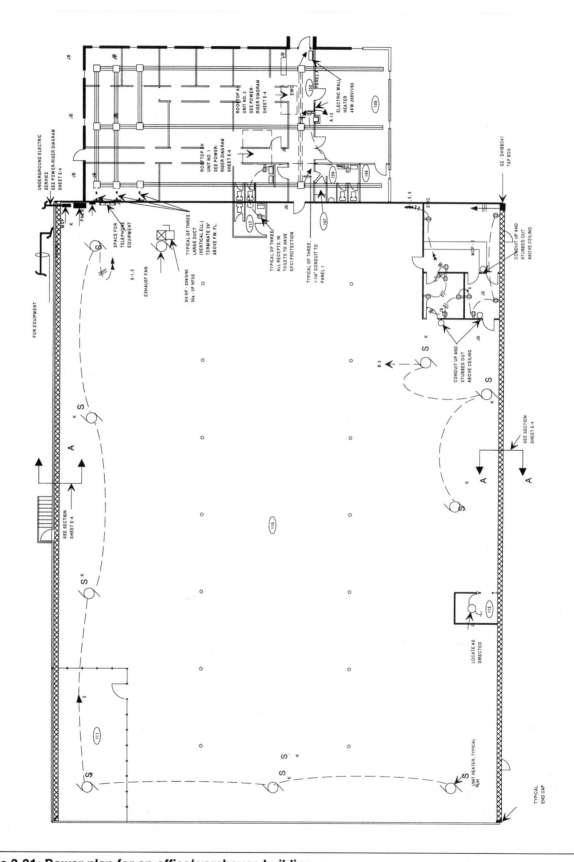

Figure 3-31: Power plan for an office/warehouse building

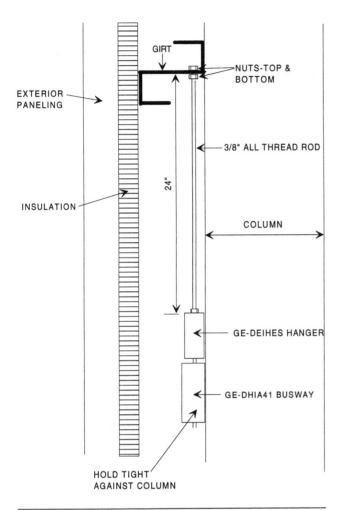

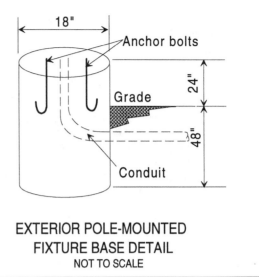

EXTERIOR POLE-MOUNTED FIXTURE BASE DETAIL
NOT TO SCALE

Figure 3-33: Pole-mounting detail for exterior lighting fixtures

Figure 3-32: Section A-A on the drawing in Figure 3-31

necessary items required to accurately describe the various areas of construction. One must rely upon reference materials — manufacturer's data, catalogs, checklists, and, best of all, a high-quality master specification.

Specification Format

For convenience in writing, speed in estimating work, and ease in reference, the most suitable organization of the specification is a series of sections dealing successively with the different trades, and in each section grouping all the work of the particular trade to which the section is devoted. All the work of each trade should be incorporated into the section devoted to that trade.

Those people who use the specifications must be able to find all information needed without taking too much time in looking for it.

The CSI Format

The Construction Specification Institute (CSI) developed the Uniform Construction Index some years ago that allowed all specifications, product information, and cost data to be arranged into a uniform system. This format is now followed on most large construction projects in North America. All construction is divided into 16 Divisions, and each division has several sections and subsections. The following outline describes the various divisions normally included in a set of specifications for building construction.

Division 1 — General Requirements. This division summarizes the work, alternatives, project meetings, submissions, quality control, temporary facilities and controls, products, and the project closeout. Every responsible person involved with the project should become familiar with this division.

Division 2 — Site Work. This division outlines work involving such items as paving, sidewalks, outside utility lines (electrical, plumbing, gas, telephone, etc.), landscaping, grading, and other items pertaining to the outside of the building.

Division 3 — Concrete. This division covers work involving footings, concrete formwork, expansion and contraction joints, cast-in-place concrete, specially finished concrete, precast concrete, concrete slabs, and the like.

Division 4 — Masonry. This division covers concrete, mortar, stone, masonry accessories, and the like.

Division 5 — Metals. Metal roofs, structural metal framing, metal joists, metal decking, ornamental metal, and expansion control normally fall under this division.

Division 6 — Carpentry. Items falling under this division include: rough carpentry, heavy timber construction, trestles, prefabricated structural wood, finish carpentry, wood treatment, architectural woodwork, and the like. Plastic fabrications may also be included in this division of the specifications.

Division 7 — Thermal and Moisture Protection. Waterproofing is the main topic discussed under this division. Other related items such as dampproofing, building insulation, shingles and roofing tiles, preformed roofing and siding, membrane roofing, sheet metal work, wall flashing, roof accessories, and sealants are also included.

Division 8 — Doors and Windows. All types of doors and frames are included under this division: metal, plastic, wood, etc. Windows and framing are also included along with hardware and other window and door accessories.

Division 9 — Finishes. Included in this division are the types, quality, and workmanship of lath and plaster, gypsum wallboard, tile, terrazzo, acoustical treatment, ceiling suspension systems, wood flooring, floor treatment, special coatings, painting, and wallcovering.

Division 10 — Specialties. Items such as chalkboards and tackboards; compartments and cubicles, louvers and vents that are not connected with the heating, ventilating, and air conditioning system; wall and corner guards; access flooring; specialty modules; pest control; fireplaces; flagpoles; identifying devices; lockers; protective covers; postal specialties; scales; storage shelving; wardrobe specialties; and the like are covered in this division of the specifications.

Division 11 — Equipment. The equipment included in this division could include central vacuum cleaning systems, bank vaults, darkrooms, food service, vending machines, laundry equipment, and many similar items.

Division 12 — Furnishing. Items such as cabinets and storage, fabrics, furniture, rugs and mats, seating, and other similar furnishing accessories are included under this division.

Division 13 — Special Construction. Such items as air-supported structures, incinerators, and other special items will fall under this division.

Division 14 — Conveying Systems. This division covers conveying apparatus such as dumbwaiters, elevators, hoists and cranes, lifts, material-handling systems, turntables, moving stairs and walks, pneumatic tube systems, and powered scaffolding.

Division 15 — Mechanical. This division includes plumbing, heating, ventilating, and air conditioning and related work. Electric heat is sometimes covered under Division 16, especially if individual baseboard heating units are used in each room or area of the building.

Division 16 — Electrical. This division covers all electrical requirements for the building including lighting, power, alarm and communication systems, special electrical systems, and related electrical equipment. This is the division that electricians will use the most. Division 16 contains the following sections:

DIVISION 16 — ELECTRICAL

16050 Electrical Contractors

16200 Power Generation

16300 Power Transmission

16400 Service and Distribution

16500 Lighting

16600 Special Systems

16700 Communications

16850 Heating and Cooling

16900 Controls and Instrumentation

DIVISION 16 - ELECTRICAL

SECTION 16A - GENERAL PROVISIONS

1. Portions of the sections of the Documents designated by the letters "A", "B" & "C" and "DIVISION ONE - GENERAL REQUIRE-MENTS" apply to this Division.

2. Consult Index to be certain that set of Documents and Specifications is complete. Report omissions or discrepancies to the Architect.

3. SCOPE OF THE WORK:

   a. The scope of the work consists of the furnishing and installing of complete electrical systems–exterior and interior–including miscellaneous systems. The Electrical Contractor shall provide all supervision, labor, materials, equipment, machinery, and any and all other items necessary to complete the systems. The Electrical Contractor shall note that all items of equipment are specified in the singular; however, the Contractor shall provide and install the number of items of equipment as indicated on the drawings and as required for complete systems.

   b. It is the intention of the Specifications and Drawings to call for finished work, tested, and ready for operation.

   c. Any apparatus, appliance, material or work not shown on drawings but mentioned in the specifications, or vice versa, or any incidental accessories necessary to make the work complete and perfect in all respects and ready for operation, even if not particularly specified, shall be furnished, delivered and installed by the Contractor without additional expense to the Owner.

   d. Minor details not usually shown or specified, but necessary for proper installation and operation, shall be included in the Contractor's estimate, the same as if herein specified or shown.

Figure 3-34: Sample page from a set of electrical specifications

The above sections are further subdivided into many subsections. For example, items covered under Section 16400 — Service and Distribution — will usually include the project's service entrance, meter-ing, grounding, service-entrance conductors, main disconnecting means and similar details. A sample page from a set of electrical specifications is shown in Figure 3-34.

Chapter 4
Electrical Load Calculations

This chapter presents rules and related calculations for determining sizes and ratings of the various conductors and equipment required by the *National Electrical Code (NEC)* to be included in each service installation — including the feeders and branch-circuit loads. The elements of the service installation and the related feeders and branch circuits are depicted in Figure 4-1.

The service-entrance conductors are normally enclosed in a service raceway such as conduit, but sometimes consist of open conductors or Type SE cable. The conductors continue through a metering device and finally terminate at the service disconnecting means, which is usually a switch or circuit breaker. The service or main overcurrent protection is a set of fuses or a circuit breaker that protects the service-entrance conductors. Each metallic part of the service must be bonded together — either by means of a metallic raceway or an equipment bonding jumper. If one conductor of the circuit, such as the neutral conductor, is grounded, a grounding electrode conductor that connects the grounded conductor to two or more grounding electrodes is required. Each feeder circuit may require an equipment grounding conductor to ground the noncurrent-carrying metal parts of equipment.

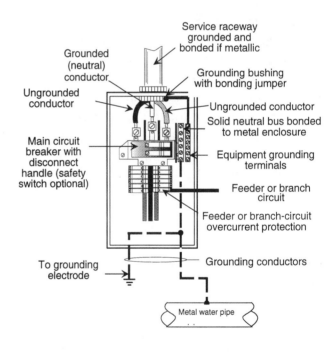

Figure 4-1: Basic elements of an electric service

Each conductor, the disconnecting means, and the overcurrent protective device must satisfy *NEC* installation requirements which specify the size or rating as appropriate for the element. These requirements form the basis for the electrical load calculation of services, feeders, and branch circuits.

BASIC CALCULATION PROCEDURES

Sometimes it is confusing just which comes first; the layout of the outlets, or the sizing of the electric service. In many cases, the service size (size of main disconnect, panelboard, service conductors, etc.) can be sized using *National Electrical Code* (*NEC*) procedures before the outlets are actually located. In other cases, the outlets will have to be laid out first. However, in either case, the service-entrance and panelboard locations will have to be determined before the circuits can be installed — so the electrician will know in which direction (and to what points) the circuit homeruns will terminate.

Traditionally, consulting engineers and electrical designers locate all outlets on the working drawings prior to sizing the service-entrance. Once the total connected load has been determined, demand factors and continuous-load factors are applied to size the branch-circuit requirements. Sub-panels are then placed at strategic locations throughout the building, and feeders are then sized according to *NEC* requirements. The sum of these feeder circuits (allowing appropriate demand factors) determines the size of the electric service. The main distribution panel, number and size of disconnect switches and overcurrent protection is then determined to finalize the load calculation. Such procedures for determining the load of any project normally surpass *NEC* requirements, but this is quite acceptable since the *NEC* specifies minimum requirements. In fact, in *NEC* Section 90-1(b), the *NEC* itself states:

> "This Code contains provisions considered necessary for safety. Compliance therewith and proper maintenance will result in an installation essentially free from hazard but not necessarily efficient, convenient, or adequate for good service or future expansion of electrical use."

Consequently, many electrical installations are designed for conditions that surpass *NEC* requirements to obtain a more efficient and convenient electrical system. However, since most electrical examinations are based on the latest *NEC* installation requirements, procedures contained in this chapter are stictly based on *NEC* regulations. To illustrate why, an electrician's examination may ask the trade size of conduit that is necessary to contain a certain number of conductors. The *NEC* may specify, say, 2-inch conduit as a minimum size for this condition. You can use, say, $2\frac{1}{2}$-inch conduit on the job which will surpass *NEC* requirements. In doing so, no electrical inspector is going to reject the installation because it surpasses *NEC* requirements. However, on most electrician's examinations, those taking the exam must give exact answers in order to get the questions right. Therefore, even though $2\frac{1}{2}$-inch conduit may work on the job, your answer will be wrong if the *NEC* minimum requirement calls for 2-inch conduit. Thus, this is the main reason for sticking with *NEC* methods of load calculations throughout this chapter.

Load Calculating Steps

In general, the type of occupancy (office, store, bank, etc.) is first determined and catergorized before the load calculation is begun. An overview of the basic steps appear in Figure 4-2. These steps proceed as follows:

Step 1. Determine the area of the building using outside dimensions, less any uninhabitable areas such as garages, patios, porches, and the like.

Step 2. Multiply the resulting area by the load-per-square-foot amount listed in *NEC* Table 220-11 for the type of occupancy.

Step 3. Determine the volt-amperes of continuous loads (if any) and multiply the volt-amperes (VA) for these loads by a factor of 1.25 (125%) as per *NEC* Section 220-3(a).

Step 4. Apply demand factors to any qualifying loads.

Step 5. Calculate the total adjusted general lighting load.

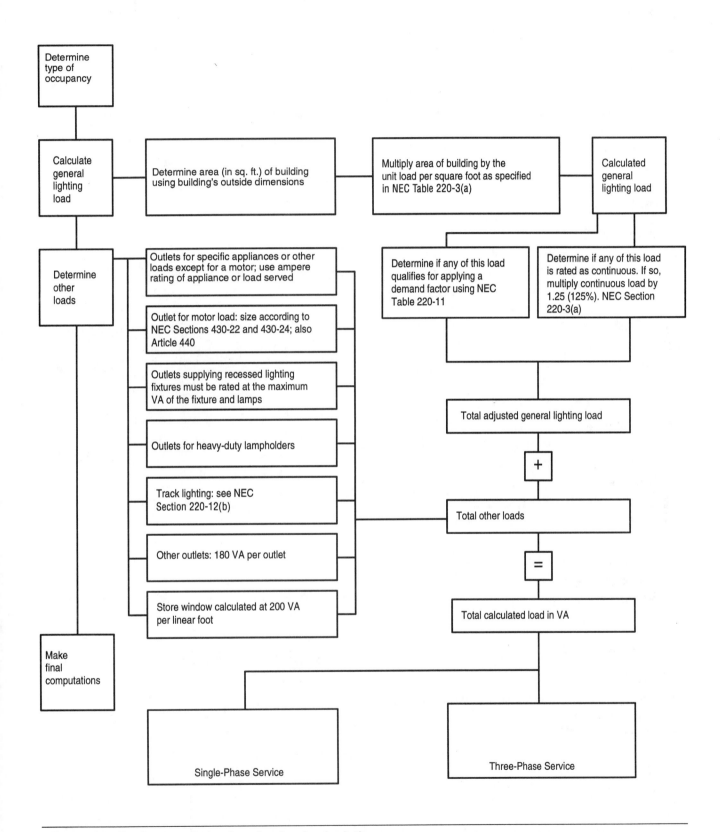

Figure 4-2: Basic steps in performing load calculations

Step 6. Determine the type and VA ratings of any other loads, such as appliance circuits, circuits for track lighting, show windows in stores and like.

Step 7. Add these "other loads" to the total adjusted general lighting load to obtain the total load (in volt-amperes) for the project.

Minimum Service Ratings

Article 230-42(b) specifies that ungrounded service conductors shall have an ampacity of not less than the minimum rating of the disconnecting means specified in Article 230-79. If you look up Article 230-79, it says that the service disconnecting means shall have a rating of not less than the load to be carried:

- (a) One-circuit installations with a single branch circuit, the service connecting means can't be less than 15 amperes.
- (b) Two-circuit installations with not more than two 2-wire branch circuits can't have a disconnecting means of less than 30 amperes.
- (c) For a one-family dwelling, the service disconnecting means must be at least 100 amperes, 3-wire.
- (d) For all others, the service disconnecting means must have a rating of not less than 60 amperes.

Rural Pump House

Figure 4-3 shows the floor plan of a small rural pump house that may be used on some farms to supply water for livestock some distances from the farmhouse or other buildings containing electricity. Therefore, a separate service must be supplied for this pump house.

The total loads for this facility consist of the following:

- Shallow well pump with a nameplate rating of 7.5 amperes at 120 volts

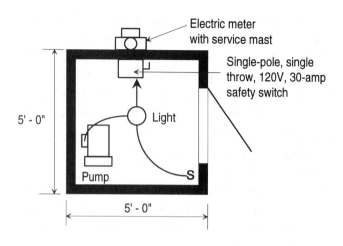

Figure 4-3: Floor plan of small pump house

- One wall-switch controlled lighting fixture containing one 60-watt lamp.

Using strict *NEC* procedures, here's how the load and service size are calculated.

Step 1. *NEC* Table 220-3(a) does not list a small pump house. Therefore, we must use *NEC* Section 220-3(b) — "Other Loads" — for our calculation.

Step 2. Determine total load in volt-amperes. Since the full load of the pump is rated at 7.5 amperes at 120 volts, the total VA is:

$$7.5 \times 120 = 900VA$$

This added to the 60-watt lamp load (60VA) gives a total connected load of 960VA.

Step 3. Determine the size of the service-entrance conductors.

$$\frac{960\,VA}{120V} = 8\ amperes$$

Step 4. Checking *NEC* Article 230-79(a) for service size. It says:

For installations to supply only limited loads of a single branch circuit, the service disconnecting means shall have a rating of not less than 15 amperes.

Since the total connected load for the pump house is only 8 amperes, the single branch circuit feeding the pump and the light fixture need only be No. 14 AWG copper (15 amperes). However a No. 12 AWG copper or No. 10 aluminum conductor would be considered a trade practice, with a 20-ampere overcurrent protection device.

Since 30 amperes is the smallest standard size safety switch, a 120-volt, single-pole, single-throw, 30-ampere with solid neutral safety switch will be selected for the disconnect. Overcurrent protection will be provided by one 15-ampere Type-S plug fuse.

Roadside Vegetable Stand

Another practical application of *NEC* Section 230-79 would be a typical roadside vegetable stand. Again, *NEC* Table 230-3(a) does not list this facility. Therefore, we must once again use *NEC* Section 220-3(b) —"Other Loads" — for our calculation.

Step 1. Determine the lighting load. Two fluorescent fixtures are used to illuminate the 9' x 12' area. Since each fixture contains two 40-watt fluorescent lamps, and the ballast is rated at 10 watts, the total connected load for each fixture is 100VA, or a total of 200VA for both fixtures. However, since this stand will be open late at night during the season, this is considered to be a continuous load, and a factor of 1.25 (125%) must be applied as follows:

$$200VA \times 1.25 = 250VA$$

Step 2. Determine remaining loads. The only other electrical outlets in the stand consist of two receptacles: one furnishes power to a refrigerator with a nameplate full-load rating of 12.2 amperes; the other furnishes power for an electric cash register rated at 300VA and an electronic calculator rated at 200VA.

Step 3. Determine if any of the receptacle loads are continuous. Since the refrigerator will more than likely operate for more than three hours during hot summer months, this load will be rated as continuous. The cash register and electronic calculator, however, will operate intermittently, and are not continuous loads.

Refrigerator load =
$$12.2 \times 120 = 1464VA \times 1.25 = 1830VA$$

Step 4. Determine total connected load with appropriate continuous-load factors applied.

Fluorescent fixtures	250VA
Receptacle for refrigerator	1830VA
Receptacle for other loads	500VA
Total calculated load	2580VA

Step 5. Determine size and rating of service-entrance conductors.

$$\frac{2580VA}{240V} = 10.75 \text{ amperes}$$

Step 6. Check *NEC* Article 230-79(b) for service size. It says:

For installations consisting of not more than two 2-wire branch circuits, the service disconnecting means shall have a rating of not less than 30 amperes.

Consequently, No. 10 copper or No. 8 aluminum is the minimum size allowed for the service entrance conductors, based on Table 310-16.

COMMERCIAL OCCUPANCY CALCULATIONS

Calculating load requirements for commercial occupancies is based on specific *NEC* requirements that relate to the loads present. The basic approach is to separate the loads as described as follows:

- Lighting
- Receptacles
- Motors
- Appliances
- Other special loads

In general, all loads for commercial occupancies should be considered continuous unless specific information is available to the contrary.

Smaller commercial establishments will utilize single-phase, 3-wire services, while the larger projects will almost always use a three-phase, 4-wire service. Furthermore, it is not uncommon to have secondary feeders supplying panelboards which, in turn, supply branch circuits operating at different voltage. In this case, the calculation of the feeder and branch circuits for each voltage is considered separately. The rating of the main service is based on the total load with the load values transformed according to the various circuit voltages if necessary.

Demand factors are also applicable to some commercial establishments. For example, the lighting load in hospitals, hotels and motels, and warehouses, is subject to the application of demand factors. In restaurants and similar establishments, the load of electric cooking equipment is subject to a demand factor if there are more than three cooking units. Optional calculation methods to determine feeder or service loads for schools and similar occupancies are also provided in the *NEC*.

Special occupancies, such as mobile homes and recreational vehicles, require the feeder or service load to be calculated in accordance with specific *NEC* requirements. The service for mobile home parks and recreational vehicle parks is also designed based on specific *NEC* requirements that apply only to those locations. The feeder or service load for receptacles supplying shore power for boats in marinas and boatyards is also specified in the *NEC*.

When transformers are not involved, a relatively simple calculation involving only one voltage results. If step-down transformers are used, the transformer itself must be protected by an overcurrent device which may also protect the circuit conductors in most cases.

Switches and panelboards used for the distribution of electricity within a commercial building are also subject to *NEC* rules. In general, a lighting and appliance panelboard cannot have more than 42 overcurrent protective devices to protect the branch circuits originating at the panelboard. This requirement could affect the number of feeders required when a large number of lighting or appliance circuits are needed.

Retail Stores With Show Windows

The drawing in Figure 4-4 shows a small store building with a show window in front. Note that the storage area has four general-purpose duplex receptacles, while the retail area has 14 wall-mounted duplex receptacles and two floor-mounted receptacles for a total of 16 in this area. These combined with the storage-area receptacles bring the total to 20 general-purpose duplex receptacles that do not supply a continuous load. What are the conductor sizes for the service-entrance if a 120/240-volt single-phase service will be used? How many branch circuits are required if 20-ampere circuits are used throughout?

Step 1. Determine the total square feet in the building by multiplying length by width.

50 × 80 = 4000 sq. ft.

Step 2. Calculate the lighting load using *NEC* Table 220-3(a); according to this table, 3 volt-amperes must be used per square foot.

4000 sq. ft. × 3 volt-amperes = 12,000 volt-amperes

Step 3. Since the lighting load for this store building is expected to continue for three hours

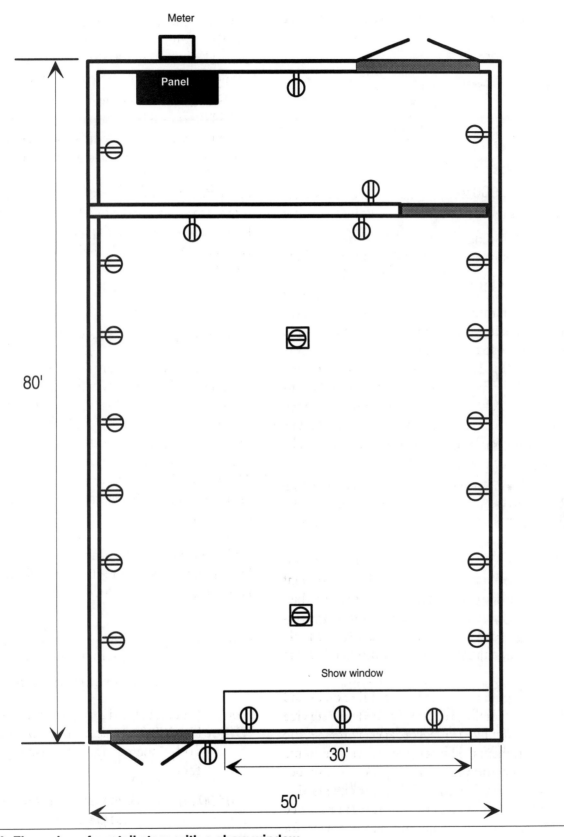

Figure 4-4: Floor plan of a retail store with a show window

or more, this will qualify as a "continuous load" and must therefore be multiplied by a factor of 1.25.

12,000 volt-amperes × 1.25 = 15,000 volt-amperes

Step 4. Determine the number of 20-ampere branch circuits required for the lighting load.

$$\frac{15000\,VA}{120\,V \times 20A} = 6.25 \; or \, 7 \; circuits$$

Step 5. Determine the total volt-amperes for the 20 general-purpose duplex receptacles.

20 × 180VA = 3600 volt-amperes

Step 6. Determine the number of 20-ampere branch circuits required for the general-purpose receptacles.

$$\frac{3600\,VA}{120\,V \times 20A} = 1.5 \; or \, 2 \; circuits$$

Therefore, the branch circuits for lighting and the 20 general-purpose receptacles require a total of nine 20-ampere circuits. These will be installed with conductors with a current-carrying capacity of 20 amperes and protected with an overcurrent device — either fuse or circuit breaker — rated at 20 amperes.

Step 7. Calculate the load for the 30-foot show window on the basis of 200 volt-amperes per linear foot which is considered a continuous load value.

30 × 200 = 6000 volt-amperes

Step 8. Determine the number of 20-ampere branch circuits required to feed the show-window outlets.

$$\frac{6000\,VA}{120\,V \times 20A} = 2.5 \; or \, 3 \; circuits$$

Step 9. Allow one additional 20-ampere circuit for the outside outlet for sign or outline lighting if the store is on the ground floor.

The feeder or service load in this example is simply the sum of the branch-circuit loads. If it is assumed that the sign circuit is to be continuously loaded, its maximum load is calculated as follows:

.8 × 120 volts × 20 amperes = 1920 volt-amperes

However, if the actual load of this circuit is not known, a 1200-volt-ampere load may be used in the calculation.

Step 10. Calculate the total load in volt-amperes.

Lighting load	15,000 volt-amperes
Receptacle load	3,600 volt-amperes
Show window	6,000 volt-amperes
Sign circuit	1,200 volt-amperes (min.)
Total calculated load	25,800 volt-amperes

Step 11. Calculate the service size in amperes.

$$\frac{25,800}{240\,volts} = 107.5 \; amperes$$

Consequently, the service-entrance conductors must be rated for no less than 107.5 amperes at 240 volts. The standard 110-ampere overcurrent protective device would be used. In actual practice, most contractors would install a 125-ampere service with disconnects, panelboard, overcurrent protection, and conductors rated for 125 amperes.

Office Building

A 20,000 square foot office building is served by a 480Y/277-volt, three-phase service. A single-line diagram of the electrical system is shown in Figure 4-5. The building contains the following loads:

- 10,000VA, 208-volt, three-phase sign

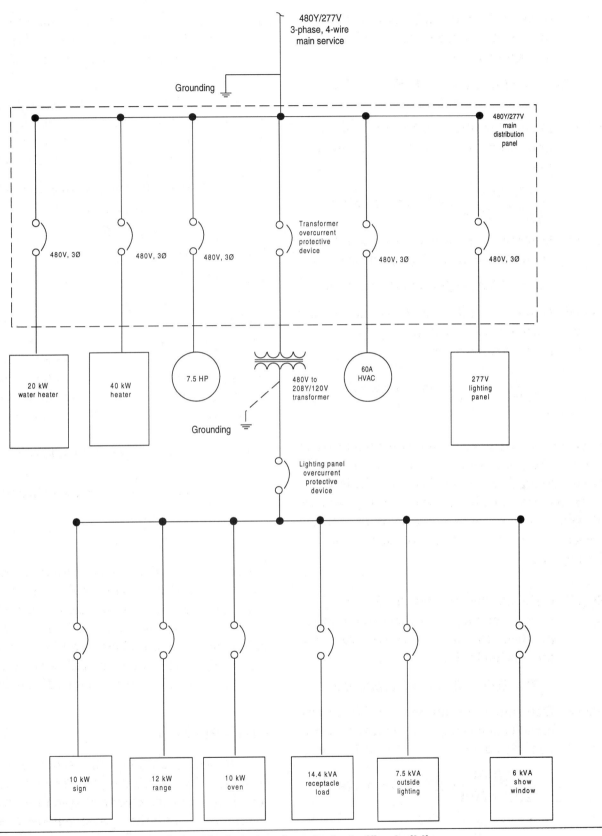

Figure 4-5: Single-line diagram of electrical system for a typical office building

- 100 duplex receptacles supplying continuous load

- 40-foot long show window

- 12-kVA, 208/120-volt, three-phase electric range

- 10-kVA, 208/120-V, three-phase electric oven

- 20-kVA, 480-volt, three-phase water heater

- Seventy-five 150-watt, 120-volt incandescent lighting fixtures

- Two hundred 200-watt, 277-volt fluorescent lighting fixtures

- 7.5 HP, 480-volt, three-phase motor for fan-coil unit

- 40-kVA, 480-volt, three-phase electric heating unit

- 60-ampere, 480-volt, three-phase air-conditioning unit

The ratings of the service epuipment, transformers, feeders, and branch circuits are to be determined, along with the required size of service grounding conductor. Circuit breakers are used to protect each circuit and THHN copper conductors are used throughut the electrical system.

Refer again to the single-line diagram of the electrical system (Figure 4-5). Note that the incoming 3-phase, 4-wire, 480Y/277V main service terminates into a main distribution panel containing six overcurrent protective devices. Since there are only six circuit breakers in this enclosure, no main circuit breaker or disconnect is required as alloved by *NEC* Section 230-71. Five of htese circuit brakers protect feeders and branch circuits to 480/277-volt equipment, while the sixth circuit breaker protects the feeder to a 480-208Y/120-volt transformer. The secondary side of this transformer feeds a 208/120-volt lighting panel with all 120-volt loads balanced and all loads on this panel are continuous. Now let's start at the loads connected to the 208/120-volt panel and perform the required calculations.

Step 1. Calculate the load for the 100 receptacles, remembering that all of these are rated as a continuous load.

100 × 180 × 1.25 = 22,500VA

Step 2. Calculate the load for the show window using 200 VA per linear foot.

200VA × 40 feet = 8,000VA

Step 3. Calculate the load for the outside lighting.

75 lamps × 150VA × 1.25
= 14,062.5 or 14,063VA

Step 4. Calculate the load for the 10 kW sign.

10,000 × 1.25 = 12,500VA

Step 5. Calculate the load for the 12 kW range.

12,000 x 1.25 = 15,000VA

Step 6. Calculate the load for the 10 kW oven.

10,000 x 1.25 = 12,500VA

Step 7. Determine the sum of the loads:

Receptacle load	22,500
Show window	8,000
Outside lighting	14,063
Electric sign	12,500
Electric range	15,000
Electric oven	12,500
Total connected load on subpanel	84,563

Step 8. Determine the feeder rating for the subpanel.

$$\frac{84563}{\sqrt{3} \times 208V} = 235 \ amperes$$

Conductors are sized according to Articles 110-14(c)(1) and (c)(2).

Step 9. Refer to *NEC* Table 310-16 and find that 250 kcmil conductor (rated at 255 amperes) is the closest conductor size that will handle the load. Since a 235- ampere circuit breaker is not standard, one rated at

250-amperes is the next standard size, and is permitted by the *NEC*.

The 208Y/120-volt circuit is a "separately derived system" from the transformer, as defined in the 1999 *NEC*, and is grounded by means of a grounding electrode conductor that must be at least a No. 2 copper conductor based on the 250 kcmil copper feeder conductors.

Step 10.   Size the transformer and the transformer overcurrent protective device. Since the transformer supplying the subpanel is rated for a continuous load, the kVA rating of the transformer need only be the actual connected load. The 1.25 continuous-rating factor does not have to be applied. Therefore, the kVA rating of the transformer need only be 69.3 kVA. A commercially available 75-kVA transformer would be selected.

Step 11.   Select the overcurrent protective device for the transformer.

$$\frac{75,000\,VA}{\sqrt{3} \times 480\,V} = 90.32 \ amperes$$

The maximum setting of the transformer overcurrent protective device is then $1.25 \times 90.32 = 112.89$ amperes; that is, it may not exceed this rating. Since 110 amperes is a standard fuse and circuit-breaker rating, this would be the maximum size normally used. No. 2 THHN copper conductors may be used for the transformer feeder.

Calculations For The Primary Feeder

Calculations for the primary feeder and other 480/277-volt circuits are based on the assumption that all of these loads are continuous, including the water heater for the building. Let's take a look at the lighting first. *NEC* Table 220-3(a) requires a minimum general lighting load for office buildings to be based on 3.5 VA per square foot. Since the building has already been sized at 20,000 sq. ft.,

the minimum VA for lighting may be determined as follows:

$$20,000 \times 3.5 = 70,000 \ VA$$

Since this load is continuous, the 70 kVA figure will have to be multiplied by a factor of 1.25 to obtain the total calculated load:

$$70,000 \times 1.25 = 87,500$$

The actual connected load, however, is as follows:

$$200 \ light \ fixtures \times 200 \ VA \ each \times 1.25 = 50,000 \ VA$$

Since this connected load is less than the *NEC* requirements, it is neglected in the calculation and the 3.5 VA per-square-foot figure is used. Therefore, the total load on the 277-volt lighting panel may be determined as follows:

$$\frac{1.25 \times 70000}{\sqrt{3} \times 480V} = 105.37 \ amperes$$

No. 2 THHN conductors will be used for the feeder supplying the 277-volt lighting panel, and the overcurrent protective device will be rated for 110 amperes — the closest standard fuse or circuit breaker size.

NEC Section 422-13 requires that the feeder or branch circuit supplying a fixed storage-type water heater having a capacity of 120 gallons or less be sized not less than 125 percent of the nameplate rating of the water heater. Consequently, the load for the water heater may be calculated as follows:

$$1.25 \times \frac{20,000\,VA}{\sqrt{3} \times 480\,V} = 30.1 \ amperes$$

The feeder for the water heater will therefore require a No. 8 AWG THHN conductor with an overcurrent protective device of either 35 or 40 amperes.

The electric-heating load is sized in a similar way. *NEC* Section 424-3(b) requires that this load be sized at not less than 125 percent of the total load of the motors and the heaters. Therefore, the load for the electric heating may be calculated as follows:

$$1.25 \times \frac{40,000}{\sqrt{3} \times 480} = 60.2 \text{ amperes}$$

This load requires a conductor size of No. 4 AWG, THHN, with an overcurrent protective device rated at 70 amperes.

An instantaneous trip circuit breaker is selected to protect the 7.5 HP motor circuit. This arrangement is allowed if the breaker is adjustable and is part of an approved controller. A full-load current of 11 amperes requires No. 14 conductors protected by a breaker set as high as ($13 \times 11 =$) 143 amperes. However, in this case, the circuit breaker will probably be set at approximately 70 to 80 amperes.

The air-conditioning circuit must have an ampacity of (1.25×60 amperes =) 75 amperes for which No. 3 THHN conductors are selected from *NEC* Table 310-16. An overcurrent protective de-vice for this circuit cannot be set higher than ($1.75 \times 60 =$) 105 amperes.

Main Service Calculations

When performing the calculations for the main service, assume that all loads are balanced and may be computed in terms of volt-amperes, but are more than likely computed in terms of amperes. Calculation of the loads in amperes also simplifies the selection of the main overcurrent protective device and service conductors. A summary of this calculation is shown in Figure 4-6.

The 84,563 VA, 208Y/120 lighting panel load represents a line load of 101.83 amperes at 480 volts. The 480/277-volt loads are then added to this for a total of 338.3 amperes. The conductors are then selected from *NEC* Table 310-16 which indicates that 500 kcmil THHN conductors have a current-carrying capacity of 380 amperes. An overcurrent protective device of the same amperage may also be selected. The load on the neutral is only 104 amperes, but the neutral conductor

Type of Load	Load in Amperes	Neutral	NEC Reference
208Y/120-volt system	101.83A	0	
277-volt lighting panel	105.37A	104	
Water heater	30.1A	0	
Electric heat (neglected)	0	0	Section 220-21
7.5 HP motor	11.0	0	
Air conditioner	75	0	
25% of largest motor = .25 × 60A	15	0	Section 430-25
Service Load	**338.3**	**104**	

Figure 4-6: Main service load calculations

cannot be smaller than the grounding electrode conductor. Referring to *NEC* Table 250-66, the grounding electrode conductor must be 1/0 copper. Consequently, the neutral cannot be smaller than 1/0 THHN copper conductor.

In actual practice, the service size for this building would probably be increased to 400 amperes to allow for some future development, or possibly larger if the situation called for it; that is, if an expansion of the system was anticipated in the near future.

MOBILE HOME AND RV PARKS

During the past couple of decades, mobile home and trailer parks have increased in number to the point where the NFPA found it necessary to include detailed wiring instructions covering such operations. *NEC* Articles 550 and 551 cover the electrical conductors and the equipment installed within or on mobile homes and recreation vehicles, as well as the means of connecting the units to an electrical supply.

Sizing Electrical Services For Mobile Homes

The electrical service for a mobile home may be installed either underground or overhead, but the point of attachment must be a pole or power pedestal located adjacent to but not mounted on or in the mobile home. The power supply to the mobile home itself is provided by a feeder assembly consisting of not more than one mobile home power cord, rated at 50 amperes or a permanently installed circuit.

The *NEC* gives specific instructions for determining the size of the supply-cord and distribution-panel load for each feeder assembly for each mobile home. The calculations are based on the size of the mobile home, the small-appliance circuits, and other electrical equipment that will be connected to the service.

Lighting loads are computed on the basis of the mobile home's area: width times length (outside dimensions exclusive of coupler) times 3 watts per square foot.

$$Length \times width \times 3 = lighting\ VA$$

Small-appliance loads are computed on the basis of the number of circuits times 1500 watts for each 20-ampere appliance receptacle circuit.

$$Number\ of\ circuits \times 1500 = small\text{-}appliance\ watts$$

The sum of the two loads gives the total load in watts. However, there is a diversity (demand) factor that may be applied to this total in sizing the service and power cord. The first 3000 watts (obtained from the previous calculation) is rated at 100 percent. The remaining watts should be multiplied by a demand factor of 0.35 (35 percent). The total wattage so obtained is divided by the feeder voltage to obtain the service size in amperes.

If other electrical loads are to be used in the mobile home, the nameplate rating of each must be determined and entered in the summation. Therefore, to determine the total load for a mobile home power supply, perform the following calculations:

Step 1. Lighting and small appliance loads, as discussed previously.

Step 2. Nameplate amperes for motors and heater loads, including exhaust fans, air conditioners, and electric heaters. Since air conditioners and heaters will not operate simultaneously, only the larger of the two needs to be included in the total load figures. Multiply the largest motor nameplate rating by 1.25 and add the answer in the calculations.

Step 3. Total of nameplate amperes for any garbage disposals, dishwashers, electric water heaters, clothes dryers, cooking units, and the like. Where there are more than three of these appliances, use 75 percent of the total load.

Step 4. The amperes for free-standing ranges (as distinguished from separate ovens and cooking units) by dividing the values shown in the table in Figure 4-7 on page 76 by the voltage between phases.

Nameplate Rating (VA or Watts)	Use
10,000 or less	80% of rating
10,001 - 12,500	8,000 watts
12,501 - 13,500	8,400 watts
13,501 - 14,500	8,800 watts
14,501 - 15,500	9,200 watts
15,501 - 16,500	9,600 watts
16,501 - 17,500	10,000 watts

Figure 4-7: Power demand factors for free-standing electric ranges in mobile homes

Step 5. The anticipated load if outlets or circuits are provided for other than factory-installed appliances.

To illustrate this procedure for determining the size of the electrical service and power cord for a mobile home, assume that a mobile home is 70 feet by 10 feet; has two portable appliance circuits; a laundry circuit; a 1200-watt air conditioner; a 200-watt, 120-volt exhaust fan; a 1000-watt water heater; and a 6000-watt electric range. The load is calculated as shown in Figure 4-8.

Lighting and Small-Appliance Load	Volt-Amperes
Lighting load: $70 \times 10 \times 3$ W/ft^2	2100
Small-appliance load:	
1500VA $\times$ 2 =	3000
Laundry	1500
First 3000 VA at 100%	3000
Remainder (6600 - 3000) at 35%	1260
Total lighting and small appliance load	4260

Figure 4-8: Summary of mobile home calculations

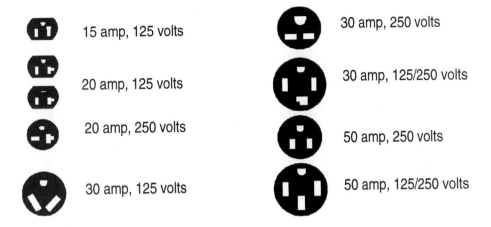

15 amp, 125 volts	30 amp, 250 volts
20 amp, 125 volts	30 amp, 125/250 volts
20 amp, 250 volts	50 amp, 250 volts
30 amp, 125 volts	50 amp, 125/250 volts

Figure 4-9: Receptacle configurations

Based on the higher current for either phase, a 50-ampere power cord should be used to furnish electric power for the mobile home. The service should be rated for a minimum of 50 amperes and be provided with overcurrent protection accordingly.

Types Of Equipment

Weatherproof electrical equipment for mobile homes, mobile home parks, and similar outdoor applications are available from many sources. The electrician should obtain catalogs and study the many types of mobile home utility power outlets and service equipment that are available. Any electrical equipment supplier should have these on hand.

Receptacle configurations used in mobile home and recreation vehicle applications are shown in Figure 4-9. Receptacles 1, 2, 4, and 8 are the most commonly used in mobile home and recreation vehicle parks.

Power units for use in mobile homes and recreation vehicles vary in design with fuse or circuit-breaker protection, attached meter sockets, mounting and junction posts, and special corrosion-resistant finish for ocean-side areas. The latter are used for boats where the outlets are located along the docks and power is leased by the dock owners on

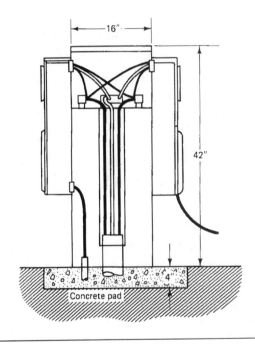

Figure 4-10: Mounting cubicle for mobile homes

a daily basis. In addition to the standard units, manufacturers build equipment to meet special requirements.

Where more than one mobile home is to be fed, a power outlet and service equipment mounting cubicle (section shown in Figure 4-10) is ideal. The busbars in this unit accommodate wire sizes to 600

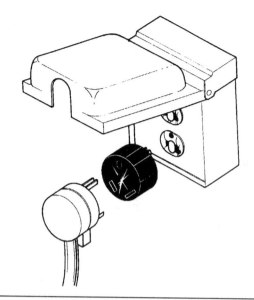

Figure 4-11: Travel-trailer adapter

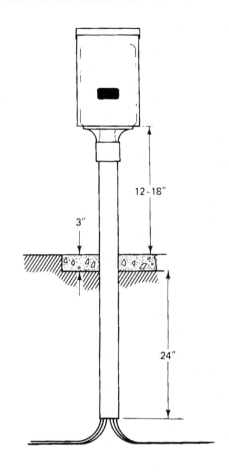

Figure 4-12: Power-outlet mounting post

kcmil for a 400-ampere capacity. Therefore, either two 200-ampere units or four 100-ampere units may be mounted on the cubicle for feeding mobile home units underground. The main service should also enter from underground.

Standard 15- or 20-ampere, 120-volt receptacles may be converted for use with standard 30-ampere, 120-volt travel trailer caps by use of the adapter shown in Figure 4-11.

The power outlet mounting post shown in Figure 4-12 is very popular for travel trailer parks and marinas. A cast-aluminum mounting base is provided on which to mount the power outlet. The installer provides a length of 2-inch rigid conduit of the desired length. This is for underground installations. Note the conductors feeding in and out of the bottom of the conduit.

Before beginning an installation, consult with the local power company for the method of serving the mobile home park and for the location of service entrance poles. Power company regulations vary from area to area, but the power company will furnish and set the pole in most cases. However, the electrician must obtain permission from the power company before performing any work on facilities on the poles.

Once a definite plan has been settled on, the electrician should obtain a piece of ½-inch-thick plywood of sufficient size to hold the service equipment (i.e., a wire trough, meter bases, and weatherproof power outlets as shown in Figure 4-13). This piece of plywood should be primed with paint, and a final coat of wood preservative then applied. Two pieces of 2- by 4-inch timbers are spiked or otherwise secured to the sheet of plywood for reinforcement before the entire assembly is spiked to the pole. The wood backing should be arranged so that the meters will be no more than 5 feet 6 inches above the ground nor less than 4 feet when they are installed.

Up to six power outlets may be fed from one service without need for a disconnect switch to shut down the entire service. However, if more than four power outlets are assembled on one piece of plywood, the arrangement shown in Figure 4-13 will not provide adequate support. Another short pole should be installed at a distance from the service pole so that the sheet of plywood can be secured to both poles (Figure 4-14 on page 80) for added support.

With the plywood backing secured in place, a wire trough, sized according to *NEC* Article 374-5, should be installed at the very top of the board as shown in Figure 4-14. The wire trough (auxiliary gutter) should not contain more than 30 current-carrying conductors nor should the sum of the cross-sectional areas of all contained conductors at any cross section exceed 20 percent of the interior cross-sectional area of the gutter. The auxiliary gutter should be approved for outdoor use.

The meter bases may usually be obtained from the power company but must be installed by the electrician. Once the entire installation is complete and has been inspected, the power company will install the meters. The connections of the meter bases to the wire trough are made with short, rigid conduit nipples using locknuts and bushings. Although straight nipples are often used for these connections, an offset nipple usually does a better job.

Weatherproof fuse or circuit-breaker disconnects are installed directly under the meter bases, again by means of conduit nipples. A weatherproof, 50-ampere mobile home power outlet with overcurrent protection is also used quite often.

The service mast comes next and should consist either of rigid metallic conduit or of EMT with weatherproof fittings. Once installed and secured to the pole with pipe straps, the service-entrance conductors may be pulled into the conduit and out into the wire trough. An approved weatherhead is

then installed on top of the mast, and at least three feet of service conductors should be left for the power company to make their connections.

With the service-entrance conductors in place, meter taps are made to the service conductors in the trough. All such splices and taps made and insulated by approved methods may be located within the gutter when the taps are accessible by means of removable covers or doors on the wire trough. The conductors, including splices and taps, must not fill more than 75 percent of the gutter area (*NEC* Article 374-8a). These taps must leave the gutter opposite their terminal connections, and conductors must not be brought in contact with uninsulated current-carrying parts of opposite polarity.

The taps in the auxiliary gutter go directly to the line side of the meter bases. Once secured, the load

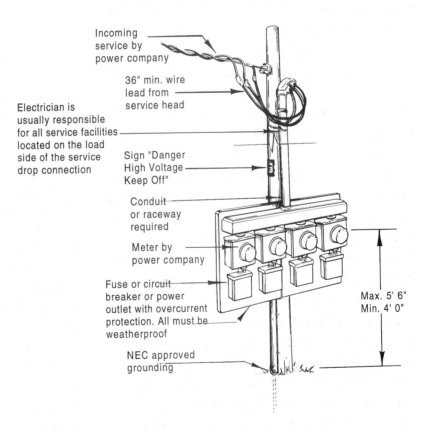

Figure 4-13: Typical mobile home park service

79

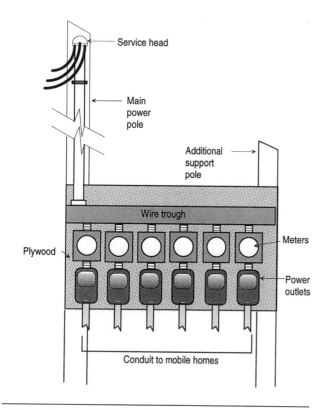

Figure 4-14: Additional pole for added support

side of the meter bases is connected to the disconnects or power outlets. All wiring should be sized according to the *NEC*.

Most water supplies for mobile homes consist of PVC (plastic) water pipe and, therefore, cannot provide an adequate ground for the service equipment. In cases like these, a grounding electrode, such as a ¾-inch by 8-foot ground rod driven in the ground near the service equipment, is used. A piece of bare copper ground wire is connected to the ground rod on one end with an approved ground clamp, and the other end is connected to the neutral wire in the auxiliary gutter. This wire must be sized according to *NEC* Table 250-66.

When all the work is complete, the service installation should be inspected by the local electrical inspector. The power company should then be notified to provide final connection of their lines.

Sizing Electrical Services And Feeders For Parks

A minimum of 75 percent of all recreation vehicle lots with electrical service equipment must be equipped with both a 20-ampere, 125-volt receptacle and a 30-ampere, 125-volt receptacle. The remainder with electrical service equipment may be equipped with only a 20-ampere, 125-volt receptacle.

Since most travel trailers and recreation vehicles built recently are equipped with 30-ampere receptacles, an acceptable arrangement is to install a power pedestal in the corner of four lots so that four different vehicles can utilize the same pedestal. Such an arrangement requires three 30-ampere receptacles and one 20-ampere receptacle to comply with *NEC* Section 551-71. A wiring diagram showing the distribution system of a park electrical system serving 20 recreation vehicle lots is shown in Figure 4-15.

Electric service and feeders must be calculated on the basis of not less than 3600 watts per lot equipped with both 20- and 30-ampere supply facilities and 2400 watts per lot equipped with only 20-ampere supply facilities. The demand factors set forth in *NEC* Table 551-73 are the minimum allowable factors that may be used in calculating load for service conductors and feeders.

Example 1:

Park area A has a capacity of 20 lots served by electricity; park B has 44. Find:

1. The diversity (demand) factor of area A
2. The diversity (demand) factor of area B
3. The total demand of area A
4. The total demand of area B

Solution:

Step 1. The diversity factor is 45 percent, read directly from Table 551-73 of the *NEC*.

Step 2. The diversity factor is 41 percent, read directly from Table 551-73 of the *NEC*.

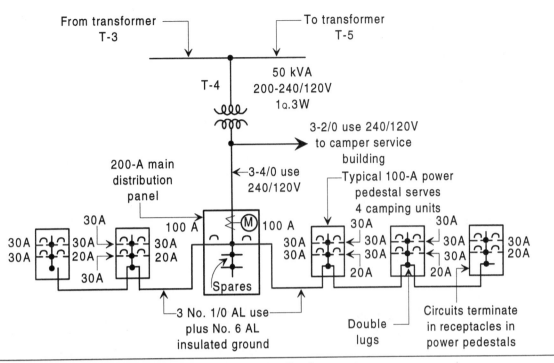

Figure 4-15: Wiring diagram for an RV trailer park distribution system

Step 3. Since each lot is calculated on the basis of 3600 watts, the total demand is $20 \times 3600 \times 0.45 = 32,400$ watts.

Step 4. The total demand is $3600 \times 44 \times 0.41 = 64,944$ watts (total demand).

RESTAURANTS

The load of three or more cooking appliances and other equipment for a commercial kitchen may be reduced in accordance with *NEC* demand factors. This provision would apply to restaurants, bakeries, and similar locations.

For example, a small restaurant is supplied by a 240/120-volt, four-wire, three-phase service. The restaurant has the following loads:

- 1000-square foot area lighted by 120-volt lamps

- Ten duplex receptacles

- 20-ampere, 240-volt, three-phase motor-compressor

- 5-horsepower, 240-volt, three-phase roof ventilation fan protected by an inverse time circuit breaker

- More than six units of kitchen equipment with a total connected load of 80 kilovolt amperes. All units are 240-volt, three-phase equipment

- Two 20-ampere sign circuits

The main service uses Type THHN copper conductors and is wired as shown in Figure 4-16 on page 82. Lighting and receptacle loads contribute 27.9 amperes to phases A and C and 25.8 amperes to the neutral. The 80-kVA kitchen equipment load is subject to the application of a 65 percent demand factor which reduces it to a demand load of $.65 \times 80 \text{ kVA} = 52 \text{ kVA}$. This load requires a minimum ampacity of 125 amperes per phase at 240 volts. The load of the three-phase motors and 25 percent of the largest motor load bring the service load total to 193.1 amperes for phases A and C and 165.2 amperes for phase B.

81

Service Loads	Line A, C	Neutral	Line B
A. 240/120-volt loads Lighting = $\dfrac{1.25 \times 2\,VA\ Per\ sq.\ ft. \times 1000\ sq.\ ft.}{240V} = 10.4$ amperes	10.4	8.3	-0-
Receptacles = $\dfrac{180\,VA \times 10}{240V} = 7.5$ amperes	7.5	7.5	-0-
B. Three-phase loads Kitchen equipment (6 or more units) = $\dfrac{80,000\,W \times .65}{\sqrt{3} \times 240V} = 125$ amperes	125	-0-	125
20-ampere three-phase motor compressor breaker setting = 1.75 x 20A = 35 amperes	20	-0-	20
5-HP three-phase motor Breaker setting = 2.5 x 15.2A = 38 amperes use 40-ampere standard size	15.2	-0-	15.2
C. 25% of largest motor load = .25 x 20A = 5 amperes	5	-0-	5
D. Sign circuit = $\dfrac{1200\,VA}{120V}$ Each	10	10	-0-
Service load	193.1 amp	25.8 amp	165.2 amp

Table 310-16, and Section 250-24(b)(1)	1. Conductors: Use No. 3/0 THHN copper for ungrounded conductors; use No. 4 THHN copper conductor for neutral (neutral based on size of grounding electrode conductor)
Sections 430-63 and 240-6	2. Overcurrent protective device: Phases A and C = 40A (largest motor device) + 10.4A + 7.5A + 125A + 20A + 10A = 212.9 amperes use standard size 200-ampere fuses. Phase B = 40A + 125A + 20A = 185 amperes
Table 250-66	3. Grounding electrode conductor required to be No. 4 copper.

Figure 4-16: Service specifications for a small restaurant

If the phase conductors are three No. 3/0 type THHN copper conductors, the grounding electrode conductor and the neutral conductor must each be at least a No. 4 copper conductor.

The fuses are selected in accordance with the *NEC* rules for motor-feeder protection. The ungrounded conductors, therefore, are protected at 200 amperes each.

SERVICES FOR HOTELS AND MOTELS

The portion of the feeder or service load contributed by general lighting in hotels and motels without provisions for cooking by tenants is subject to the application of demand factors. In addition, the receptacle load in the guest rooms is included in the general lighting load at 2 watts per square foot. The demand factors, however, do not apply to any area where the entire lighting is likely to be used at one time, such as the dining room or a ballroom. All other loads for hotels or motels are calculated as shown previously.

For example, let's determine the 120/240-volt feeder load contributed by general lighting in a 100-unit motel. Each guest room is 240 square feet in area. The general lighting load is:

$$2 \text{ VA/ft}^2 \times 240 \text{ ft}^2/\text{unit} \times 100 \text{ units} = 48,000 \text{ VA}$$

but the reduced lighting load is

First 20,000 at 50%	= 10,000 volt-amperes
Remainder (48,000 - 20,000) at 40%	= 11,200 volt-amperes
Total	= 21,200 volt-amperes

This load would be added to any other loads on the feeder or service to compute the total capacity required.

OPTIONAL CALCULATION FOR SCHOOLS

The *NEC* provides an optional method for determining the feeder or service load of a school equipped with electric space heating or air conditioning, or both. This optional method applies to the building load, not to feeders within the building.

The optional method for schools basically involves determining the total connected load in volt-amperes, converting the load to volt-amperes/square foot, and applying the demand factors from the *NEC* table. If both air-conditioning and electric space-heating loads are present, only the larger of the loads is to be included in the calculation.

Let's take one example. A school building has 200,000 square feet of floor area. The electrical loads are as follows:

- Interior lighting at 3 volt-amperes per square foot
- 300 kVA power load
- 100 kVA water heating load
- 100 kVA cooking load
- 100 kVA miscellaneous loads
- 200 kVA air-conditioning load
- 300 kVA heating load

The service load in volt-amperes is to be determined by the optional calculation method for schools.

The combined *connected* load is 1500 kVA. Based on the 200,000 square feet of floor area, the load per square foot is

$$\frac{1,500,000 \text{ VA}}{200,000 \text{ sq. ft.}} = 7.5 \text{ VA per sq. ft.}$$

The demand factor for the portion of the load up to and including 3 volt-amperes/square foot is 100 percent. The remaining 4.5 volt-amperes/square foot in the example is added at a 75 percent demand factor for a total load of 1,275,000 volt-amperes.

MARINAS AND BOATYARDS

The wiring system for marinas and boatyards is designed by using the same *NEC* rules as for other commercial occupancies except for the application of several special rules dealing primarily with the design of circuits supplying power to boats.

The smallest sized receptacle that may be used to provide shore power for boats is 20 amperes. Each single receptacle that supplies power to boats must be supplied by an individual branch circuit with a rating corresponding to the rating of the receptacle.

The feeder or service ampacity required to supply the receptacles depends on the number of receptacles and their rating, but demand factors may be applied that will reduce the load of five or more receptacles. For example, a feeder supplying ten 30-ampere shore power receptacles in a marina requires a minimum ampacity of

$$10 \times 30A \times .8 = 240 \; \textit{amperes}$$

Although this computed feeder ampacity might seem rather large, this is the minimum required by *NEC* Section 555-6.

Summary

Electrical load calculations are necessary for determining sizes and ratings of conductors, equipment, and overcurrent protection required by the *NEC* to be included in each electrical installation — including the service, feeders, branch circuts and branch-circuit loads. These calculations are necessary in every electrical installation from the smallest roadside vegetable stand to the largest high-rise office building or industrial establishment. Therefore, every electrician must know how to calculate services, feeders and branch circuits for any given installation or siatuation, and also know which *NEC* requirements apply.

Chapter 5
Electric Services

This chapter is designed to cover most electric service applications that will be encountered by electricians working on commercial projects. Detailed installation techniques are also presented for secondary systems up to 600 volts, including the various connections for outdoor distribution and interior dry-type transformers; the latter is used mostly on 480/277V, three-phase, wye-connected systems where lower voltage is required to operate 120/208V outlets and equipment. A review of service grounding requirements is also presented, along with the installation of main distribution panels, multiple disconnects, subpanels, current transformers, and other service equipment.

Electric services can range in size from a small 120-volt, single-phase, 15-ampere service (the minimum allowed by the *National Electrical Code (NEC)* Section 230-79(a) for a roadside vegetable stand to huge industrial installations involving substations dealing with thousands of volts and amperes. Regardless of the size, all electric services are provided for the same purpose: for delivering electrical energy from the supply system to the wiring system on the premises served. Consequently, all establishments containing equipment that utilizes electricity require an electric service.

Figure 5-1 on page 86 shows the basic sections of a typical commercial electric service. In this illustration, note that the high-voltage lines terminate on a power pole near the building being served. A bank of transformers is mounted on the pole to reduce the transmission voltage to a usable level (120/208V, three-phase, wye-connected in this case). The remaining sections are described as follows:

- Service drop: The overhead conductors, through which electrical service is supplied, between the last power company pole and the point of their connection to the service facilities located at the building or other support used for the purpose.

- Service entrance: All components between the point of termination of the overhead service drop or underground service lateral and the building's main disconnecting device, except for metering equipment.

- Service-entrance conductors: The conductors between the point of termination of the overhead service drop or underground service lateral and the main disconnecting device in the building or on the premises.

- Service-entrance equipment: Provides overcurrent protection to the feeder and service conductors, a means of disconnecting the feeders from energized service conductors, and a means of measuring the energy used by the use of metering equipment.

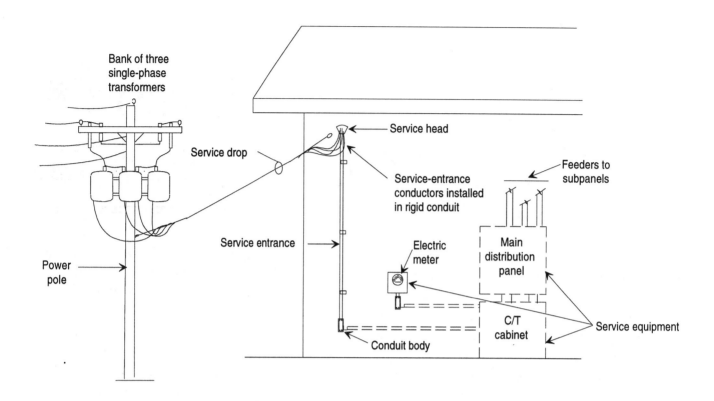

Figure 5-1: Typical three-phase overhead service

When the service conductors to the building are routed underground, as shown in Figure 5-2, these conductors are known as the *service lateral,* defined as follows:

- Service lateral: The underground conductors through which service is supplied between the power company's distribution facilities and the first point of their connection to the building or area service facilities located at the building or other support used for the purpose.

ELECTRICAL DISTRIBUTION

A review of electrical distribution systems is the best foundation for understanding alternating current and the purpose of electric services.

The essential elements of an ac electrical system capable of producing useful power include generating stations, transformers, substations, transmission lines and distribution lines. Figure 5-3 on page 88 shows these elements and their relationships.

Generation

Electricity is produced at the generating plant at voltages varying from 2,400 volts to 13,200 volts. Transformers are also located at the generating plant to step up the voltage to hundreds of thousands of volts for transmission — a kind of wholesale block technique for economically moving massive amounts of power from the generation point to key locations.

Electricity is transported from one part of the system to another by metal conductors, cables made up of many strands of wire. A continuous

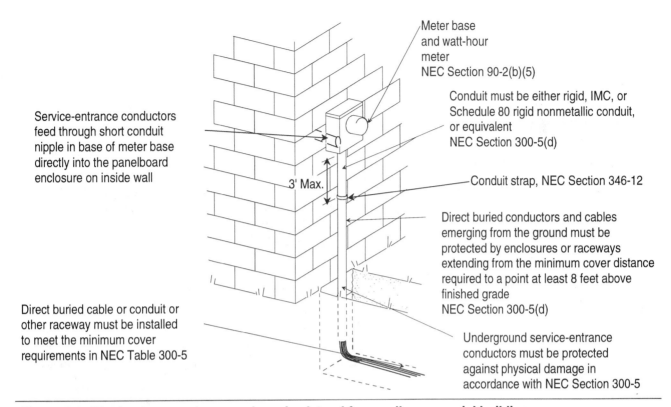

Service-entrance conductors feed through short conduit nipple in base of meter base directly into the panelboard enclosure on inside wall

Meter base and watt-hour meter NEC Section 90-2(b)(5)

Conduit must be either rigid, IMC, or Schedule 80 rigid nonmetallic conduit, or equivalent NEC Section 300-5(d)

3' Max.

Conduit strap, NEC Section 346-12

Direct buried conductors and cables emerging from the ground must be protected by enclosures or raceways extending from the minimum cover distance required to a point at least 8 feet above finished grade NEC Section 300-5(d)

Direct buried cable or conduit or other raceway must be installed to meet the minimum cover requirements in NEC Table 300-5

Underground service-entrance conductors must be protected against physical damage in accordance with NEC Section 300-5

Figure 5-2: Single-phase underground service lateral for small commercial building

system of conductors through which electricity flows is called the distribution circuit.

Transmission

The system for moving high voltage electricity is called the transmission system. Transmission lines are interconnected to form a network of lines. Should one line fail, another will take over the load. Such interconnections provide a reliable system for transporting power from generating plants to communities for use in residential, commercial, and industrial establishments.

Most transmission lines installed by power companies utilize three-phase current — three separate streams of electricity, traveling on separate conductors. This is an efficient way to transport large quantities of electricity. At various points along the way, transformers step down the transmission voltage at facilities known as substations. This is usually the first step in conditioning the voltage for utilization by consumers.

Substations

Substations can be small buildings or fenced-in yards containing switches, transformers, and other electrical equipment and structures. Substations are convenient places to monitor the system and adjust circuits. Devices called regulators, which maintain system voltage as the demand for electricity changes, are also installed in substations. Another device, which momentarily stores energy, is called a capacitor, and is sometimes installed in substations; this device reduces energy losses and improves voltage regulation. Within the substation, rigid tubular or rectangular bars, called busbars or buses, are used as conductors.

At the substation, the transmission voltage is stepped down to voltages below 69,000 volts which feed into the distribution system.

The distribution system delivers electrical energy to user's energy-consuming equipment — such as lighting, motors, machines and appliances from residential to industrial establishments.

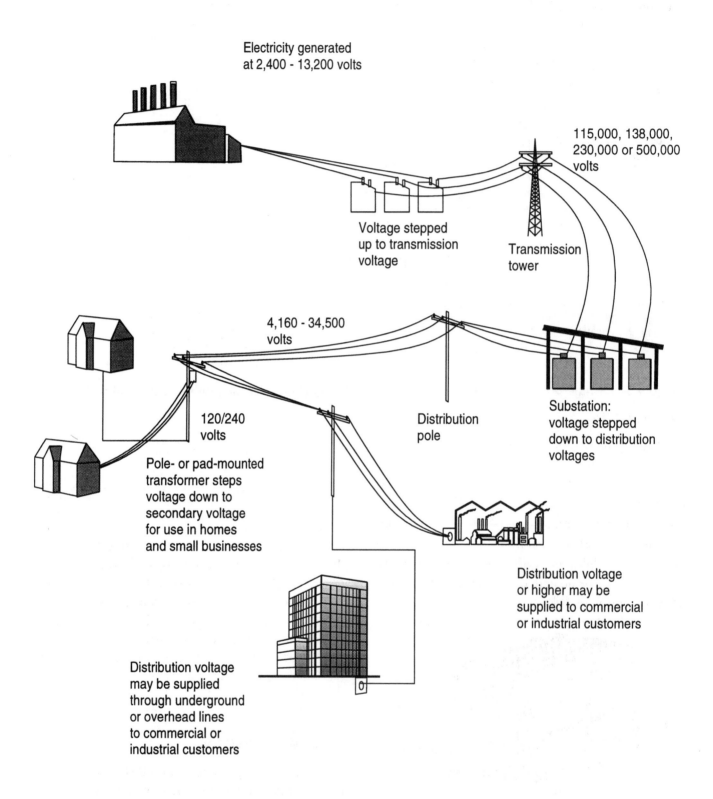

Electricity generated
at 2,400 - 13,200 volts

115,000, 138,000,
230,000 or 500,000
volts

Voltage stepped
up to transmission
voltage

Transmission
tower

4,160 - 34,500
volts

120/240
volts

Pole- or pad-mounted
transformer steps
voltage down to
secondary voltage
for use in homes
and small businesses

Distribution
pole

Substation:
voltage stepped
down to distribution
voltages

Distribution voltage
or higher may be
supplied to commercial
or industrial customers

Distribution voltage
may be supplied
through underground
or overhead lines
to commercial or
industrial customers

Figure 5-3: Parts of a typical electrical distribution system

Conductors called feeders, radiating in all directions from the substation, carry the power from the substation to various distribution centers. At key locations in the distribution system, the voltage is stepped down by transformers to the level needed by the customer. Distribution conductors on the high-voltage side of a transformer are called primary conductors (primaries); those on the low-voltage side are called secondary conductors (secondaries).

Transformers are actually smaller versions of substation distribution transformers that are installed on poles, on concrete pads, or in transformer vaults throughout the distribution system.

Distribution lines carry either three-phase or single-phase current. Single-phase power is normally used for residential and small commercial occupancies, while three-phase power serves most of the other users.

Underground

Most power companies now utilize transmission systems that include both overhead and underground installations. In general, the terms and the devices are the same for both. In the case of the underground system, distribution transformers are installed at or below ground level. Those mounted on concrete slabs are called padmounts (Figure 5-4), while those installed in underground vaults are called submersibles.

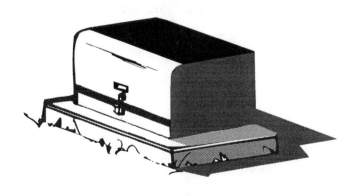

Figure 5-4: Padmount transformer

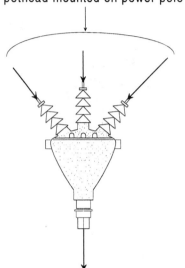

Overhead high-voltage lines connect to these terminals on pothead mounted on power pole

High-voltage cable down power pole to either manhole, padmount or submersible transformer

Figure 5-5: Typical pole-mounted pothead

Buried conductors (cables) are insulated to protect them from soil chemicals and moisture. Many overhead conductors do not require such protective insulation.

When underground transmission or distribution cables terminate and connect with overhead conductors at buses or on the tops of poles, special devices called *potheads* or cable terminators are employed. These devices prevent moisture from entering the insulation of the cable and also serve to separate the conductors sufficiently to prevent arcing between them. The cable installation along the length of the pole is known as the cable riser. See Figure 5-5.

Secondary Systems

From a practical standpoint, most of the electricians' work will be with the power supply on the secondary (usage) side of the transformer.

Two general arrangements of transformers and secondaries are in common use. The first arrangement is the sectional form, in which a unit of load, such as one city street or city block, is served by secondary conductors, with the transformer located in the middle. The second arrangement is the continuous form in which the primary is installed in one long continuous run, with transformers spaced along it at the most suitable points to form the secondaries. As the load grows or shifts, the transformers spaced along it can be moved or rearranged, if desired. In sectional arrangement, such a load can be cared for only by changing to a larger size of transformer or installing an additional unit in the same section.

One of the greatest advantages of the secondary bank is that the starting currents of motors are divided among transformers, reducing voltage drop and also diminishing the resulting lamp flicker at the various outlets.

Power companies all over the United States and Canada are now trying to incorporate networks into their secondary power systems, especially in areas where a high degree of service reliability is necessary. Around cities and industrial applications, most secondary circuits are three-phase, either 120/208V or 277/480V and wye-connected.

Usually, two to four primary feeders are run into the area, and transformers are connected alternately to them. The feeders are interconnected in a grid, or network, so that if any feeder goes out of service the load is still carried by the remaining feeders.

The primary feeders supplying networks are run from substations at the usual primary voltage for the system, such as 4160, 4800, 6900, or 13,200 volts. Higher voltages are practicable if the loads are large enough to warrant them.

Common Power Supplies

The most common power supply used for residential and small commercial applications is the 120/240 V, single-phase service; it is used primarily for light and power, including single-phase motors up to about $7\frac{1}{2}$ horsepower (hp). A diagram of this service is shown in Figure 5-6, depicting all of the major components.

Four-wire wye-connected secondaries (Figure 5-7) and four-wire, delta-connected secondaries (Figure 5-8) are common for commercial applications.

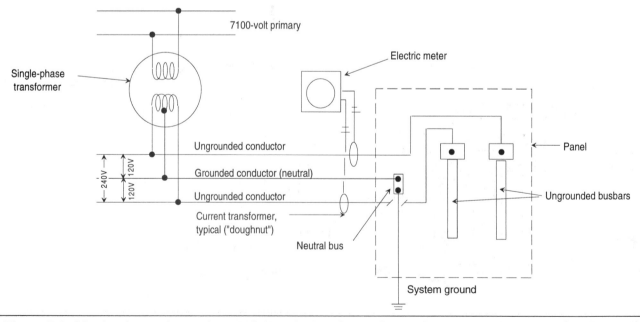

Figure 5-6: Single-phase, three-wire, 120/240-volt electric service

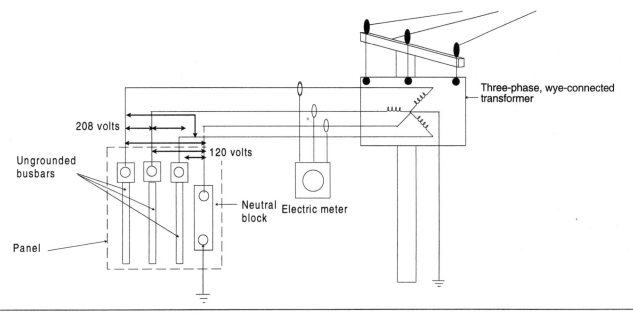

Figure 5-7: Three-phase, four-wire wye-connected secondary

The characteristics of the electric service and the equipment connected to the service must match; also, the characteristics of an electric service will often dictate those for the electrical equipment or vice versa.

Referring again to the three-phase, wye-connected service in Figure 5-7, note that the voltage between any one of the three-phase conductors and the grounded (neutral) conductor is 120 volts. Consequently, one would probably assume that the

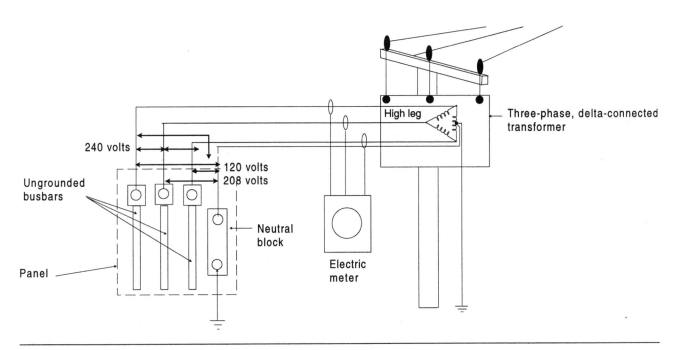

Figure 5-8: Four-wire delta-connected secondary

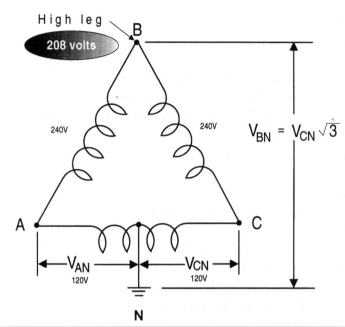

On a 3-phase, 4-wire 120/240V delta-connected system, the midpoint of one phase winding is grounded to provide 120V between phase A and ground; also between phase C and ground. Between phase B and ground, however, the voltage is higher and may be calculated by multiplying the voltage between C and ground (120V) by the square root of 3 or 1.73. Consequently, the voltage between phase B and ground is approximately 208 volts. Thus, the name "high leg."

The NEC requires that conductors connected to the high leg of a 4-wire delta system be color coded with orange insulation or tape.

Figure 5-9: Characteristics of a three-phase delta system

voltage between any two of the phase conductors would be 240 volts. However, this is not the case. When dealing with any three-phase system, a factor — the square root of 3 ($\sqrt{3}$) — enters the picture. Therefore, to find the voltage between any two-phase conductors, multiply the voltage of one phase conductor to ground (120 volts) by the square root of 3. Thus,

$$120 \times \sqrt{3} =$$

$$120 \times 1.73(2050808) = 207.84$$

Rounds off to 208 volts

Therefore, feeder and branch circuits connected to 120/208-volt, three-phase, four-wire systems can supply the following loads:

- 120-volt, single-phase, two-wire.
- 208-volt, single-phase, two-wire.
- 208-volt, three-phase, three-wire.
- 120/208-volt, three-phase, four-wire.

The 120/208-volt, three-phase, four-wire system yields an electrical supply for loads rated 120/208-volt requiring only three wires. These loads usually include such items as HVAC equipment, cooking units, washers and dryers used in high-rise apartments.

Another popular wye-connected system is the three-phase, four-wire, 277/480-volt system.

Feeder and branch circuits connected to the 277/480-volt, three-phase, four-wire systems can supply loads rated:

- 277-volt, single-phase, two-wire.
- 480-volt, single-phase, two-wire
- 480-volt, three-phase, three-wire.
- 277/480-volt, three-phase, four-wire.

The delta-connected system in Figure 5-8 operates a little differently. While the wye-connected system is formed by connecting one terminal from three equal voltage transformer windings together to make a common terminal, the delta-connected system has its windings connected in series, forming a triangle or the Greek delta symbol Δ. Note in Figure 5-9 that a center-tap terminal is used on one winding to ground the system. On a 240/120-volt system, there are 120 volts between the center-tap terminal and each ungrounded terminal on either side. There are 240 volts across the full winding of each phase.

Refer again to Figure 5-9. Note that a high leg results at point "B." This is known in the trade as the "high leg," "red leg," or "wild leg." This high leg has a higher voltage to ground than the other two phases. The voltage of the high leg can be determined by multiplying the voltage to ground of either of the other two legs by the square root of 3. Therefore, if the voltage between phase A to ground is 120 volts, the voltage between phase B to ground may be determined as follows:

$$120 \times \sqrt{3} \quad = \quad 207.84 = 208 \ volts$$

From this, it should be obvious that no single-pole breakers should be connected to the high leg of a center-tapped, four-wire delta-connected system. In fact, *NEC* Section 215-8 requires that the

phase busbar or conductor having the higher voltage to ground to be permanently marked by an outer finish that is orange in color. By doing so, this will prevent future workers from connecting 120-volt single-phase loads to this high leg which will probably result in damaging any equipment connected to the circuit. Remember the color *orange*; no 120-volt loads are to be connected to this phase.

SERVICE COMPONENTS

Electricians working mostly on commercial projects will be involved with the service installation from the power company's point of attachment to the building to the service equipment, including all wiring and components in between, with the possible exception of the electric meter.

To understand the function of each part of an electric service, let's take an actual installation — a commercial retail store (Figure 5-10). Let's assume that you are in charge of the project and this is your first day on the job. The contractor's super-

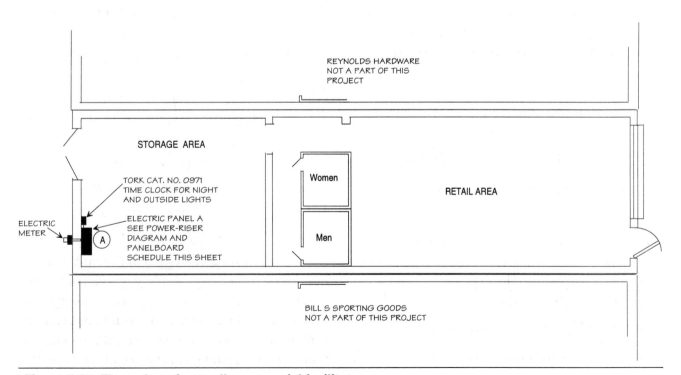

Figure 5-10: Floor plan of a small commercial facility

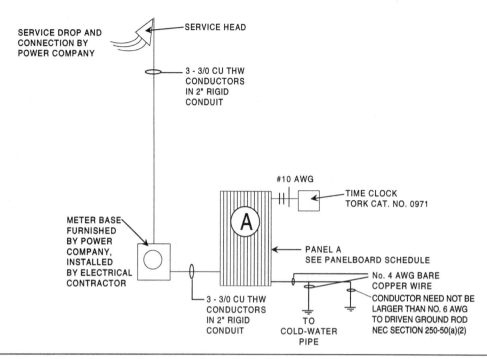

SERVICE DROP AND
CONNECTION BY
POWER COMPANY

SERVICE HEAD

3 - 3/0 CU THW
CONDUCTORS
IN 2" RIGID
CONDUIT

#10 AWG

TIME CLOCK
TORK CAT. NO. 0971

METER BASE
FURNISHED
BY POWER
COMPANY,
INSTALLED
BY ELECTRICAL
CONTRACTOR

PANEL A
SEE PANELBOARD SCHEDULE

No. 4 AWG BARE
COPPER WIRE

3 - 3/0 CU THW
CONDUCTORS
IN 2" RIGID
CONDUIT

TO
COLD-WATER
PIPE

CONDUCTOR NEED NOT BE
LARGER THAN NO. 6 AWG
TO DRIVEN GROUND ROD
NEC SECTION 250-50(a)(2)

Figure 5-11: Power-riser diagram corresponding to the floor plan in Figure 5-10

intendent stops by the project site and hands you a set of working drawings and written specifications. It is up to you to determine how the service is to be installed. Furthermore, you will be required to compile a material list and order all necessary items to complete the service installation.

This particular project consists of a rectangular building that is a part of a shopping center complex. The concrete block walls have been erected and the building is under roof. The concrete floor will not be poured until all electrical and plumbing work has been installed. However, the permanent electric service is to be installed immediately to provide temporary power for the workers. The remaining wiring in the building will be installed later.

Consulting The Construction Documents

The first order of business is to consult the working drawings and then perhaps read the appropriate sections in the written specifications. In doing so, the floor plan of the building appears as shown in Figure 5-10. Note that the standard panel

symbol is used (a solid rectangle) to indicate the location of Panel "A" — the only power panel used on this project. The panel symbol indicates that the panel is to be surface mounted on the inside rear wall of the building in the storage area. The electric meter, and a time clock for controlling night and outside lighting are also shown on this floor plan; the meter is installed on the outside rear wall, while the time clock is installed on the inside rear wall, next to panel "A."

Notes and call-out arrows on this floor plan refer to a power-riser diagram and also a panelboard schedule on the same drawing sheet; these appear in Figures 5-11 and 5-12 respectively. This drawing sheet — showing the floor plan, power-riser diagram, and panelboard schedule — provides most of the required information so that the service can be installed to the project specifications. In most cases, electrical workers are not required to design electrical sytems; rather, they are required to interpret engineer's designs. Consequently, panelboard schedules will vary with each designer. However, once you have a "feel" for interpreting electrical working drawings, you should have little

PANEL No.	CABINET TYPE	PANEL MAINS			BRANCHES					ITEMS FED OR REMARKS
		AMPS	VOLTS	PHASE	1P	2P	3P	PROT.	FRAME	
A	SURFACE	200A	120/240	1φ,3-W	12	-	-	20A	70A	LTS., RECEPTS, W.C.
SQUARE "D" TYPE NQO 200A MAIN CIRCUIT BREAKER					-	1	-	60A	100A	CONDENSING UNIT
					-	1	-	30A	70A	WATER HEATER
					-	1	-	20A		AIR-HANDLING UNIT
					-	2	-	20A		TOILET HEATERS
					8	-	-	-	↓	PROVISIONS ONLY

Figure 5-12: Panelboard schedule corresponding to the floor plan in Figure 5-10

difficulty in "reading" any schedules that will be encountered.

The written specifications should also be read just to make certain that no conflicts exist, and to further verify the information found on the working drawings. A sample specification appears in Figure 5-13, beginning on page 96.

From the information obtained from the drawings and specifications, we know that the service for this project is single-phase, three-wire, 120/240-volt, 200 amperes. The main panel (panel "A") is a surface-mounted Square D Type NQO (or equivalent) with a 200-ampere main circuit breaker. Furthermore, we can determine the number of spaces required in the panel by totaling the number of circuit breakers listed in the panelboard schedule as follows:

12	1-pole, 20A breakers =	12 spaces
1	2-pole, 60A breaker =	2 spaces
1	2-pole, 30A breaker =	2 spaces
3	2-pole, 20A breakers =	6 spaces
8	Provisions only =	8 spaces
	Total	**30 spaces**

Therefore, a surface-mounted panel — Square D and type NQO with 200 ampere main circuit breaker and provisions for 30 spaces — can be ordered. The required circuit breakers should also be ordered and installed at the same time. This will meet with project specifications.

Service Head

Referring again to Figure 5-11, let's start at the top of the service riser. The first item shown is the service head, sometimes called "weatherhead." Since the service raceway in our example consists of 2-inch rigid conduit, *NEC* Section 230-54(a) requires this conduit (raceway) to be equipped with a raintight service head at the point of connection to the service-drop conductors.

A service head (Figure 5-14 on page 98) is a fitting that prevents water from entering the service raceway. This is accomplished by bending the service conductors (contained in the raceway) downward as they exit from the service head so that any water or moisture will drip from the outside conductors before entering the service head. These conductors are also protected by a plastic or fiber strain insulator or bushing —

PANELBOARDS—CIRCUIT BREAKER

A. GENERAL:

Furnish and install circuit-breaker panelboards as indicated in the panelboard schedule and where shown on the drawings. The panelboard shall be dead front safety type equipped with molded case circuit breakers and shall be the type as listed in the panelboard schedule: Service entrance panelboards shall include a full capacity box bonding strap and be approved for service entrance. The acceptable manufacturers of the panelboards are ITE, General Electric, Cutler-Hammer, and Square D, provided that they are fully equal to the type listed on the drawings. The panelboard shall be listed by Underwriters' Laboratories and bear the UL Label.

B. CIRCUIT BREAKERS:

Provide molded case circuit breakers of frame, trip rating and interrupting capacity as shown on the schedule. Also, provide the number of spaces for future circuit breakers as shown in the schedule. The circuit breakers shall be quick-make, quick-break, thermal-magnetic, trip indicating and have common trip on all multipole breakers with internal tie mechanism.

C. WIRING TERMINALS:

Terminals for feeder conductors to the panelboard mains and neutral shall be suitable for the type of conductor specified. Terminals for branch circuit wiring, both breaker and neutral, shall be suitable for the type of conductor specified.

D. CABINETS AND FRONTS:

The panelboard bus assembly shall be enclosed in a steel cabinet. The size of the wiring gutters and gauge of steel shall be in accordance with NEMA Standards. The box shall be fabricated from galvanized steel or equivalent rust resistant steel. Fronts shall include door and have flush, brushed stainless steel, spring-loaded door pulls. The flush lock shall not protrude beyond the front of the door. All panelboard locks shall be keyed alike. Fronts shall not be removable with door in the locked position.

Figure 5-13: Sample panelboard specifications

E. DIRECTORY:

On the inside of the door of each cabinet, provide a typewritten directory which will indicate the location of the equipment or outlets supplied by each circuit. The directory shall be mounted in a metal frame with a nonbreakable transparent cover. The panelboard designation shall be typed on the directory card and panel designation stenciled in $1\frac{1}{2}$" high letters on the inside of the door.

F. PANELBOARD INSTALLATION:

(1) Before installing panelboards check all of the architectural drawings for possible conflict of space and adjust the location of the panelboard to prevent such conflict with other items.

(2) When the panelboard is recessed into a wall serving an area with accessible ceiling space, provide and install an empty conduit system for future wiring. All $1\frac{1}{4}$" conduit shall be stubbed into the ceiling space above the panelboard and under the panelboard if such accessible ceiling space exists.

(3) The panelboards shall be mounted in accordance with Article 373 of the NEC. The Electrical Contractor shall furnish all material for mounting the panelboards.

Figure 5-13: Sample panelboard specifications (Cont.)

placed at the entrance of the service head — to separate the service conductors as required by *NEC* Section 230-54(e). Two types of service heads are in common use: one type has internal threads that enable the service head to be screwed directly onto the conduit; the other type utilizes a clamp with retaining screws. In this latter type, the service head is placed on top of the service raceways and the clamp tightened with the retaining screws.

Further protection from water and moisture is provided by drip loops (*NEC* Section 230-54(f). Service heads are required to be located above the service-drop attachment. Drip loops are then formed where the service-drop conductors are connected to the service conductors and these drip loops must be located below the service head.

Figure 5-14 shows how drip loops prevent water from entering the service raceway; that is, water will not flow uphill into the service head, so the water drips from the conductors at the lowest point of the drip loop.

The service-entrance conductors must have a minimum length of $3\frac{1}{2}$ feet after they leave the service head. This is to ensure a good drip loop and to give adequate length for splicing onto the service drop.

SERVICE-ENTRANCE CONDUCTORS

The size 3/0 AWG, Type THW conductors shown in the power-riser diagram in Figure 5-11 are service-entrance conductors. These conductors

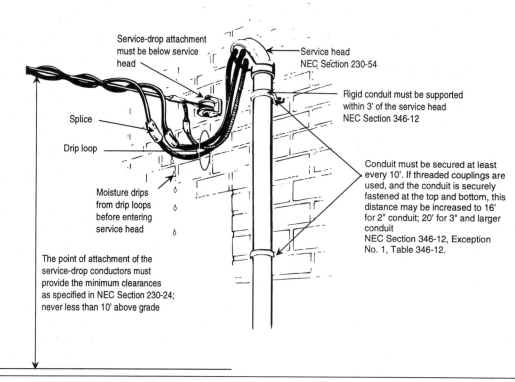

Service-drop attachment must be below service head

Service head NEC Section 230-54

Rigid conduit must be supported within 3' of the service head NEC Section 346-12

Splice

Drip loop

Conduit must be secured at least every 10'. If threaded couplings are used, and the conduit is securely fastened at the top and bottom, this distance may be increased to 16' for 2" conduit; 20' for 3" and larger conduit NEC Section 346-12, Exception No. 1, Table 346-12.

Moisture drips from drip loops before entering service head

The point of attachment of the service-drop conductors must provide the minimum clearances as specified in NEC Section 230-24; never less than 10' above grade

Figure 5-14: Service head and related components

are run from the main disconnect breaker, through the meter, to the service head, and terminate with splices onto the service drop. The conductors must not be spliced at any place between these points except for the following:

- Clamped or bolted connections in metering equipment.

- Where service-entrance conductors are tapped to supply two to six disconnecting means that are grouped at a common location.

- Where service conductors are extended from a service drop to an outside meter location and returned to connect to the service-entrance conductors of an existing installation.

- Where the service-entrance conductors consist of busway, connections are

permitted as required to assemble the various section and fittings.

Service conductors are normally installed in two different ways: in a raceway system or in a cable assembly. In our sample, the conductors are installed in 2-inch rigid conduit and extend from the service head down to the threaded weatherproof hub on top of the meter base. The *NEC* permits the conductors to be spliced at this point; that is, connected to the bolted terminals on the meter base. However, no splices are permitted from the service head to the meter base.

Service Equipment

Equipment and components falling under this heading include the main disconnect switch or breaker, circuit breakers, fuses, and other necessary items to meter, control, and cut off the power supply.

Metering Equipment

An electric meter is used by the power company to determine the amount of electrical energy consumed by the customer.

Energy is the product of power (kilowatts) and time (hours). The type of meter connected to most residential and small commercial occupancies provides a reading in kilowatt-hours. For example, if the meter reads a usage of 500 watts for a period of six hours, it would register ($.5 \times 6 =$) 3 kilowatt-hours.

There are several different types of metering devices in use. The type used on our sample building is known as the feed-through type. This type of meter is used mostly for services up to 200 amperes, although feed-through meters up to 400 amperes are not uncommon in many locations. Services rated above 400 amperes will almost always use separate current transformers enclosed in a current-transformer cabinet (CT cabinet). Current transformers are discussed in greater detail later in this chapter.

A typical watt-hour meter consists of a combination of coils, conductors, and gears — all encased in a housing as shown in Figure 5-15. The coils are constructed on the same principle as a split-phase induction motor, in that the stationary current coil and the voltage coil are placed so that they produce a rotating magnetic field. The disc near the center of the meter is exposed to the rotating magnetic field. The torque applied to the disc is proportional to the power in the circuit, and the braking action of the eddy currents in the disc makes the speed of the rotation proportional to the rate at which the power is consumed. The disc, through a train of gears, moves the pointers on the register dials to record the amount of power used directly in kilowatt hours (kWh).

Most watt-hour meters utilize five dials; again, see Figure 5-15. The dial farthest to the right on the meter counts the kilowatt hours singly. The second dial from the right counts by tens, the third dial by hundreds, the fourth dial from the right by thousands, and the left-hand dial by ten-thousands. The dials may seem a little strange at first, but are actually very simple to read. The number which the dial has passed is the reading. For example, look at the dial on the very left in Figure 5-15. Note that the pointer is about halfway between the number 2 and the number 3. Since it has passed the number 2, but has not yet reached number 3, the dial reading is "2." The same is true of the second dial from the left; that is the pointer is between the number 2 and 3. Consequently, the reading of this dial is also "2." Following this same procedure, the reading in the illustration (Figure 5-15) is 2, 2, 1, 7, 9, or 22,179 kilowatt-hours.

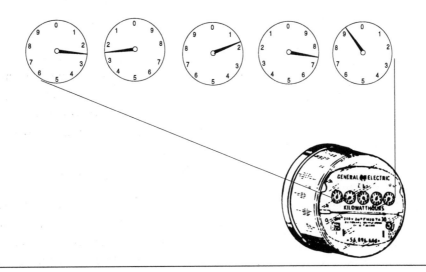

Figure 5-15: A typical watt-hour meter

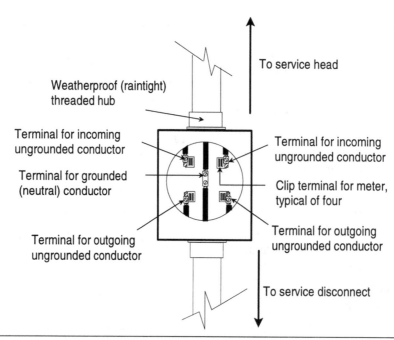

Figure 5-16: Arrangement of conductors in a typical single-phase meter base

Although knowing how to read an electric meter is interesting, most electricians will be involved only with installing the meter base and making the connections therein. Once these connections are made and inspected, the local power company will install and seal the meter.

Meter bases should be installed securely with anchors sufficient to hold the weight of the meter as well as the raceway system resting upon the meter base. In our example, this base must support the 2-inch conduit, service head, and the copper conductors. Although the conduit will be supported with conduit straps, most of the weight will rest upon the meter base; the straps are used mainly to keep the conduit from moving sideways in this example.

Most single-phase, feed-through meter bases are arranged as shown in Figure 5-16. The ungrounded service conductors from the service drop terminate in the top terminals. These conductors are once again picked up from the bottom terminals. Clips are provided on these terminals to clamp-in the meter itself, allowing current from the ungrounded conductors to pass through the meter for a reading. Since the grounded conductor

(neutral) is not metered, one terminal is provided for both the incoming and outgoing conductors.

Service Disconnecting Means

A service disconnecting means is a device that enables the electric service from the power company to be disconnected from the building or premises. Several different configurations are possible. In our sample building, a single panel with a 200-ampere main circuit breaker acts as the disconnecting means. This arrangement and other possible service configurations for this same project appear in Figure 5-17.

Service switches, load centers, or main distribution panelboards are normally installed at a point immediately where the service-entrance conductors enter the building. Branch circuits and feeder panelboards (when required in addition to the main service panel) are usually grouped together at one or more centralized locations to keep the length of the feeders and branch-circuit conductors at a practical minimum of operating efficiency (minimum voltage drop) and to lower the initial installation costs.

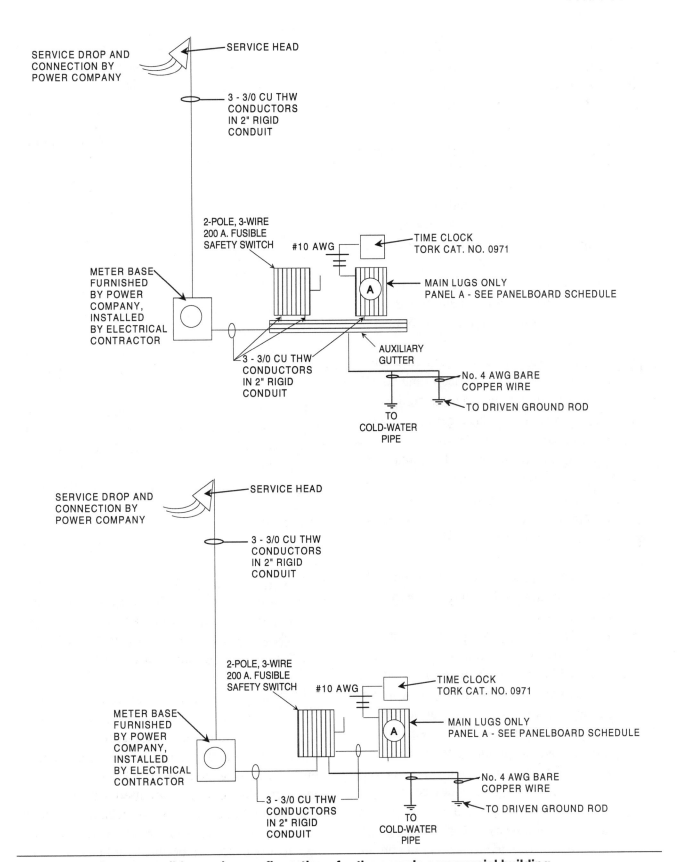

Figure 5-17: Some possible service configurations for the sample commercial building

Distribution equipment and panelboards are generally intended to carry and control electrical current, but are not intended to dissipate or utilize energy. Eight basic factors influence the selection of distribution equipment:

1. Codes and Standards: Suitability for installation and use, in conformity with the provisions of the *NEC* and all local codes, must be considered. Suitability of equipment may be evidenced by listing or labeling.

2. Mechanical Protection: Mechanical strength and durability, including the adequacy of the protection provided must be considered.

3. Wiring Space: Wire bending and connection space is provided according to UL standards in all distribution equipment. When unusual wire arrangements or connections are to be made, then extra wire bending space, gutters, and terminal cabinets should be investigated for use.

4. Electrical Insulation: All distribution equipment carries labels showing the maximum voltage level that should be applied. The electrical supply voltage should always be equal to, or less than, the voltage rating of distribution equipment; never more.

5. Heat: Heating effects under normal conditions of use and also under abnormal conditions likely to arise in service must be constantly considered. Ambient heat conditions, as well as wire insulation ratings, along with the heat rise of the equipment, must be evaluated during selection.

6. Arcing Effects: The normal arcing effects of overcurrent protective devices must be considered when the application is in or near combustible materials or vapors. Enclosures are selected to prevent or contain fires created by normal operation of the equipment. Selected locations of equipment must be made when another location may cause a hazardous condition.

7. Classification: Classification according to type, size, voltage, current capacity, interrupting capacity and specific use must be considered when selecting distribution equipment. Loads may be continuous or noncontinuous and the demand factor must be determined before distribution equipment can be selected.

8. Personal Protection: Other factors that contribute to the practical safeguarding of a person using or likely to come in contact with the equipment must be considered. The equipment selected for use by only qualified persons may be different from equipment used or applied where unqualified people may come in contact with it.

In electrical wiring installations, overcurrent protective devices, consisting of fuses or circuit breakers, are sometimes factory assembled in a metal cabinet, the entire assembly commonly being called a *panelboard*. At other times, the panelboards will be delivered unassembled, consisting of an enclosure ("can"), the interior busbars, and the trim. Circuit breakers are then installed as the project dictates.

Sometimes the main service-disconnecting means will be made up on the job by assembling individually enclosed fused switches or circuit breakers on a length of metal auxiliary gutter as shown in the top view in Figure 5-17.

Grounding

NEC Article 250 covers general requirements for grounding and bonding electric services. In general, the *NEC* requires a premises wiring system, supplied by an alternating-current service to be grounded by a grounding electrode conductor connected to a grounding electrode. The grounding electrode conductor must be bonded to the grounded service conductor (neutral) at any accessible point from the load end of the service drop or

Size of Largest Service-Entrance Conductor or Equivalent Area for Parallel Conductors		Size of Grounding Electrode Conductor	
Copper	Aluminum or Copper-Clad Aluminum	Copper	Aluminum or Copper-Clad Aluminum
2 or smaller	1/0 or smaller	8	6
1 or 1/0	2/0 or 3/0	6	4
2/0 or 3/0	4/0 or 250 kcmil	4	2
Over 3/0 through 350 kcmil	Over 250 kcmil through 500 kcmil	2	1/0
Over 350 kcmil through 600 kcmil	Over 500 kcmil through 900 kcmil	1/0	3/0
Over 600 kcmil through 1100 kcmil	Over 900 kcmil through 1750 kcmil	2/0	4/0
Over 1100 kcmil	Over 1750 kcmil	3/0	250 kcmil

Figure 5-18: Sizes of grounding electrode conductor for ac systems (*NEC* Table 250-66)

service lateral to, and including, the terminal bus to which the grounded service conductor is connected at the service disconnecting means. A grounding connection must not be made to any grounded circuit conductor on the load side of the service disconnecting means.

Most applications require the grounded service conductor to be bonded to at least two grounding electrodes according to *NEC* Section 250-50.

The table in Figure 5-18 gives the required sizes of grounding conductors for various sizes of electric services. Referring to this table, since the service in our sample building is 200 amperes, requiring 3/0 THW copper conductors, a #4 AWG copper or #2 AWG aluminum wire must be used for the grounding electrode conductor.

THREE-PHASE SERVICES

Most services encountered on commercial and industrial projects will consist of three-phase systems with voltage ratings as low as 120/208 volts to as high as perhaps 4160 volts or more.

In single-phase current, only one voltage curve is generated, while three-phase indicates that there are three voltage curves present on the system simultaneously. There is very little difference between installing a single-phase service (just de-

scribed) and a three-phase service. The main difference being an extra service conductor for a three-phase, four-wire system, requiring that the service head have four openings instead of three. Of course, the panelboard will also have to be arranged for a three-phase system — requiring an extra ungrounded busbar. Other than these changes, the installation process is essentially the same as described for the single-phase system.

To illustrate, Figure 5-19 on page 104 shows a power-riser diagram of a 1200-ampere, three-phase, wye-connected service. Note that the service lateral consists of three $3\frac{1}{2}$-inch empty conduits to be installed from the C/T cabinet to 4 feet beyond the building's footings. These are provided for the power company's service lateral.

The service conductors continue through the C/T cabinet and terminate in an auxiliary gutter. Three service-equipment taps are made in this gutter to feed the following:

- An 800-ampere fusible safety switch which, in turn, feeds a motor-control center.
- One 200-ampere power panel with main circuit breaker.
- One 200-ampere lighting panel with main circuit breaker.

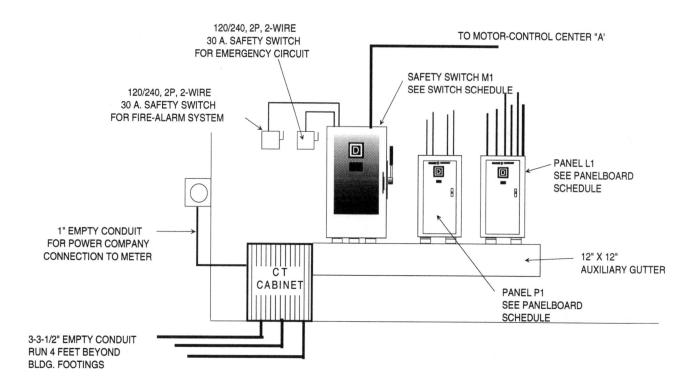

120/240, 2P, 2-WIRE
30 A. SAFETY SWITCH
FOR EMERGENCY CIRCUIT

TO MOTOR-CONTROL CENTER "A'

120/240, 2P, 2-WIRE
30 A. SAFETY SWITCH
FOR FIRE-ALARM SYSTEM

SAFETY SWITCH M1
SEE SWITCH SCHEDULE

PANEL L1
SEE PANELBOARD
SCHEDULE

1" EMPTY CONDUIT
FOR POWER COMPANY
CONNECTION TO METER

CT
CABINET

12" X 12"
AUXILIARY GUTTER

PANEL P1
SEE PANELBOARD
SCHEDULE

3-3-1/2" EMPTY CONDUIT
RUN 4 FEET BEYOND
BLDG. FOOTINGS

Figure 5-19: Power-riser diagram for a three-phase, four-wire, 1200-ampere service

Current Transformers

Meters used by power companies to record the amount of current used by customers usually respond to a current which varies from zero to five amperes. To respond to the actual current of the service, each meter is provided with current transformers. If the peak demand of the service is 100 amperes, a 100:5 current transformer is used. If the peak current demand is expected to be 200 amperes, a 200:5 current transformer is used.

Services above 400 amperes usually utilize a group of current transformers — one for each ungrounded conductor in the service. There are two basic types of current transformers: the busbar type and the "doughnut" type. These latter current transformers encircle the ungrounded conductors in the system to read the current flow, much the same as a clamp-on ammeter. See Figure 5-20.

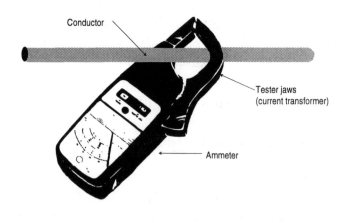

Conductor

Tester jaws
(current transformer)

Ammeter

Figure 5-20: A clamp-on ammeter operates on the same principle as service current transformers

They are sometimes called "doughnuts" due to their appearance. The busbar type current transformer has

104

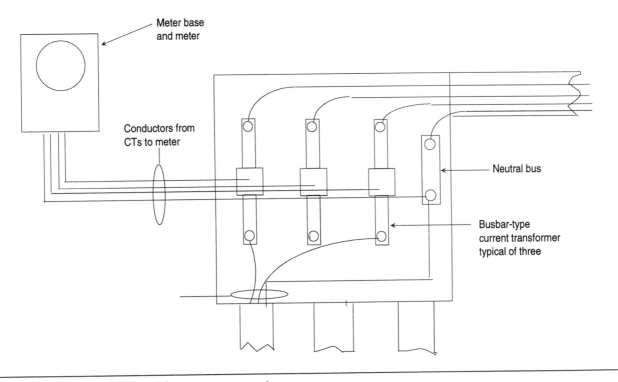

Meter base and meter

Conductors from CTs to meter

Neutral bus

Busbar-type current transformer typical of three

Figure 5-21: Typical C/T-cabinet arrangement

each transformer connected in series with a busbar, and does not encircle the conductor.

Current transformers are normally enclosed in an enclosure called a "C/T cabinet." The letters "C/T," of course stand for "current transformer." Figure 5-21 shows a typical C/T cabinet with current transformers and their related wiring. In some rare cases, the current transformers may be mounted exposed on overhead conductors, but this is more the exception than the rule.

Power companies have different requirements for sizes of C/T cabinets but the dimensions shown in Figure 5-22 on page 106 are typical for several service sizes.

Power companies also have different specifications for the location and wiring of C/Ts and C/T cabinets, depending on the locale. The following are the requirements of one power company. However, always check locally to verify their requirements.

1. The meter base and meter may be located on either side or top of the current transformer cabinet, or it may be located at a distance away, if approved by power company and the conduit containing the instrument wiring is run exposed.

2. In no instance shall more than one set of conductors terminate in the instrument transformer cabinet. Sub-feeders and branch circuits are to terminate at the customer's distribution panel. The instrument transformer cabinet shall not be used as a junction box.

3. When service-entrance conduits enter or leave through the back of the current-transformer cabinet, the size of the C/T cabinet must be increased to provide additional working space.

4. For services at higher voltages, additional space must be provided in the transformer cabinet for mounting potential transformers. Consult the local power company for exact dimensions for cabinets containing current and potential transformers..

5. If kilovar metering is required, increase the width of meter mounting from 18 inches to 36 inches.

Phase	Service Characteristics	Cabinet Size
Single-phase	120/240 volts, 3-wire	10″ × 24″ × 32″
Three-phase	120/240 volts, 4-wire	10″ × 36″ × 42″
Three-phase	120/208 volts, 4-wire	10″ × 36″ × 42″
Three-phase	480 volts, 3-wire	10″ × 36″ × 42″
Three-phase	480/277 volts, 4-wire	10″ × 36″ × 48″

Figure 5-22: Typical C/T cabinet sizes

6. If recording demand instruments are required, increase the height of meter mounting from 36 inches to 48 inches.

Gutters

The auxiliary gutter shown in Figure 5-19 is used to route the service conductors and also to provide an enclosure for tapping these conductors for the safety switch and two 200-ampere panels. In this type of arrangement, appropriate connectors are normally used to make the taps. However, many electrical contractors have found that a bussed gutter saves labor and provides for a neater installation.

A bussed gutter is an assembly of busbars in an enclosure. The enclosure may be rated for outdoor (weatherproof) or indoor installations. Busbars installed in the gutter may be made of aluminum or copper and must have an ampacity rating for the application; that is, if the service conductors are rated for 1200 amperes (in our sample) the busbars must be rated for at least the same ampacity. Furthermore, it must be UL listed.

From an installation or a maintenance/modification viewpoint, bussed gutters are one of the favorite types of wiring methods for use with multiswitch services. An advantage of a bussed gutter is the ease of installation and modification. Adding disconnect switches or changing switches is relatively easy. No connectors have to be untaped and reconnected as with systems using wire connectors on the conductors for taps.

In our example — using service conductors with an ampacity of 1200 amperes — the rating of the busbars in bussed gutters must also be rated at 1200 amperes, if bussed gutters were to be used. See Figure 5-23. Furthermore, the bussed gutter must have an AIC rating sufficient for the available fault current and must have sufficient wire-bending space per *NEC* Section 110-3.

Bus Bracing

One characteristic of ground-faults is an induced torque in conductors carrying the fault. Because of this torque, the busbars in a bussed gutter must be attached to the enclosure in such a manner as to prevent their being dislodged and/or making contact with the gutter frame during the fault. When busbars are attached in such a manner as to withstand the torque created by the available amount of ground-fault current, they are said to be "braced" for that amount of current. For instance, busbars may be braced for 20,000 amperes, 30,000 amperes or whatever level of ground-fault current

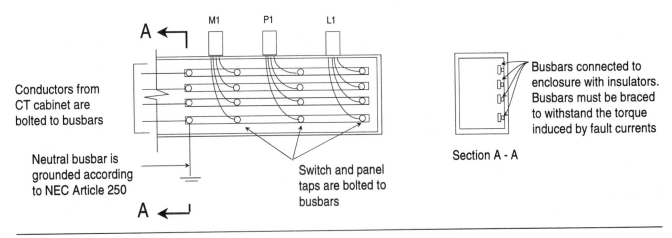

Figure 5-23: Three-phase bussed gutter

required up to 200,000 amperes. The bussed gutter must be labeled by the manufacturer for the amount of fault current the busses are braced to withstand.

480/277-VOLT SERVICES

Wye-connected, 480/277-volt services are common in medium to large commercial projects. This voltage is also frequently used in industrial applications to power motors for driving machinery and other apparatus. Such installations frequently utilize switchgear enclosures such as the one shown in Figure 5-24. Service equipment of this type is made up of "vertical sections" that connect to a common bus system within the enclosure. These sections contain fusible switches or circuit breakers, metering equipment, or other devices related to the electric service.

NEC Section 230-71 allows a maximum of six service disconnecting means per service grouped in any one location. If there are more than six

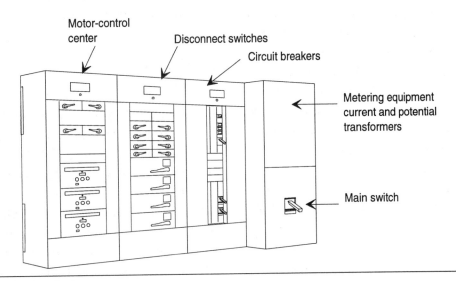

Figure 5-24: Typical switchgear

switches or circuit breakers in the switchboard, then a main switch or circuit breaker must be provided to disconnect all service conductors in the building or structure from the power supply.

Typically, the metering equipment (CTs, potential transformers, etc.) are installed in the same enclosure as the main disconnecting means; additional space is usually provided in the switchgear for this equipment. Furthermore, taps are normally provided in the main bus in a barriered section for connection to emergency switches, such as for fire-alarm systems, emergency lighting, etc., as allowed by *NEC* Section 230-82, Exception 4.

Practical Applications

Figure 5-25 shows a plot plan of a shopping center facility that utilizes a 480/277-volt, wye-connected, three-phase service to supply numerous tenant areas. In general, a pad-mounted transformer installed on the property perimeter reduces the distribution voltages to 480/277 volts. An underground service is installed from this pad-mounted transformer to a switchgear room in one section of the shopping center. A single-line diagram of the electrical system for this project is shown in Figure 5-26 on page 110. Note that there are only six fusible switches in the main switchgear so no main disconnecting means is necessary. Each of these six feeders supplies a bussed-gutter system (discussed previously), each of which contains six meter bases, a fused safety switch that feeds a 480/277-volt panel for lighting and HVAC equipment. This latter panel also feeds a 480-120/208-volt dry transformer which, in turn, furnishes power to a 120/208-volt panel for feeding tenant receptacles and display lighting.

In general, this system records the amount of power used by each tenant so that tenants may be billed accordingly. A 277-volt fluorescent lighting system provides general illumination which is fed from the 480/277-volt panel in each tenant space. However, some 120-volt display lighting is employed which is fed from the 120/208-volt panel.

SWITCHES, PANELBOARDS, AND LOAD CENTERS

Panelboards consist of assemblies of overcurrent protective devices, with or without disconnecting devices, placed in a metal cabinet. The cabinet includes a cover or trim with one or two doors to allow access to the overcurrent and disconnecting devices and, in some types, access to the wiring space in the panelboard.

There is some confusion concerning the definition of "load center" and "panelboard." It's almost like the statement, "All Cognacs are brandy, but not all brandies are Cognac." Typically, load centers are fuse or circuit-breaker cabinets used on residential or small commercial projects. They are preassembled units with the interior busses installed at the factory. Upon installation, the required number of plug-in circuit breakers or fuse holders are installed, the circuit conductors terminated, and the front cover installed.

Many electricians classify "panelboard" as an enclosure for overcurrent protective devices used on larger commercial and industrial installations. Furthermore, the "can" or housing usually consists of "raw" unpainted galvanized metal. Frequently, the circuit breakers are factory installed using bolt-in circuit breakers.

A person would probably be correct in calling all load centers a *panelboard*, but not all panelboards are load centers!

Panelboards fall into two mounting classifications: (1) flush mounting (Figure 5-27, page 111), wherein the trim extends beyond the outside edges of the cabinet to provide a neat finish with the wall surface, and (2) surface mounting (Figure 5-28, page 111), wherein the edge of the trim is flush with the edge of the cabinet.

Panelboards fall into two general classifications with regard to overcurrent devices: (1) circuit breaker and (2) fused. Small circuit-breaker and fusible panelboards commonly referred to as *load centers* are manufactured for use in residential and small commercial and industrial occupancies.

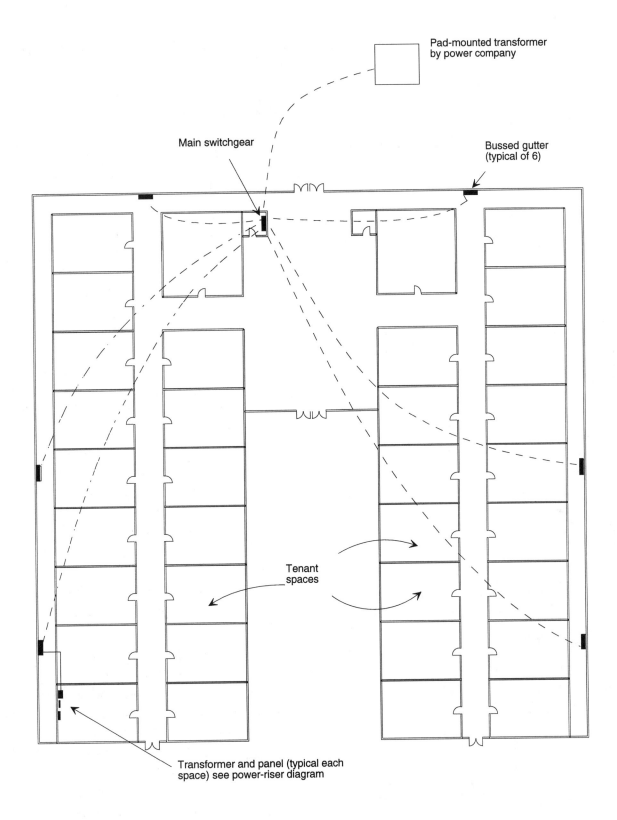

Figure 5-25: Plot plan of shopping center installation

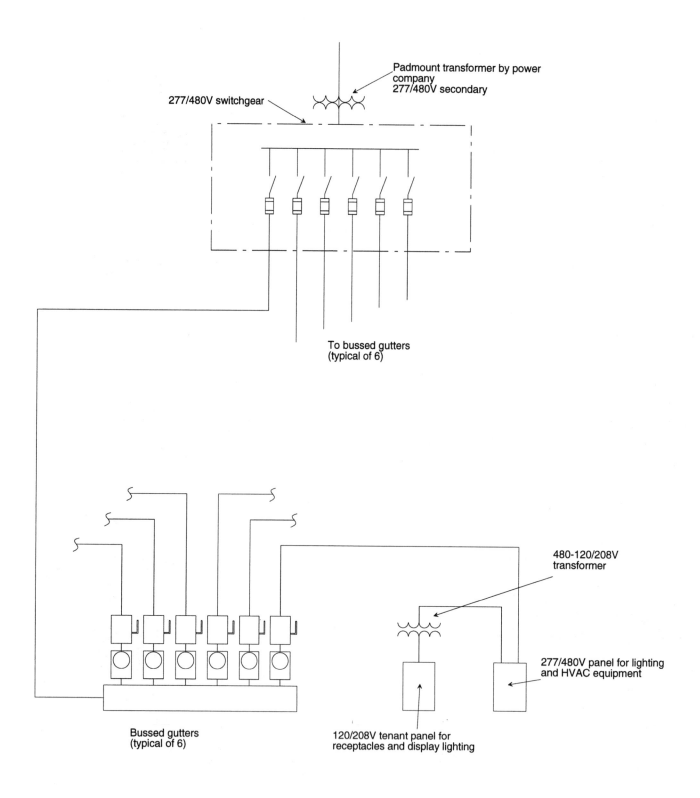

Figure 5-26: Power-riser diagram of shopping center in Figure 5-25

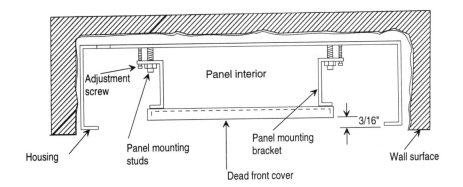

Figure 5-27: Flush-mounted panelboard

Panel Installation

Prior to installing a panel, the selected location must be examined to verify that proper clearances exist and that the environment is proper for the panel installation.

In general, all panelboards must have a rating not less than the minimum feeder capacity required for the load computed in accordance with *NEC* Article 220. Panelboards must be durably marked by the manufacturer with the voltage and the current rating and the number of phases for which they are designed and with the manufacturer's name or trademark in such a manner as to be visible after installation, without disturbing the interior parts or wiring. All panelboard circuits and circuit modifications

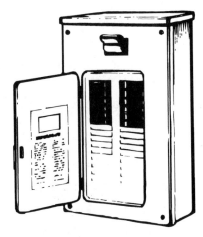

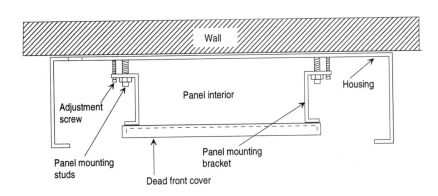

Figure 5-28: Surface-mounted panelboard

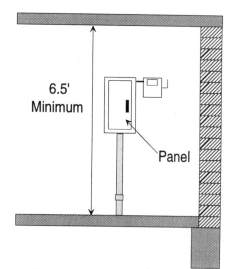

The minimum headroom of working spaces about service equipment has increased from 6.25 feet to 6.5 feet *NEC* Section 110-26(e)

Exception No. 1 to this requirement remains the same; that is, service equipment under 200 amperes in existing dwellings does not require this much headroom

Figure 5-29: *NEC* headroom requirements

shall be legibly identified as to purpose or use on a circuit directory located on the face or inside of the panel doors. *NEC* Section 384-13.

The working height about panelboards has increased with the 1999 *NEC*. Formerly 6.25 feet, the new requirement is 6.5 feet. See Figure 5-29.

Once a proper location has been determined, the panel is removed from its packing boxes, assembled, and installed. When removing the panel from its packing, verify that all necessary components have been delivered and make sure that any stray packing material has been removed from the panel. Check to make sure that the right panel is to be installed. A checklist might include the following items:

- Is the panel to be top fed or bottom fed? This information should be obtained from the drawings or from the project supervisor.

- Check to verify that the voltage rating of the panel is as specified on the drawings.

- Check to verify that the ampacity of the panel is as specified on the drawings.

- Check to verify that the phase and number of conductors is as specified on the drawings.

- Verify that the panel was not damaged during shipping.

Installing Flush-Mounted Enclosures

Flush-mounted enclosures installed in noncombustible material must be mounted so that the front edge of the enclosure is not set back further than 1/4 inch from the finished surface. If installed in other than noncombustible walls, the panel edge must be flush with the finished wall. *NEC* Section 373-3.

Installing Surface-Mounted Enclosures

Surface-mounted enclosures must be securely fastened in place. If the wall structure offers little structural support, as in the case of 1/4-inch wood paneling or 1/2-inch gypsum board, the enclosure must be located so that it may be attached to framing members inside the wall covering. In some cases, a framing structure will have to be built to support the panel.

Installing the Panel Interior

Prior to installing the panel interior, check to verify that the enclosure is securely fastened in place and is free of all foreign material. Obtain and study the specifications and instructions that are

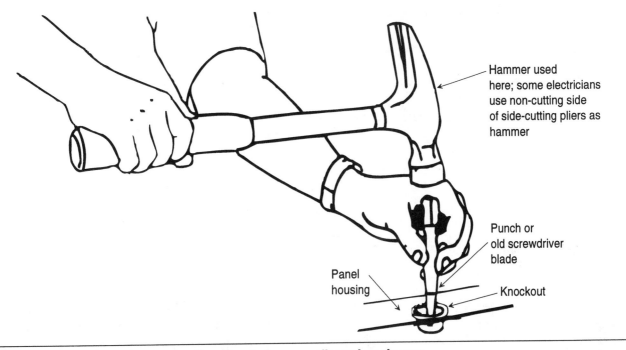

Figure 5-30: Method of removing knockouts in a panelboard enclosure

included with the panel. If no instructions are available, the following is a general installation procedure that may be used.

- Mount the interior to the enclosure using the four mounting studs installed on the enclosure back.

- Adjust the depth of the interior with the adjustment screws. The dead front cover should be no further than $3/16$ inch from the wall surface for a flush panel, or the same distance from the enclosure face for a surface-mounted panel.

Knockouts

A series of concentric or eccentric circular partial openings are usually cut in the top, bottom, and sides of both load center and panelboard housings; some may also be cut in the back. These openings are cut in such a manner that they may be removed by tapping (knocking) them out — usually with a punch or screwdriver blade and hammer as shown in Figure 5-30.

The direction from which the knockouts can be removed alternates from inside the enclosure to outside the enclosure; that is, $1/2$-inch knockouts are knocked outward from inside the enclosure; $3/4$-inch knockouts are knocked inward from outside the enclosure; 1-inch knockouts are knocked outward from inside the enclosure, etc. See Figure 5-31.

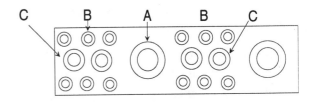

A	=	2 - 2 or 2½″ knockouts; space for 2 - 3½″ knockouts
B	=	12 - ½″ or ¾″ knockouts
C	=	4 - ¾″ or 1″ knockouts

Figure 5-31: Typical knockouts in panelboard

113

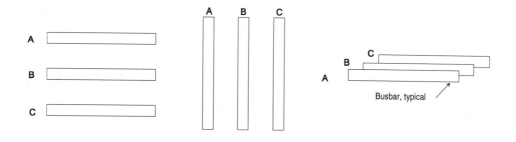

Figure 5-32: *NEC* **approved phase arrangement**

In most cases, raceways connected to panelboards using the concentric knockout openings have poor equipment grounding connections. Consequently, *NEC* Section 250-97 requires bonding jumpers to be used around concentric or eccentric knockouts that are punched or otherwise formed so as to impair the electrical connection to ground. This is accomplished by using a grounding locknut or bushing and then connecting a bonding jumper to either another grounding locknut or bushing, or else to the equipment grounding terminal inside the panelboard.

Panel Connections

Electrical connections in a panelboard fall under two categories:

- Line connections, which include termination and routing of the service and feeder conductors.

- Load connections, which include termination and routing of the branch-circuit and feeder conductors.

When installing the line connections, verify that the lugs are stamped "CU/AL" or a label is inside the panel which states that the connection of aluminum conductors is permitted prior to terminating aluminum conductors.

When installing the load connections, again verify that the lugs are suitable for both copper and aluminum if aluminum conductors are used. Due to the difficulty in keeping aluminum conductors tight in their termination lugs, copper is usually specified for most industrial installations.

NEC Section 384-3(f) requires the phase arrangement on three-phase busses to be A, B, C, from front to back, top to bottom, or left to right, as viewed from the front of the panel. The B phase must be the phase with the highest voltage to ground on a three-phase, four-wire delta-connected system. See Figure 5-32.

Enclosures

The majority of overcurrent devices (fuses and circuit breakers) are used in some type of enclosure; that is, panelboards, switchboards, motor-control centers, individual enclosures, etc.

NEMA has established enclosure designations because individually enclosed overcurrent-protective devices are used in so many different types of locations, weather and water conditions, dust and other contaminating conditions, etc. A designation such as "NEMA 12" indicates an enclosure type to fulfill requirements for a particular application. The NEMA designations were recently revised to obtain a clearer and more precise definition of the enclosure needed to meet various standard requirements.

Some of the revisions in the NEMA designations are: The NEMA Type 1A (semi-dusttight) has been dropped. The NEMA 12 enclosure now can be substituted in many installations in place of the NEMA 5. The advantage of this substitution is that the NEMA 12 enclosure is much less expensive than the NEMA 5 enclosure. NEMA Type 3R as applied to circuit breaker enclosures is a lighter weight, less expensive rainproof enclosure than the other "Weather Resistant" enclosure types. The table in Figure 5-33 gives a brief explanation of the NEMA enclosure specifications:

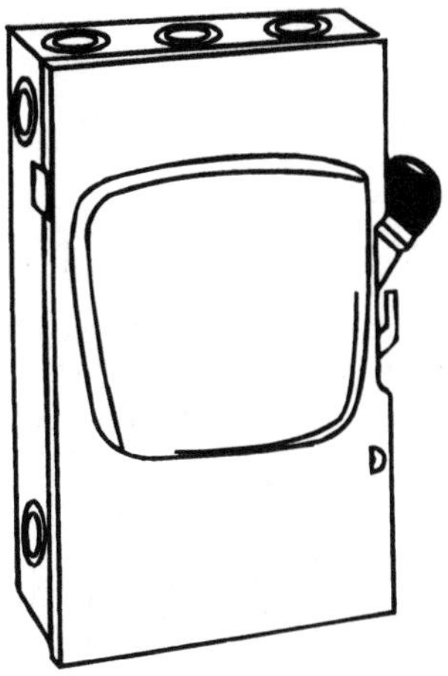

General-duty safety switch

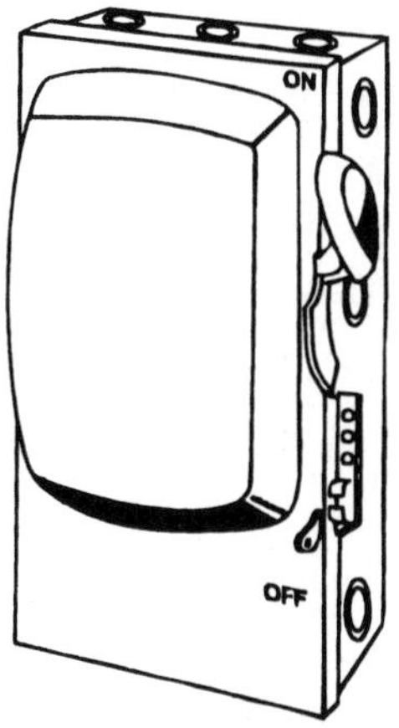

Heavy-duty safety switch

Enclosure	Explanation
NEMA Type 1 General Purpose	To prevent accidental contact with enclosed apparatus. Suitable for application indoors where not exposed to unusual service conditions
NEMA Type 3 Weatherproof (Weather Resistant)	Protection against specified weather hazards. Suitable for use outdoors
NEMA Type 3R Raintight	Protects against entrance of water from a rain. Suitable for general outdoor application not requiring sleetproof
NEMA Type 4 Watertight	Designed to exclude water applied in form of hose stream. To protect against stream of water during cleaning operations
NEMA Type 5 Dusttight	Constructed so that dust will not enter the enclosed area. Being replaced in some equipment by NEMA 12 Type
NEMA Type 7 Hazardous Locations A, B, C or D Class I—letter or letters following type number indicates particular groups of hazardous locations per NEC	Designed to meet application requirements of NEC for Class I, hazardous locations (explosive atmospheres). Circuit interruption occurs in air
NEMA Type 9 Hazardous Locations E, F or G Class II Letter or letters following type number indicates particular groups of hazardous locations per NEC	Designed to meet application requirements of NEC for Class II hazardous locations (combustible dusts, etc.)
NEMA Type 12 Industrial Use	For use in those industries where it is desired to exclude dust, fibers and filings, or oil or coolant seepage

Figure 5-33: NEMA classifications of enclosures

Safety Switches

Most manufacturers of safety switches have at least two complete lines to meet industrial, commercial and residential requirements. Both types usually have visible blades and safety handles. With visible blades, the contact blades are in full view so you can clearly see you're safe. Safety handles are always in complete control of the switch blades, so whether the cover is open or closed, when the handle is in the "OFF" position the switch is always "OFF"; that is, on the load side of the switch. The feeder or line side of the switch is still "hot" (energized) so when working with safety switches keep this in mind. See Figure 5-34.

WARNING! Even though a safety-switch handle is in the OFF position, the line side of the switch is still energized.

Heavy-duty switches are intended for applications where price is secondary to safety and continued performance. This type of switch is usually subjected to frequent operation and rough handling. Heavy-duty switches are also used in atmospheres where a general-duty switch would be unsuitable. Heavy-duty switches are widely used by most heavy industrial applications; motors and HVAC equipment will also be controlled by such switches. Most heavy-duty switches are rated 30 through 1200 amperes, 240 to 600 volts (ac-dc). The switches with horsepower ratings are able to interrupt approximately six times the full-load, motor-current ratings. When equipped with Class J or Class R fuses, many heavy-duty safety switches are UL listed for use on systems with up to 200,000 amperes available fault current.

Heavy-duty switches are available with NEMA 1, 3R, 4, 4X, 5, 7, 9 and 12 enclosures.

Switch Contacts: There are two types of switch contacts used in today's safety switches. One is the "butt" contact similar to those used in circuit-breaker devices; the other is a knife-blade and jaw type. The knife-blade types are considered to be superior to other types on the market.

All current-carrying parts of safety switches are usually plated with tin, cadmium or nickel to reduce

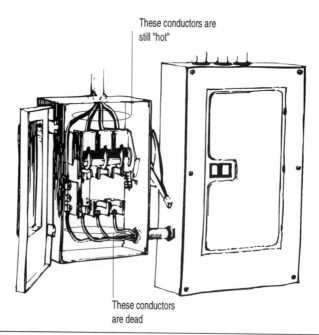

Even though this safety switch handle is in the OFF position, and the fuses have been removed, the line side of the switch, as well as the line-side conductors, are still energized.

These conductors are still "hot"

These conductors are dead

Figure 5-34: Safety switch acting as disconnect for a panelboard

heating by keeping metal oxidation at a minimum. Switch blade and jaws are made of copper for high conductivity. With knife-blade construction, the jaws distribute a uniform clamping pressure over the entire blade-to-jaw contact surface. In the event of high-current fault, the electromagnetic forces which develop tend to squeeze the jaws tightly against the blade. In the butt-type contact, these forces tend to force the contacts apart, causing them to burn severely.

Fuse clips are also plated to control corrosion and keep heating to a minimum. All heavy duty fuse clips have steel reinforcing springs to increase their mechanical strength and give a firmer contact pressure. As a result, fuses will not work loose due to vibration or rough handling.

Insulating Materials: As the voltage rating of switches is increased, arc suppression becomes more difficult and the choice of insulation material becomes a more critical problem. Arc suppressors used by many manufacturers consist of a housing made of insulation material and one or more magnetic suppressor plates. All arc suppressors are tested to assure proper control and extinguishing of arcing.

Operating Mechanism: Heavy-duty safety switches have spring-driven, quick-make, quick-break mechanisms. A quick-breaking action is necessary if a switch is to be safely switched "OFF" under a heavy load. The spring action, in addition to making the operation quick-make, quick-break, firmly holds the switch blades in an "ON" or "OFF" position. The operating handle is an integral part of the switching mechanism, so if the springs should fail the switch can still be operated. When the handle is in the "OFF" position the switch is always "OFF."

A one-piece cross bar is usually employed to offer direct control over all blades simultaneously. The one-piece cross bar means stability and strength, plus proper alignment for uniform blade operation.

Dual cover interlocks are also standard on all heavy-duty switches. The dual interlock prevents the enclosure door from being opened when the switch is "ON" and also keeps the switch from being turned "ON" while the door is open.

General-Duty Safety Switches: General-duty switches are for residential and light commercial applications where the price of the device is a limiting factor. General-duty switches are meant to be used where operation and handling are moderate and where the available fault current is less than 10,000 amperes. Some examples of general-duty switch applications would be: residential HVAC equipment, light duty fan-coil circuit disconnects for commercial projects, and the like.

General-duty switches are rated up to 600 amperes at 240 volts (ac only) in general purpose (NEMA 1) and rainproof (NEMA 3R) enclosures. These switches are horsepower rated and capable of opening a circuit with approximately six times a motor's full-load current rating.

All current-carrying parts of general-duty switches are plated with either tin or cadmium to reduce heating. Switch jaws and blades are made of plated copper for high conductivity. A steel reinforcing spring increases the mechanical strength of the jaws and assures a firm contact pressure between blade and jaw.

Double-Throw Safety Switches: Double-throw switches are used as transfer switches and are not intended as motor circuit switches; therefore, they are not horsepower rated.

Safety switches are available as either fused or unfused devices. These switches have quick-make, quick-break action, plated current-carrying parts, a key controlled interlock mechanism and screw-type lugs. Arc suppressors are supplied on all switches rated above 250 volts.

General-duty safety switches are manually operable and not quick-make, quick-break. They are available as either fused or unfused devices in NEMA 1 enclosures only.

Panels

There are many different types of load centers and panelboards used in commercial wiring installations. Such equipment will vary in size from a small 30-ampere, single-pole switch to huge switchgear and motor-control centers. If you have been used to residential wiring, one of the major differences that you will encounter with commercial load centers and panelboards is that the equipment used on commercial projects — for the most part — is designed for heavier duty than those used on residential projects. For example, a load center for a residence may have a thin sheet metal housing that is protected from the elements by a coat of enamel. The circuit breakers will usually be of the plug-in type. Panelboard housing for commercial installations are usually of a heavier gauge galvanized steel, and the circuit breakers will be of the bolt-in type. There are, of course, exceptions in both cases.

The following are some of the most popular types of load centers and panelboards.

Main Lugs Only: Main lugs load centers provide distribution of electrical power where a main disconnect with overcurrent protection is provided separately from the load center. All terminals are suitable for aluminum or copper conductors.

Main Breaker Load Centers: These load centers are similar to the main lugs only panels except the main circuit breaker is factory installed. Furthermore, the main circuit breaker and neutral terminals are usually located at the same end of the load center and are adjacent to one another.

Split-Bus Load Centers: The *NEC* allows the use of a maximum of six main disconnects in a common enclosure. Split-bus QO load centers have the bus split or divided into sections which are insulated from each other to provide an economical service entrance device in applications not requiring a single main disconnect.

The line or main section of the load center has provisions for up to six main disconnects for the heavier 240V appliances, subfeeders and lighting main disconnects. The lower section contains provisions for lighting and 120V appliance circuits and is fed by the lighting main disconnect which is located in the main bus section. All split-bus devices are provided with factory installed wires connected to the lower section. These wires can then be field connected to the lighting main disconnect in the main section.

Riser Panels: Riser panels, consisting of a main lugs only load center with an extended gutter of over 6 inches, are ideally suited for high-rise office buildings and for apartment complexes. They are available with 6, 8 and 12 circuit load centers. The box, interior and covers are sold separately so they can be installed at the most convenient time during construction.

Panelboards: Switchgear manufacturers offer complete lines of lighting and distribution panelboards, most of which are available either unassembled from distributor stock or factory assembled. All types should be UL listed and meet Federal Specification WP-115a.

Exact descriptions and installation instructions are provided in later chapters of this book, under the appropriate section and application.

Chapter 6
Grounding

The grounding system is a major part of the electrical system. Its purpose is to protect life and equipment against the various electrical faults that can occur. It is sometimes possible for higher-than-normal voltages to appear at certain points in an electrical system or in the electrical equipment connected to the system. Proper grounding ensures that the high electrical charges that cause these high voltages are channeled to earth or ground before damaging equipment or causing danger to human life.

When we refer to *ground*, we are talking about ground potential or earth ground. If a conductor is connected to the earth or to some conducting body that serves in place of the earth, such as a driven ground rod (electrode) or cold-water pipe, the conductor is said to be *grounded*. The neutral conductor in a three- or four-wire service, for example, is intentionally grounded and therefore becomes a *grounded conductor*. However, a wire used to connect this neutral conductor to a grounding electrode or electrodes is referred to a *grounding conductor*. Note the difference in the two meanings; one is ground**ed**, while the other is ground**ing**.

TYPES OF GROUNDING SYSTEMS

There are two general classifications of protective grounding:

- System grounding
- Equipment grounding

The system ground relates to the service-entrance equipment and its interrelated and bonded components. That is, system and circuit conductors are grounded to limit voltages due to lightning, line surges, or unintentional contact with higher voltage lines, and to stabilize the voltage to ground during normal operation.

Equipment grounding conductors are used to connect the noncurrent-carrying metal parts of equipment, conduit, outlet boxes, and other enclosures to the system grounded conductor, the grounding electrode conductor, or both, at the service equipment or at the source of a separately derived system. Equipment grounding conductors are bonded to the system grounded conductor to provide a low impedance path for fault current that will facilitate the operation of overcurrent devices under ground-fault conditions.

Article 250 of the 1999 *National Electrical Code* (*NEC*) covers general requirements for grounding and bonding. Nearly 75 changes or additions have been made to this Article since the last edition of the *NEC* was printed. This should be reason enough to carefully read all parts of this Article over several times until you have a thorough understanding of its contents.

Single-Phase Systems

To better understand a complete grounding system, let's take a look at a small commercial system

119

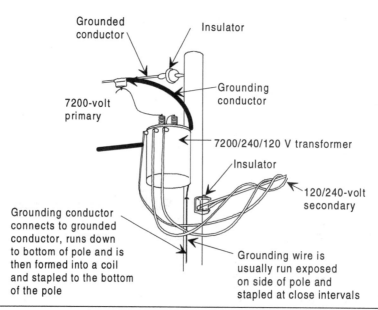

Figure 6-1: Pole-mounted transformer reducing transmission voltage to usable consumer current

beginning at the power company's high-voltage lines and transformer as shown in Figure 6-1. The pole-mounted transformer is fed with a two-wire 7200-volt system which is transformed and stepped down to a three-wire, 120/240-volt, single-phase electric service suitable for residential use. A wiring diagram of the transformer connections is shown in Figure 6-2. Note that the voltage

between phase A and phase B is 240 volts. However, by connecting a third wire (neutral) on the secondary winding of the transformer — between the other two — the 240 volts are split in half, giving 120 volts between either phase A or phase B and the neutral conductor. Consequently, 240 volts are available for household appliances such as ranges, hot-water heaters, clothes dryers, and

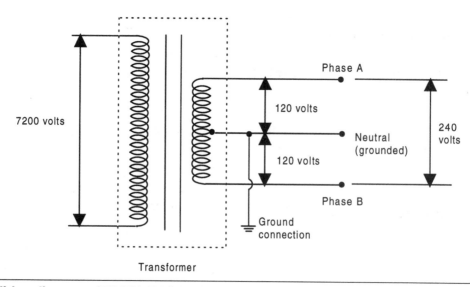

Figure 6-2: Wiring diagram of 7200V-120/240V, single-phase transformer connection

the like, while 120 volts are available for lights, small appliances, TVs, and similar electrical appliances.

Referring again to the diagram in Figure 6-2, conductors A and B are ungrounded conductors, while the neutral is a grounded conductor. If only 240-volt loads were connected, the neutral (grounded conductor) would carry no current. However, since 120-volt loads are present, the neutral will carry the unbalanced load and becomes a current-carrying conductor. For example, if phase A carries 60 amperes and phase B carries 50 amperes, the neutral conductor would carry only (60 - 50 =) 10 amperes. This is why the *NEC* allows the neutral conductor in an electric service to be smaller than the ungrounded conductors.

The typical pole-mounted service-entrance is normally routed by messenger cable from a point on the pole to a point on the building being served, terminating in a meter housing. Another service conductor is installed between the meter housing and the main service switch or panelboard. This is the point where most systems are grounded — the neutral bus in the main panelboard. However, some localities require electric services to be grounded at the meter base.

OSHA AND *NEC* REQUIREMENTS

The grounding equipment requirements established by Underwriters' Laboratories, Inc., have served as the basis for approval for grounding of the *NEC*. The *NEC*, in turn, provides the grounding premises of the Occupational Safety and Health Act (OSHA).

All electrical systems must be grounded in a manner prescribed by the *NEC* to protect personnel and valuable equipment. To be totally effective, a grounding system must limit the voltage on the electrical system and protect it from:

- Exposure to lightning.
- Voltage surges higher than that for which the circuit is designed.

- An increase in the maximum potential to ground due to abnormal voltages.

Grounding Methods

Methods of grounding an electric service are covered in *NEC* Section 250-50 and are shown in Figure 6-3 on page 122. In general, all of the following (if available) and any made electrodes must be bonded together to form the grounding electrode system:

- An underground water pipe in direct contact with the earth for no less than 10 feet.

- The metal frame of a building where effectively grounded.

- An electrode encased by at least 2 inches of concrete, located within and near the bottom of a concrete foundation or footing that is in direct contact with the earth. Furthermore, this electrode must be at least 20 feet long and must be made of electrically conductive coated steel reinforcing bars or rods of not less than $\frac{1}{2}$-inch diameter, or consisting of at least 20 feet of bare copper conductor not smaller than No. 2 AWG wire size.

- A ground ring encircling the building or structure, in direct contact with the earth at a depth below grade not less than $2\frac{1}{2}$ feet. This ring must consist of at least 20 feet of bare copper conductor not smaller than No. 2 AWG wire size. See Figure 6-3 on page 122.

Grounding systems used in large commercial buildings will frequently use all of the methods shown in Figure 6-3, and the methods used will often surpass the *NEC*, depending upon the manufacturing process, and the calculated requirements made by consulting engineers. Figure 6-4 on page 123 shows a floor plan of a typical commercial grounding system.

In some structures, only the water pipe will be available, and this water pipe must be supplemented by an additional electrode as specified in *NEC* Sections 250-50(a)(2) and 250-52. With these facts in mind, let's take a look at a typical electric service, and the available grounding electrodes. See Figure 6-5 on page 124.

The building in Figure 6-5 has a metal underground water pipe that is in direct contact with the earth for more than 10 feet, so this is one valid grounding source. The building also has a metal underground gas-piping system, but this may not be used as a grounding electrode (*NEC* Section 250-52(a)). *NEC* Section 250-50(a)(2) further states that the underground water pipe must be supplemented by an additional electrode of a type specified in Section 250-50 or in Section 250-52. If a grounded metal building frame, concrete-encased electrode, or a ground ring is not available, *NEC* Section 250-52 — *Made and Other Electrodes* —

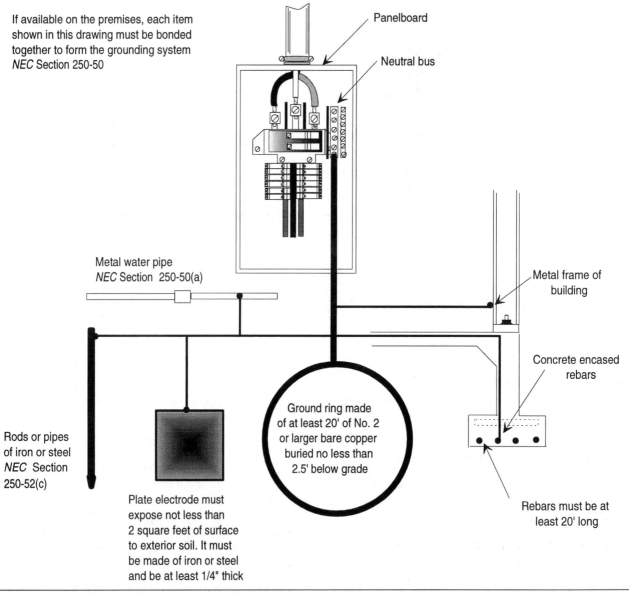

If available on the premises, each item shown in this drawing must be bonded together to form the grounding system *NEC* Section 250-50

Panelboard

Neutral bus

Metal water pipe
NEC Section 250-50(a)

Metal frame of building

Concrete encased rebars

Rods or pipes of iron or steel
NEC Section 250-52(c)

Ground ring made of at least 20' of No. 2 or larger bare copper buried no less than 2.5' below grade

Plate electrode must expose not less than 2 square feet of surface to exterior soil. It must be made of iron or steel and be at least 1/4" thick

Rebars must be at least 20' long

Figure 6-3: *NEC* approved grounding electrodes

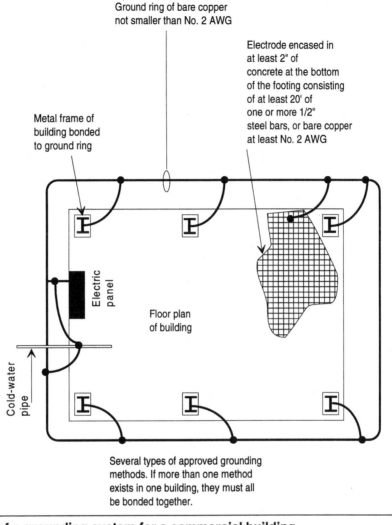

Ground ring of bare copper
not smaller than No. 2 AWG

Electrode encased in
at least 2" of
concrete at the bottom
of the footing consisting
of at least 20' of
one or more 1/2"
steel bars, or bare copper
at least No. 2 AWG

Metal frame of
building bonded
to ground ring

Electric
panel

Floor plan
of building

Cold-water
pipe

Several types of approved grounding
methods. If more than one method
exists in one building, they must all
be bonded together.

Figure 6-4: Floor plan of a grounding system for a commercial building

must be used in determining the supplemental electrode. In most cases, this supplemental electrode will consist of either a driven rod or pipe electrode, specifications for which are shown in Figure 6-6 on page 125. However, if any of the other specified electrodes are available anywhere on the premises, then all must be bonded together.

In general, the *NEC* requirements for grounding electrodes include the following:

- Provide a low-impedance path to ground for personnel and equipment protection and effective circuit relaying.

- Withstand and dissipate repeated fault and surge circuits.

- Provide corrosion resistance to various soil chemistries to ensure continuous performance for the life of the equipment being protected.

- Provide rugged mechanical properties for easy driving with minimum effort and rod damage.

An alternate method to the pipe or rod method is a plate electrode. Each plate electrode must expose not less than 2 square feet of surface to the surrounding earth. Plates made of iron or steel must be at least $\frac{1}{4}$ inch thick, while plates of nonferrous metal like copper need only be .06 inch thick.

123

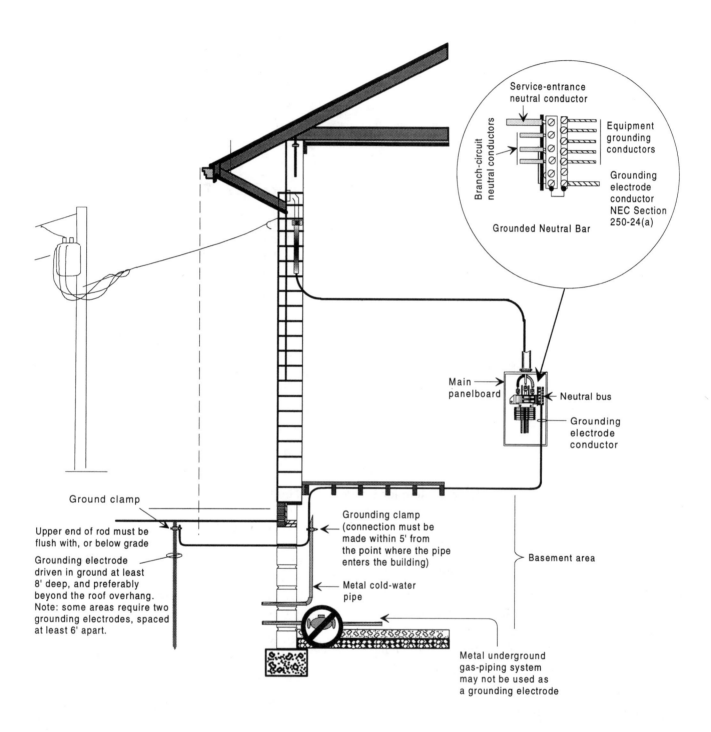

Figure 6-5: Summary of *NEC* grounding requirements

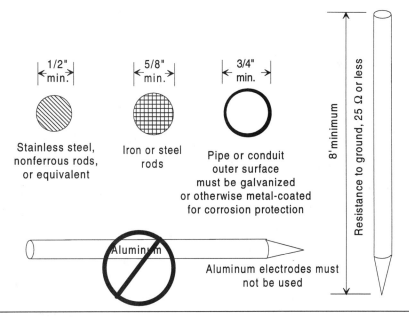

Figure 6-6: *NEC* requirements for ground rods

Either type of electrode must have a resistance to ground of 25 ohms or less. If not, they must be augmented by an additional electrode spaced not less than 6 feet from each other. In fact, many locations require two electrodes regardless of the resistance to ground. This, of course, is not an *NEC* requirement, but is required by some power companies and local ordinances in some cities and counties. Always check with the local inspection authority for such rules that surpass the requirements of the *NEC*.

EARTH ELECTRODES

The area of contact between the earth and ground rod must be sufficient so that the resistance of the current path into and through the earth will be within the allowable limits of the particular application. The resistance of this earth path must be relatively low and must remain reasonably constant through the seasons of the year.

To understand why earth resistance must be low, see Figure 6-7 on page 126 and then apply Ohm's law, $E = I \times R$. (E is volts, I is the current in amperes, and R is the resistance in ohms.) For example, assume a 4000-volt supply (2300 volts to "ground"), with a resistance of 13 ohms. Now assume an exposed ungrounded wire in this system touches a motor frame that is connected to a grounding system that has a 10-ohm resistance to earth.

According to Ohm's law, there will be a current of 100 amperes through the fault, from the motor frame to the earth. If a person touches the motor frame and is solidly grounded to earth, the person could be subjected to 1000 volts (10 ohms times 100 amperes). This is more than enough for a fatality.

Nature Of An Earth Electrode

Resistance to current through an earth electrode system has three components:

- Resistance of the ground rod itself and connections to it.
- Contract resistance between the ground rod and the earth adjacent to it.
- Resistance of the surrounding earth.

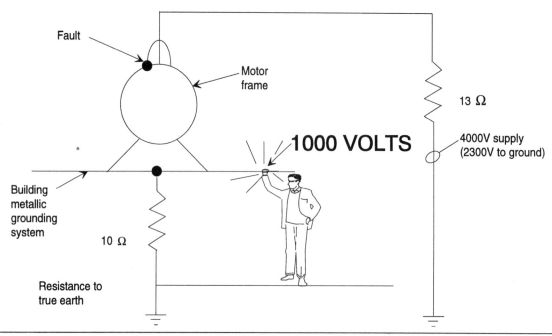

Figure 6-7: An electric circuit with too high an earth resistance

Ground rods, masses of metal, structures, and other devices are commonly used for earth electrodes. These electrodes are usually adequate in size or cross section so that their resistance is a negligible part of the total resistance.

The resistance between the electrode and the earth is much less than normally suspected. If the electrode is free of paint or grease, and the earth is firmly packed, the ANSI standards have shown that contact resistance is negligible.

A ground rod driven into earth of uniform resistivity conducts current in all directions. Think of the electrode as being surrounded by shells of earth, all of equal thickness (Figure 6-8). The earth shell closest to the ground rod has the smallest surface area and consequently offers the greatest resistance. The next earth shell is somewhat larger in area and offers less resistance. Finally, a distance from the ground rod will be reached where additional earth shells do not add significantly to the earth resistance surrounding the ground rod. This distance is approximately 8 to 10 feet, and is known as the *effective resistance area*; it is mainly dependent on the depth of the ground rod.

Of the three components involved in resistance, the resistivity of the earth is most critical and most difficult to calculate and overcome.

The equations for earth resistance from various systems of electrodes are quite complicated and in some cases may be expressed only as approximations. All such equations are derived from the general equation $R = pL/A$ and are based on the assumption of uniform earth resistivity throughout the entire soil volume being considered, although this is seldom the case. This is the reason that in critical situations, a two- or three-point megger ground test should be made.

The following is a commonly-used resistance-to-earth equation for single ground rods:

$$R = \frac{p}{2\mu L} \ln \frac{4L}{a} - 1$$

where

p = average soil resistivity, ohms-centimeters

L = ground rod length, centimeters

a = ground rod radius, centimeters

R = resistance of rod to earth, ohms

126

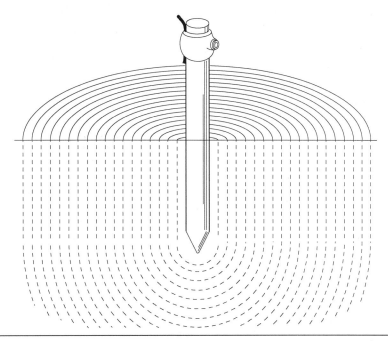

Figure 6-8: Resistive components of earth electrode

Effect Of Rod Size On Resistance

Whenever a ground rod is driven deeper into the earth, its resistance is substantially reduced. Generally, doubling the rod length reduces resistance by an additional 40 percent. Increasing the diameter of the rod, however, does not materially reduce its resistance. Doubling the diameter, for instance, reduces resistance by less than 10 percent.

Effects Of Soil Resistivity On Earth Electrode Resistance

The resistance to earth of grounding electrodes depends not only on the depth and surface area of grounding electrodes but on soil resistivity as well.

Soil resistivity is the key factor that determines what the resistance of a grounding electrode will be and to what depth it must be driven to obtain low ground resistance. The resistivity of the soil varies widely throughout the world and changes seasonally. Soil resistivity is determined largely by its electrolyte content, consisting of moisture, minerals, and dissolved salts. A dry soil has high resistivity, but a wet soil may also have a high resistivity if it contains no soluble salts. Some values found for earth resistivity are given in Figure 6-9 on page 128.

GROUNDING CONDUCTORS

The grounding conductor, connecting the grounded conductor and panelboard neutral bus to ground and grounding electrodes, must be of either copper, aluminum, or copper-clad aluminum. Furthermore, the material selected must be resistant to any corrosive condition existing at the installation or it must be suitably protected against corrosion. The grounding conductor may be either solid or stranded, covered or bare, but it must be in one continuous length without a splice or joint — except for the following conditions:

- Splices in busbars are permitted.

- Where a service consists of more than one single enclosure, it is permissible to connect taps to the grounding electrode conductor provided the taps are made within the enclosures.

Soil	Resistivity, Ohm-cm (Range)
Surface soils, loam, etc.	100 — 5,000
Clay	200 — 10,000
Sand and gravel	5,000 — 100,000
Surface limestone	10,000 — 1,000,000
Limestones	500 — 400,000
Shales	500 — 400,000
Sandstone	2,000 — 200,000
Granites, basalts, etc.	100,000

Figure 6-9: Resistivities of different soils

- Grounding electrode conductors may also be spliced at any location by means of irreversible compression-type connectors listed for the purpose or the exothermic welding process.

The size of grounding conductors depends on service-entrance size; that is, the size of the largest service-entrance conductor or equivalent for parallel conductors. The table in Figure 6-10 gives the proper sizes of grounding conductors for various sizes of electric services. Also refer to Figure 6-11 on page 130 for a summary of *NEC* requirements pertaining to system grounding.

EQUIPMENT GROUNDING

The general *NEC* regulations concerning equipment grounding apply to all installations except for specific equipment (special applications) as indicated in Article 250-114. The *NEC* also lists specific equipment that is to be grounded regardless of voltage.

In all occupancies, major appliances and many hand-held appliances and tools are required to be grounded. The appliances include refrigerators, freezers, air conditioners, clothes dryers, washing machines, dishwashing machines, sump pumps, and electrical aquarium equipment. Other tools likely to be used outdoors and in wet or damp locations must be grounded or have a system of double insulation.

Although most appliance circuits require an equipment grounding conductor, the frames of electric ranges, clothes dryers and similar appliances that utilize both 120 and 240 volts may be grounded via the grounded circuit conductor (neutral) under most conditions. In addition, however, the grounding contacts of any receptacles on the equipment must be bonded to the equipment. If these specified conditions are met, it is not necessary to provide a separate equipment grounding conductor, either for the frames or any outlet or junction boxes which are part of the circuit for these applications. However, in wiring old, existing installations — without grounded outlets — a separate ground must be provided.

Size of Largest Service-Entrance Conductor or Equivalent Area for Parallel Conductors		Size of Grounding Electrode Conductor	
Copper	Aluminum or Copper-Clad Aluminum	Copper	Aluminum or Copper-Clad Aluminum
2 or smaller	1/0 or smaller	8	6
1 or 1/0	2/0 or 3/0	6	4
2/0 or 3/0	4/0 or 250 kcmil	4	2
Over 3/0 through 350 kcmil	Over 250 kcmil through 500 kcmil	2	1/0
Over 350 kcmil through 600 kcmil	Over 500 kcmil through 900 kcmil	1/0	3/0
Over 600 kcmil through 1100 kcmil	Over 900 kcmil through 1750 kcmil	2/0	4/0
Over 1100 kcmil	Over 1750 kcmil	3/0	250 kcmil

Figure 6-10: Grounding conductor sizes for various sizes of service conductors (*NEC* Table 250-66)

Bonding

Bonding means the permanent joining of metal parts to form a conductive path for electrical current. The purpose of the conductive path is to ensure electrical continuity and the ability of the grounding circuit to safely conduct any current that is likely to be imposed.

Bonding is accomplished in electrical systems by either installing an additional grounding wire in the cable or raceway system, or else using metallic conduit or tubing as the bonding conductor. When flexible metal conduit or tubing is used as part of the grounding system, there are certain *NEC* requirements that must be followed. In general, where a proper or complete path to ground is questionable, a bonding jumper is utilized to ensure electrical conductivity between metal parts. When the jumper is installed to connect two or more portions of the equipment grounding conductor, the jumper is referred to as an *equipment bonding jumper*. Some specific cases in which a bonding jumper is required are also listed below:

- Metal raceways, cable armor, and other metal noncurrent-carrying parts that serve as grounding conductors must be bonded whenever necessary in order to assure electrical continuity.

- When flexible metal conduit is used for equipment grounding, an equipment bonding jumper is required if the length of the ground return path exceeds 6 feet or the circuit enclosed is rated over 20 amperes. When the path exceeds 6 feet, the circuit must contain an equipment grounding conductor and the bonding may be accomplished by approved fittings.

- A short length of flexible metal conduit that contains a circuit rated over 20 amperes may not service as a grounding conductor itself, but a separate bonding jumper can be provided in place of a separate equipment grounding conductor for the circuit. This bonding jumper may be installed inside or outside the conduit, but an outside jumper cannot exceed 6 feet in length.

The electrical continuity can be further assured by using special connectors and bushings. For example, a grounding bushing has a threaded point

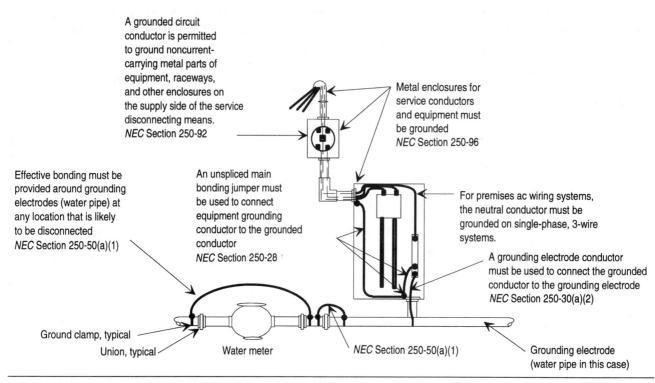

A grounded circuit conductor is permitted to ground noncurrent-carrying metal parts of equipment, raceways, and other enclosures on the supply side of the service disconnecting means. *NEC* Section 250-92

Metal enclosures for service conductors and equipment must be grounded *NEC* Section 250-96

Effective bonding must be provided around grounding electrodes (water pipe) at any location that is likely to be disconnected *NEC* Section 250-50(a)(1)

An unspliced main bonding jumper must be used to connect equipment grounding conductor to the grounded conductor *NEC* Section 250-28

For premises ac wiring systems, the neutral conductor must be grounded on single-phase, 3-wire systems.

A grounding electrode conductor must be used to connect the grounded conductor to the grounding electrode *NEC* Section 250-30(a)(2)

Ground clamp, typical

Union, typical

Water meter

NEC Section 250-50(a)(1)

Grounding electrode (water pipe in this case)

Figure 6-11: *NEC* **requirements for system grounding**

that "bites" into metal enclosures to ensure a better bond. However, if the conduit is to be connected to an enclosure with concentric knockouts, then a bonding jumper is required to ensure proper connection. In this case, the bush-ing is provided with a set-screw terminal in which the bonding jumper is secured, and a threaded point to "bite" into the metal enclosure. Two examples of bonding jumpers are shown in Figure 6-12.

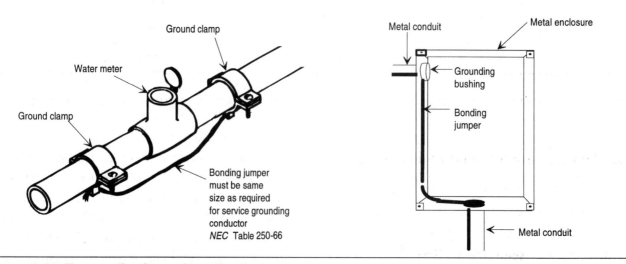

Ground clamp

Water meter

Ground clamp

Bonding jumper must be same size as required for service grounding conductor *NEC* Table 250-66

Metal conduit

Metal enclosure

Grounding bushing

Bonding jumper

Metal conduit

Figure 6-12: Two applications of bonding jumpers

Maximum Rating of Overcurrent Device	Copper Wire No.	Aluminum or Copper-Clad Aluminum Wire No.
15	14	12
20	12	10
30	10	8
40	10	8
60	10	8
100	8	6
200	6	4
300	4	2
400	3	1
500	2	1/0
600	1	2/0
800	1/0	3/0
1000	2/0	4/0
1200	3/0	250 kcmil
1600	4/0	350 kcmil
2000	250 kcmil	400 kcmil
2500	350 kcmil	600 kcmil
3000	400 kcmil	600 kcmil
4000	500 kcmil	800 kcmil
5000	700 kcmil	1200 kcmil
6000	800 kcmil	1200 kcmil

Figure 6-13: Minimum size equipment grounding conductors for grounding raceways and equipment

Size Of Equipment Grounding Conductors

Copper, aluminum, or copper-clad aluminum equipment grounding conductors must be capable of safely conducting any ground-fault current likely to be imposed on them. The size must conform to *NEC* Table 250-122. See Figure 6-13.

When current-carrying conductors are increased in size to compensate for voltage drop, equipment grounding conductors must be adjusted proportionately. For example, if a circuit is rated at 47 amperes, and a conductor with a current-carrying capacity of 60 amperes is installed, a No. 10 AWG equipment grounding conductor is required.

However, due to the length of the circuit run, a conductor with a current-carrying capacity of 100 amperes is installed to compensate for voltage drop, the equipment grounding conductor must be increased to No. 8 AWG.

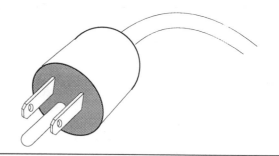

Figure 6-14: Receptacles designed to accept a three-wire grounding plug are required in all occupancies

GROUNDING ENCLOSURES

We previously mentioned that equipment grounding covers the metallic noncurrent-carrying parts of an electrical system. Such parts include metallic conduit, outlet boxes, enclosures, and frames on motors, controllers, and other electrically-operated equipment. These items are bonded together for the following purposes:

- To limit the voltage to ground on metallic enclosures and conduit.

- To assure operation of overcurrent devices in case of ground faults.

A bare or green-insulated grounding wire is attached to the metal frame or cabinet of the equipment. When connected to the circuit, this grounding wire is attached to the equipment grounding system which, in turn, was originally connected to the system's neutral busbar at the service-entrance equipment. When properly connected, if a "live" ungrounded wire should make contact with the frame or cabinet of a motor, appliance, or other metallic object in the system, a ground fault will occur and open the overcurrent protective device that is protecting the circuit. Consequently, equipment grounding is one of the best methods of protecting life and property should a ground fault occur. The best method, however, is to use a ground-fault circuit interrupter.

Grounding Outlet Boxes And Devices

All receptacles used in residential, commercial, or industrial applications must be of the grounded type which means that one of the receptacle openings connects the equipment ground to the appli-

ance, tool, or other apparatus that is connected to the receptacles. See Figure 6-14.

Ensuring continuity of the equipment grounding system to each receptacle on any electrical system is handled in different ways — depending upon the wiring method used. For example, in some small commercial wiring systems, type NM (Romex) cable will be used. If metallic device boxes are used, the bare or green insulated grounding wire must be attached to the box. This is accomplished by using either a grounding clip as shown in Figure 6-15, or else a screw designed for the purpose and that is secured to the outlet box.

Figure 6-16 shows methods of providing a grounding conductor to duplex receptacles using several wiring methods. For example, if nonmetallic boxes are used with type NM cable, no connection to the box itself is required. However,

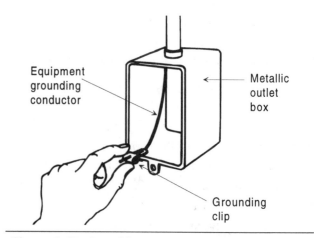

Equipment grounding conductor

Metallic outlet box

Grounding clip

Figure 6-15: Attaching grounding conductor to metallic outlet box with a grounding clip

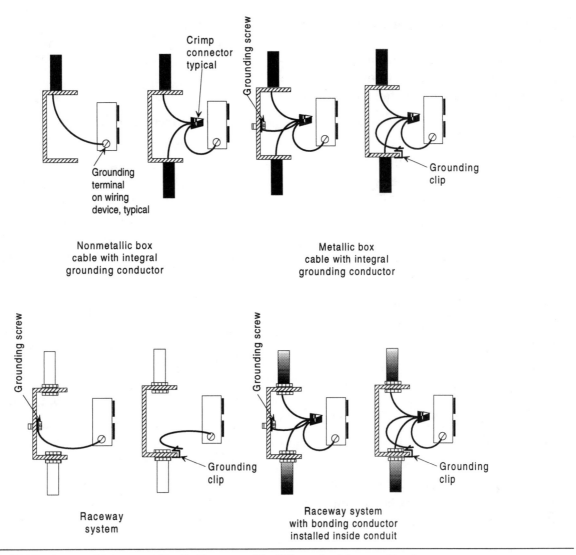

Figure 6-16: Methods used to ground outlet boxes with different wiring methods

the equipment grounding conductor must be connected to the grounding terminal on the receptacle. If two or more cables enter the nonmetallic box, each of the equipment grounding wires must be connected with an approved connector, and one wire then attaches to the grounding terminal on the receptacle. The terminal itself may not contain more than one grounding wire. Therefore, if more than one enters the box, they must be spliced independently from the device, and then only one conductor may be attached to the device.

When metallic boxes are used, these equipment grounding wires must be attached to the box, along with the wiring device (receptacle). As mentioned previously, this may be accomplished with either a grounding clip or a grounding screw. Again, refer to Figure 6-15.

Any of the following are recognized by the *NEC* as being adequate for use as an equipment grounding conductor:

- Metallic raceways, cable trays, cablebus framework, or cable armor or sheath.

- All metallic enclosures containing conductors, including meter fittings, boxes, or the like.

GROUND-FAULT CIRCUIT-INTERRUPTERS

Under certain conditions, the amount of current needed and the time it takes to open an overcurrent-protective device can be fatal. Because of this fact, the *NEC* requires ground-fault circuit-interrupter (GFCIs) to be installed on the following:

- Circuits feeding impedance heating units operating at voltages greater than 30 volts. Such units are frequently used for de-icing and snow-melting equipment.

- Circuits for electrically operated pool covers.

- Power or lighting circuits for swimming pools, fountains, and similar locations.

- Receptacles in both commercial and residential garages.

- Receptacles in residential bathrooms.

- Receptacles installed outdoors where there is direct grade level access.

- Receptacles installed in residential crawl spaces or unfinished basements.

- Countertop receptacles mounted within 6 feet of a sink.

- Receptacles installed in bathhouses.

- Receptacles installed in bathrooms of commercial, industrial, or any other building.

- Receptacles installed on roofs of any building except dwellings.

- Branch circuits derived from autotransformers.

GFCI circuit breakers require the same mounting space as standard single-pole circuit breakers and provide the same branch circuit wiring protection as standard circuit breakers. They also provide Class A ground-fault protection.

GFCI breakers that are UL listed are available in single and two-pole construction; 15, 20, 25 and 30 ampere ratings; and have a 10,000 ampere interrupting capacity. Single-pole units are rated 120V ac; two-pole units 120/240V ac.

GFCI circuit breakers not only can be used in load centers and panelboards, but they are also available factory-installed in meter pedestals and power outlet panels for RV parks and construction sites.

The GFCI sensor continuously monitors the current balance in the ungrounded "hot" load conductor and the neutral load conductor. If the current in the neutral load wire becomes less than the current in the "hot" load wire, then a ground fault exists, since a portion of the current is returning to the source by some means other than the neutral load wire. When an imbalance in current occurs, the sensor, which is a differential current transformer, sends a signal to the solid state circuitry which activates the ground trip solenoid mechanism and breaks the "hot" load connection. A current imbalance as low as 6 milliamperes (ma) will cause the circuit breaker to interrupt the circuit. This will be indicated by the trip indicator as well as the position of the operating handle centered between "OFF" and "ON." See Figure 6-17.

The two-pole GFCI breaker (Figure 6-18) continuously monitors the current balance between the two "hot" conductors and the neutral conductor. As long as the sum of these three currents is zero, the device will not trip; that is, if there was 10 amps current in the A load wire, 5 amps in the neutral, and 5 amps in the B load wire, then the sensor is balanced and will not produce a signal. A current imbalance from a ground-fault condition as low as 6 ma will cause the sensor to produce a signal of sufficient magnitude to trip the device.

Single-Pole GFCI Circuit Breakers

The single-pole breaker has two load lugs and a white wire "pigtail" in addition to the line side plug-on or bolt-on connector. The line side "hot" connection is made by installing the GFCI breaker in the panel the same as you would install any

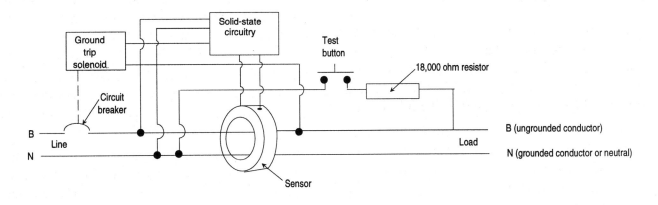

Figure 6-17: Operating circuitry of a typical GFCI

other circuit breaker. The white wire "pigtail" is attached to the panel neutral (S/N) assembly. Both the neutral and "hot" wires of the branch circuit being protected are terminated in the GFCI breaker. These two load lugs are clearly marked "LOAD POWER" and "LOAD NEUTRAL" by moldings in the breaker case. Also molded in the case is the identifying marking for the "pigtail," "PANEL NEUTRAL."

Single-pole GFCI circuit breakers must be installed on independent circuits. Circuits which employ a neutral common to more than one

"hot" conductor cannot be protected against ground faults by a single-pole breaker because a common neutral cannot be split and retain the necessary "hot" wire/neutral wire balance under normal use to prevent the GFCI breaker from tripping.

Care should be exercised when installing GFCI breakers in existing panels to be sure the neutral wire for the branch circuit corresponds with the "hot" wire of the same circuit.

Always remember that unless the current in the neutral wire is equal to that in the "hot" wire

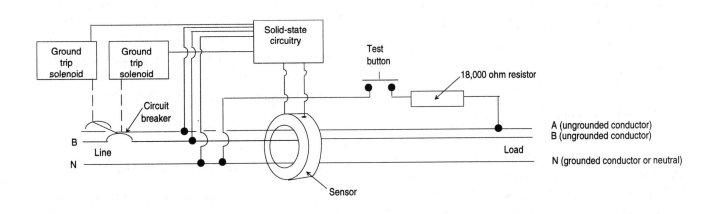

Figure 6-18: Operating characteristics of a two-pole GFCI

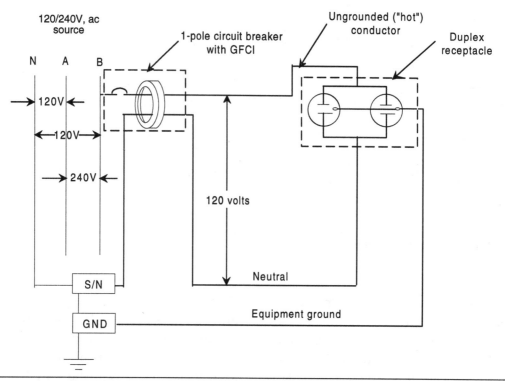

Figure 6-19: Operating characteristics of single-pole circuit breaker with GFCI

(within 6 ma), the GFCI breaker senses this as being a possible ground-fault. See Figure 6-19.

Two-Pole GFCI Circuit Breakers

A two-pole GFCI circuit breaker can be installed on a 120/240V ac 1φ-3W system, the 120/240V ac portion of a 120/240V ac three-phase, four-wire system, or two phases and neutral of a 120/208V ac three-phase, four-wire system. Regardless of the application, the installation of the breaker is the same — connections made to two "hot" busses and the panel neutral assembly. When installed on these systems, protection is provided for two-wire 240V ac or 208V ac circuits; three-wire 120/240V ac or 120/208V ac circuits and 120V ac multi-wire circuits.

The circuit in Figure 6-20 can be used to illustrate the problems that would be encountered if a common load neutral were to be used for two single-pole GFCI breakers. Either or both breakers

would trip when a load is applied at #2 duplex receptacle. The neutral current from #2 duplex receptacle would be flowing through breaker #1; this increase in neutral current through breaker #1 causes an imbalance in its sensor, thus causing it to produce a fault signal. At the same time, there is no neutral current flowing through breaker #2; therefore, it also senses a current imbalance. What happens if a load is applied at #1 duplex receptacle? As long as there is no load at #2 duplex receptacle, then neither breaker will trip because neither breaker will sense a current imbalance.

Wiring practices often used in junction boxes can also present problems when the junction box is used for taps for more than one branch circuit. Even though the circuits are not wired using a common neutral, sometimes all neutral conductors are connected together. Thus, parallel neutral paths would be established, producing an imbalance in each GFCI breaker sensor, causing them to trip.

The two-pole GFCI circuit breaker eliminates the problems encountered when trying to use two

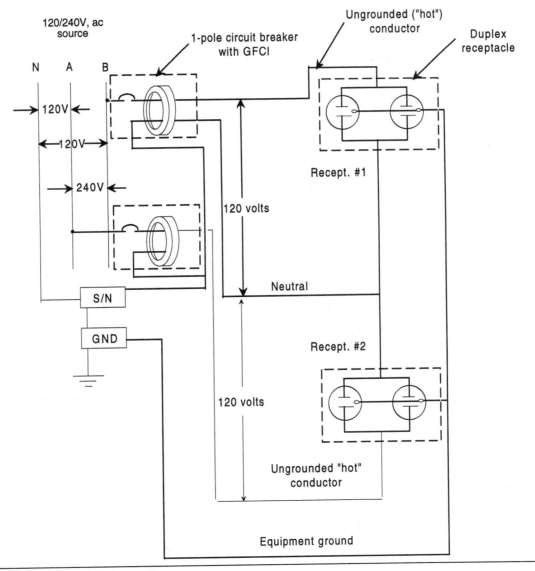

Figure 6-20: Circuit depicting common load neutral

single-pole GFCI breakers with a common neutral. Because both "hot" currents and the neutral current pass through the same sensor, under normal load condition, no imbalance in current occurs between the three currents and the breaker will not trip.

Direct-Wired GFCI Receptacles

Direct-wired GFCI receptacles provide Class A ground fault protection on 120V ac circuits.

They are available in both 15- and 20-ampere arrangements. The 15-ampere unit has a NEMA 5-15R receptacle configuration for use with 15-ampere plugs only. The 20-ampere device has a NEMA 5-20R receptacle configuration for use with 15- or 20-ampere plugs. Both 15- and 20-ampere units have a 120V ac, 20-ampere circuit rating. This is to comply with *NEC* Table 210-24 which requires that 15-ampere circuits use 15-ampere rated receptacles but permits the use of either 15- or 20-ampere rated receptacles on 20-ampere circuits. Therefore, the GFCI receptacle units which contain a 15 ampere receptacle may be used on 20-ampere circuits.

These receptacles have terminals for the "hot" neutral and ground wires. In addition, they have feed-through terminals which can be used to provide ground fault protection for other receptacles electrically "downstream" on the same branch circuit. All terminals will accept #14-#10 AWG copper wire.

GFCI receptacles have a two pole tripping mechanism which breaks both the "hot" and the neutral load connections.

When tripped, the "RESET" button extends, making visible a red indicating band. The unit is reset by pushing this button.

GFCI receptacles have the additional benefit of being noise-suppressed. Noise suppression minimizes false tripping due to spurious line voltages or RF signals between 10 and 500 Mhz.

The GFCI receptacle can be mounted, without adapters, in wall outlet boxes that are at least 1½ inches deep.

Plug-in GFCI Receptacles

The plug-in GFCI receptacle is a plug-in ground fault protection adapter for use in either two- or three-wire 120V ac receptacles. This device has a unique retractable ground pin which makes it possible to provide ground fault protection at existing two-wire polarized receptacles as well as on three-wire receptacles. This unit provides two Class A GFCI protected receptacles.

To use the unit on three-wire receptacles, lock the ground pin on the back of the unit. For two-wire receptacles, unlock the ground pin. The ground pin will retract automatically as the unit stabs are inserted into the receptacle. A yellow indicator pin on the front shows when the ground pin has been retracted.

When tripped, the plug-in GFCI receptacle has a red fault light which illuminates. To reset the unit, just push the blue reset button.

Summary

No other phase of the electrical industry is more important than grounding. It is the chief means of protecting life and property from electrical hazards and also ensures proper operation of the system as well as helping other protective devices (fuses, circuit breakers, overload relays, and the like) to function properly.

The term grounded means connecting to earth by a conductor or to some conducting body that serves in place of the earth. The earth as a whole is properly classed as a conductor. For convenience, its electric potential is assumed to be zero. When a metal object is grounded, it, too, is forced to take the same zero potential as the earth. Therefore, the main purpose of grounding is to ensure that the grounded object cannot take on a potential differing sufficiently from earth potential to be hazardous. In short, make absolutely certain that all electrical systems are properly grounded prior to energizing them or otherwise putting them into operation.

WARNING!   It should be impressed on all personnel that a lethal potential can exist between the system ground and a remote ground if a system fault involving the system ground occurs while the ground is being lifted or while ground tests are being made.

Chapter 7
Overcurrent Protection

All electrical circuits, and their related components, are subject to destructive overcurrents. Harsh environments, general deterioration, accidental damage or damage from natural causes, excessive expansion or overloading of the electrical system are factors that contribute to the occurrence of such overcurrents. Reliable protective devices prevent or minimize costly damage to transformers, conductors, motors, equipment, and the many other components and loads that make up the complete electrical system. Therefore, reliable circuit protection is essential to avoid the severe monetary losses which can result from power blackouts and prolonged downtime of facilities. To protect electrical conductors and equipment against abnormal operating conditions and their consequences, protective devices are used in the circuits.

Two types of automatic overload devices normally used in electrical circuits to prevent fires or the destruction of the circuit and its associated equipment are fuses and circuit breakers.

Basically, a circuit breaker is a device for closing and interrupting a circuit between separable contacts under both normal and abnormal conditions. This is done manually (normal condition) by use of its "handle" by switching to the ON or OFF positions. However, the circuit breaker also is designed to open a circuit automatically on a predetermined overload or ground-fault current without damage to itself or its associated equipment. As long as a circuit breaker is applied within its rating, it will automatically interrupt any "fault" and therefore must be classified as an inherently safe overcurrent protective device.

The internal arrangement of a circuit breaker is shown in Figure 7-1 while its external operating characteristics are shown in Figure 7-2 on page 140. Note that the handle on a circuit breaker resembles an ordinary toggle switch. On an overload, the circuit breaker opens itself or *trips*. In a tripped position, the handle jumps to the middle position (Figure 7-2). To reset, turn the handle to the OFF position and then turn it as far as it will go beyond this position; finally, turn it to the ON position.

A standard molded-case circuit breaker usually contains:

- A set of contacts
- A magnetic trip element
- A thermal trip element
- Line and load terminals
- Bussing used to connect these individual parts

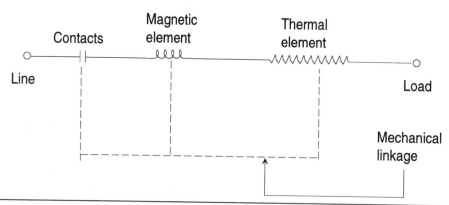

Figure 7-1: Internal arrangement of a circuit breaker

• An enclosing housing of insulating material

The circuit breaker handle manually opens and closes the contacts and resets the automatic trip units after an interruption. Some circuit breakers also contain a manually operated "push-to-trip" testing mechanism.

Circuit breakers are grouped for identification according to given current ranges. Each group is classified by the largest ampere rating of its range. These groups are:

• 15-100 amperes

• 125-225 amperes

• 250-400 amperes

• 500-1,000 amperes

• 1,200-2,000 amperes

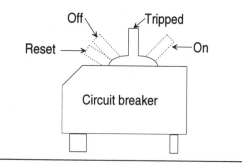

Figure 7-2: Operating characteristics of a circuit breaker

Therefore, they are classified as 100, 225, 400, 1,000 and 2,000-ampere frames. These numbers are commonly referred to as "frame classification" or "frame sizes" and are terms applied to groups of molded case circuit breakers which are physically interchangeable with each other.

CIRCUIT BREAKER RATINGS

The established voltage rating of a circuit breaker is based on its clearances or space, both through air and over surfaces between components of the electrical circuit and between the electrical components and ground. Circuit breaker voltage ratings indicate the maximum electrical system voltage on which they can be applied. Underwriters' Laboratories, Inc. recognizes only the ratings listed in Figure 7-3.

A circuit breaker can be rated for either alternating current (ac) or direct current (dc) system applications or for both. Single-pole circuit breakers, rated at 120/240 volts ac or 125/250 volts dc can be used singly and in pairs on three-wire circuits having a neutral connected to the mid-point of the load. Single-pole circuit breakers rated at 120/240 volts ac or 125/250 volts dc also can be used in pairs on a two-wire circuit connected to the un-grounded conductors of a three-wire system. Two-pole or three-pole circuit breakers rated at 120/240 volts ac or 125/250 volts dc can be used only on a three-wire, direct current, or single-phase, alter-

For Alternating Current	For Direct Current
120 volts	125 volts
120/240 volts	125 volts
240 volts	250 volts
277 volts	600 volts
277/480 volts	
480 volts	
600 volts	

Figure 7-3: Circuit breaker ratings recognized by UL

nating current system having a grounded neutral. Circuit breaker voltage ratings must be equal to or greater than voltage of the electrical system on which they are used.

Circuit breakers have two types of current ratings. The first — and the one that is used most often — is the *continuous current rating*. The second is the *ground-fault current interrupting capacity.*

Current Rating

The rated continuous current of a device is the maximum current in amperes which it will carry continuously without exceeding the specified limits of observable temperature rise. Continuous current ratings of circuit breakers are established based on standard UL ampere ratings. These are 15, 20, 25, 30, 35, 40, 45, 50, 60, 70, 80, 90, 100, 110, 125, 150, 175, 200, 225, 250, 300, 350, 400, 450, 500, 600, 700, 800, 1000, 1200, 1600, 2000, 2500, 4000, 5000, and 6000 amperes. The ampere rating of a circuit breaker is located on the handle of the device and the numerical value alone is shown. See Figure 7-4.

General application requires that the circuit breaker current rating must be equal to or less than the load circuit conductor current-carrying capacity (ampacity). Consequently, in most cases, the conductor size dictates the circuit-breaker ampacity and rating.

Most overcurrent protective devices are labeled with the following current ratings:

- Normal current rating
- Interrupting rating

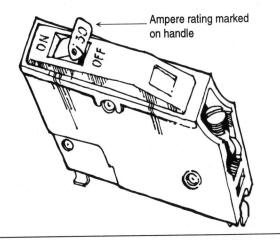

Figure 7-4: The current rating of a circuit breaker is located on the handle

141

INTERRUPTING CAPACITY RATING

The interrupting capacity (AIC) rating of a circuit breaker is the maximum short circuit current which the breaker will interrupt safely. This AIC rating is at rated voltage and frequency.

Section 110-9 of the *National Electrical Code (NEC)* clearly states that

- "Equipment intended to break current at fault levels (fuses and circuit breakers) shall have an interrupting rating sufficient for the nominal circuit voltage and the current that is available at the line terminals of the equipment."

Equipment intended to break current at other than fault levels must have an interrupting rating at nominal circuit voltage sufficient for the current that must be interrupted.

These *NEC* statements mean that fuses and circuit breakers (and their related components) designed to break fault or operating currents (open the circuit) must have a rating sufficient to withstand such currents. This *NEC* section emphasizes the difference between clearing fault level currents and clearing operating currents. Protective devices such as fuses and circuit breakers are designed to clear fault currents and therefore must have short-circuit interrupting ratings sufficient for fault levels. Equipment such as contactors and safety switches have interrupting ratings for currents at other than fault levels. Thus, the interrupting rating of electrical equipment is now divided into two parts:

- Current at fault (short-circuit) levels
- Current at operating levels

Most people are familiar with the normal current-carrying ampere rating of fuses and circuit breakers. For example, if an overcurrent protective device is designed to open a circuit when the circuit load exceeds 20 amperes for a given time period, as the current approaches 20 amperes, the overcurrent protective device begins to overheat. If the current barely exceeds 20 amperes, the circuit breaker will open normally or a fuse link will melt after a given period of time with little, if any, arcing. If, say, 40 amperes of current were instantaneously applied to the circuit, the overcurrent protective device will open instantly, but again with very little arcing. However, if a ground fault occurs on the circuit that ran the amperage up to, say, 5000 amperes, an explosion effect would occur within the protective device. One simple indication of this "explosion effect" is blackened windows of plug fuses.

If this fault current exceeds the interrupting rating of a fuse or circuit breaker, the protective device can be damaged or destroyed; such current can also cause severe damage to equipment and injure personnel. Therefore, selecting overcurrent protective devices with the proper interrupting capacity is extremely important in all electrical systems.

For better understanding of interrupting rating, consider the following analogy. Let's use a dammed stream (Figure 7-5) as an example. Consider the reservoir capacity to be the available fault current in an electrical circuit; the flood gates (located downstream from the dam) to be the overcurrent protective device in the circuit rated at 10,000 gallons per minute (10,000 AIC) and the stream of water coming through the discharge pipes in the dam to be the normal load current. Our drawing shows a normal flow of 100 gallons per minute. Also note the bridge crossing the stream, downstream from the flood gates. This bridge will represent downstream circuit components or equipment connected to the circuit.

Figure 7-6 shows this same diagram with a fault in the dam — creating a water "short-circuit" that allows 50,000 gallons per minute to flow (50,000 fault-circuit amperes). Such a situation destroys the flood gates because of inadequate interrupting rating. The overcurrent protective device in the circuit will also be destroyed. With the flood gates damaged, this surge of water continues downstream, wrecking the bridge. The downstream components may not be able to withstand the let-through current in an electrical circuit either.

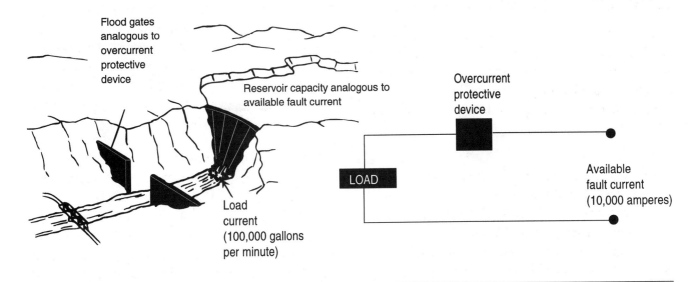

Figure 7-5: Normal current operation

Figure 7-7 on page 144 shows the same situation, but with adequate interrupting capacity. Note that the flood gates have adequately contained the surge of water and restricted the let-through current to an amount that can be withstood by the bridge, or the components downstream.

There are several factors that must be considered when calculating the required interrupting capacity of an overcurrent protective device. *NEC* Section 110-10 states the following:

- "The overcurrent protective devices, the total impedance, the component short-circuit withstand ratings, and other characteristics of the circuit to be

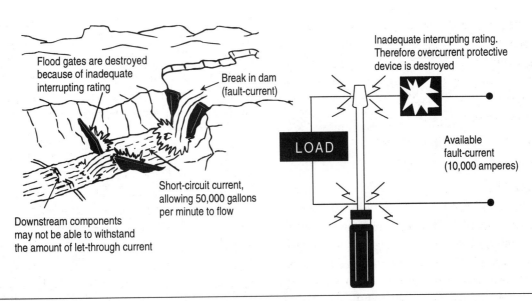

Figure 7-6: Short-circuit operation with inadequate interrupting rating

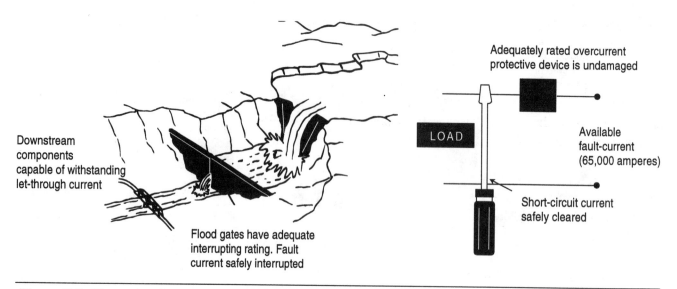

Figure 7-7: Short-circuit operation with adequate interrupting rating

protected shall be so selected and coordinated as to permit the circuit protective devices that are used to clear a fault without the occurrence of extensive damage to the electrical components of the circuit. This fault shall be assumed to be either between two or more of the circuit conductors, or between any circuit conductor or the grounding conductor or enclosing metal raceway."

Component short-circuit rating is a current rating given to conductors, switches, circuit breakers and other electrical components, which, if exceeded by fault currents, will result in "extensive" damage to the component. The rating is expressed in terms of time intervals and/or current values. Short-circuit damage can be the result of heat generated or the electro-mechanical force of high-intensity, magnetic field.

The *NEC*'s intent is that the design of a system must be such that short-circuit currents cannot exceed the withstand ratings of the components selected as part of the system. Given specific system components, and level of "available" short-circuit currents that could occur, overcurrent protective devices (mainly fuses and/or circuit break-

ers) must be used which will limit the energy let-through of fault currents to levels within the withstand ratings of the system components.

In most large commercial and industrial installations, it is necessary to calculate available short-circuit currents at various points in a system to determine if the equipment meets the requirements of *NEC* Sections 110-9 and 110-10. There are a number of methods used to determine the short-circuit requirements in an electrical system. Some give approximate values; some require extensive computations and are quite exacting. A simple, yet quite accurate method is the point-by-point method which will be discussed in more detail in this chapter under fuses.

An overcurrent protective device must be selected with interrupting capacity equal to or greater than the available short-circuit current at the point where the circuit breaker or fuse is applied in the system. The breaker interrupting capacity is based on tests to which the breaker is subjected. There are two such tests; one set up by UL and the other by NEMA. The NEMA tests are self-certification while UL tests are certified by unbiased witnesses. UL tests have been limited to a maximum of 10,000 amperes in the past, so the emphasis was placed on NEMA tests with higher ratings. UL tests now include the NEMA tests plus other ratings. Consequently, the emphasis is now being placed on UL tests.

The interrupting capacity of a circuit breaker is based on its rated voltage. Where the circuit breaker can be used on more than one voltage, the interrupting capacity will be shown for each voltage level. For example, the LA type circuit breaker has 42,000 amperes, symmetrical interrupting capacity at 240 volts, 30,000 amperes symmetrical at 480 volts, and 22,000 amperes symmetrical at 600 volts.

Standard Interrupting Capacity

Standard interrupting capacity rated circuit breakers can be identified by the black operating handle and black printed interrupting rating labels.

The interrupting rating of a circuit breaker is as important in application as the voltage and current ratings and should be considered each time a breaker is applied.

In residential applications, the available ground-fault circuit current is seldom higher than 10,000 amperes or even near this value. The QO and A1 breakers (15 -150 amperes) with the interrupting rating of 10,000 amperes are used in these applications. Higher interrupting ratings are available when required up to 65,000 AIC.

High Interrupting Capacity

Where still higher interrupting capacity than the standard ratings discussed previously is required in the 15-100 ampere FA type circuit breaker, the FH ("H" for high interrupting capacity) is available. The FH type circuit breaker has an interrupting rating of 65,000 amperes, symmetrical at 240 volts ac. The continuous current ratings are duplicated in these breakers (15-100 amperes) but the interrupting capacity has been increased to satisfy the need for greater interrupting capacity. This type of breaker is applied in installations where the higher ground-fault currents are available (such as large industrial plants). FA and FH circuit breakers are identified quickly by their gray colored handles and red interrupting rating labels. This is in contrast to the black handles on breakers with standard interrupting ratings. These breakers are dimensionally identical to the lower interrupting capacity rated breakers, but physically they are built with a case material that will withstand the higher shocks from heat and interrupting forces and is extremely flame resistant. Circuit breakers with high interrupting capacity are available in the FH, KH, LH, MH, NH, and PH types.

Three ampere ratings (15, 20, and 30) are also available in the QH high interrupting capacity breakers. These breakers satisfy conditions of lighting circuits supplied from these high available ground-fault circuit breakers.

Current-Limiting Circuit Breakers

The need for current limitation is the result of increasingly higher available ground-fault currents associated with the growth and interconnection of modern power systems. To meet this need, current-limiting circuit breakers have been developed. This type of breaker operates extremely fast to provide downstream protection for other types of fuses with as little as 10,000 AIC rating on systems with 100,000 ampere available fault current.

There are circuit breakers available for almost every need in any electrical system. For example, in circuits where the breaker must be highly sensitive to amperage changes, in an environment where the temperature changes are great, an ambient-compensating circuit breaker may be used. Its design permits it to compensate for temperature variation.

Figure 7-8 on page 146 shows the internal parts of a current-limiting circuit breaker. This is basically a conventional, common-trip, thermal magnetic circuit breaker with an independently acting limiter section in series with each pole. For purposes of explanation, each limiter can be represented by a set of contacts shunted by a "transformable" resistor.

At current below a threshold of about 1,000 amperes, the current-limiting circuit breaker per-

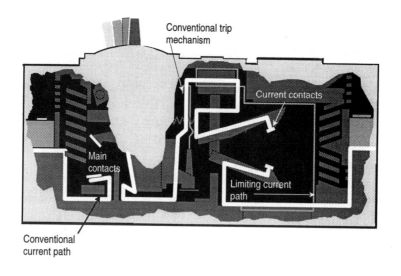

Figure 7-8: Current-limiting circuit breaker

forms in a manner similar to conventional thermal magnetic circuit breakers. In the event of an overload or minor fault current, only the breaker section contacts open to interrupt the circuit. The limiter contacts remain closed. See Figure 7-9.

For 100-ampere frame current-limiting circuit breakers, the threshold of current limitation occurs in the range of 1,000 to 2,000 amperes. Above this point, the limiter contacts open first — starting the limiting action within $1/4$ of a millisecond of fault

initiation. A substantial arc voltage is generated across the limiter contacts as shown in Figure 7-10. Rapid generation of this arc voltage is essential to drastically limit the peak let-through current in the faulted circuit.

The rapid rising equivalent resistance of the arc across the limiter contacts forces the current to transfer to the alternate path provided by the limiter resistor. This special resistor has a high positive temperature coefficient of resistance. As

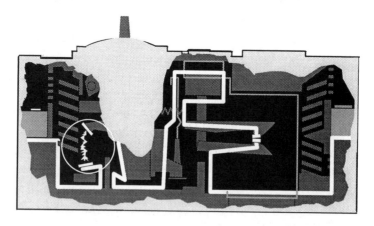

Figure 7-9: Conventional operation

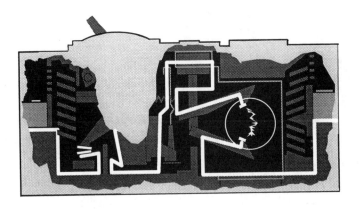

Figure 7-10: Limiting action begins

current is transferred to it, the temperature of the resistor increases rapidly with a consequent rapid rise in resistance. The resistor dissipates and limits let-through energy and increases the short-circuit power factor of the circuit nearly to unity, resulting in an easier, more reliable, transient voltage free interruption. See Figure 7-11.

During the extremely short period of operation of the limiter section, the breaker section tripping mechanism is activated and its contacts begin to open. After the arc between the limiter contacts has been extinguished and all the current is flowing through the limiter resistor, the breaker contacts interrupt the already drastically limited fault current. Because the fault current and system voltage are by then nearly in phase, complete interruption of the circuit takes place near the first voltage zero of the ac wave. See Figure 7-12 on page 148.

When a current-limiting circuit breaker trips to clear a faulted circuit, it can be reset in the same manner as a conventional molded case circuit breaker. This is true whether tripping was due to

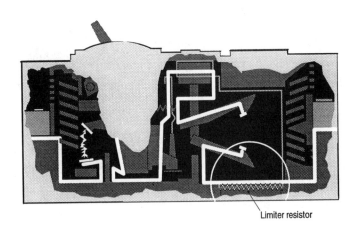

Limiter resistor

Figure 7-11: The limiter resistor

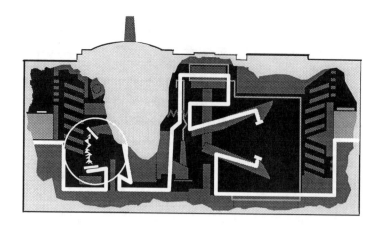

Figure 7-12: Operation of the breaker contacts

low level overload currents or major faults involving the breaker's current limiting mode of operation. The limiter contacts are automatically closed after circuit interruption. Breaker contacts are re-closed by moving the handle to the extreme OFF position and then to ON. There are no fusible elements in this type of circuit breaker to replace. See Figure 7-13.

More detailed information on current-limiting circuit breakers may be obtained from manufacturers' data, usually at no charge. Check with your local electrical-equipment supplier.

Installing And Maintaining Circuit Breakers

After circuit breakers have been installed, make sure all connections are tight.

Good maintenance of circuit breakers and their enclosures is very important and necessary to obtain the best service and performance. In order to keep circuit breakers in proper operating condition, the scheduling of maintenance inspections is important.

Since every circuit-breaker failure represents a potential hazard to other equipment on the system,

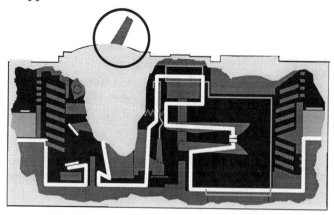

Figure 7-13: Resetting is done with breaker handle

it is difficult to calculate risks involved in prolonging maintenance inspections. One of the chief causes of circuit-breaker failure is high heat caused by loose connections at the load side of the breaker. Another cause of circuit-breaker failure is defective loads; that is, electrical equipment that cycle too frequently which in turn causes the circuit breaker to overheat. Therefore, in the majority of circuit-breaker failures, heat will be found to be the biggest culprit.

LABELING CONSIDERATIONS

NEC Sections 110-22 and 240-83 require special marking for UL recognized series rated systems. For example, *NEC* Section 110-22 states:

- "Where circuit breakers or fuses are applied in compliance with the series combination ratings marked on the equipment by the manufacturer, the equipment enclosure(s) shall be legibly marked in the field to indicate the equipment has been applied with a series combination rating. The marking shall be readily visible and state Caution — Series Rated System. . ."

On UL recognized series rated systems, the downstream equipment will be marked with a field installed label per *NEC* Section 110-22.

FUSES

A fuse is the simplest device for opening an electric circuit when excessive current flows because of an overload or such fault conditions as grounds and short circuits. A "fusible" link or links encapsulated in a tube and connected to contact terminals comprise the fundamental elements of the basic fuse. Electrical resistance of the link is so low that it simply acts as a conductor, and every fuse is intended to be connected in series with each phase conductor so that current flowing through the conductor to any load must also pass through the fuse. The continuous current rating of the fuse in amperes establishes the maximum amount of current the fuse will carry without opening. When circuit current flow exceeds this value, an internal element (link) in the fuse melts due to the heat of the current flow and opens the circuit. Fuses are manufactured in a wide variety of types and sizes with different current ratings, different abilities to interrupt fault currents, various speeds of operation (either quick-opening or time-delay opening), different internal and external constructions, and voltage ratings for both low-voltage (600 volts and below) and medium-voltage (over 600 volts) circuits.

Voltage Rating: Most low voltage power distribution fuses have 250V or 600V ratings (other ratings are 125V and 300V). The voltage rating of a fuse must be at least equal to the circuit voltage. It can be higher but never lower. For instance, a 600V fuse can be used in a 208V circuit. The voltage rating of a fuse is a function of or depends upon its capability to open a circuit under an overcurrent condition. Specifically, the voltage rating determines the ability of the fuse to suppress the internal arcing that occurs after a fuse link melts and an arc is produced. If a fuse is used with a voltage rating lower than the circuit voltage, arc suppression will be impaired and, under some fault current conditions, the fuse may not safely clear the overcurrent.

Ampere Rating: Every fuse has a specific ampere rating. In selecting the ampacity of a fuse, consideration must be given to the type of load and code requirements. The ampere rating of a fuse should normally not exceed current-carrying capacity of the circuit. For instance, if a conductor is rated to carry 20A, a 20A fuse is the largest that should be used in the conductor circuit. However, there are some specific circumstances when the ampere rating is permitted to be greater than the current-carrying capacity of the circuit. A typical example is the motor circuit; dual-element fuses generally are permitted to be sized up to 175 percent and non-time-delay fuses up to 300 percent of the motor full-load amperes. Generally, the ampere rating of a fuse and switch combination should be selected at 125 percent of the load current. There are exceptions, such as when the fuse-switch combination is approved for continuous operation at 100 percent of its rating.

Interrupting Rating: A protective device must be able to withstand the destructive energy of short-circuited currents. If a fault current exceeds a level beyond the capability of the protective device, the device may actually rupture and cause severe damage. Thus, it is important in applying a fuse or circuit breaker to use one which can sustain the largest potential short-circuit currents. The rating that defines the capacity of a protective device to maintain its integrity when reacting to fault currents is termed its interrupting rating. The interrupting rating of most branch-circuit, molded case, circuit breakers typically used in residential service entrance boxes is 10,000A. The rating is usually expressed as "10,000 amperes interrupting capacity (AIC)." Larger, more expensive circuit breakers may have AIC's of 14,000A or higher. In contrast, most modern, current-limiting fuses have an interrupting capacity of 200,000A and are commonly used to protect the lower rated circuit breakers. The 1999 *NEC*, Section 110-9, requires equipment intended to break current at fault levels to have an interrupting rating sufficient for the current that must be interrupted.

Time Delay: The time-delay rating of a fuse is established by standard UL tests. All fuses have an inverse time current characteristic. That is, the fuse will open quickly on high currents and after a period of time delay, on low overcurrents. Specific types of fuses are made to have specially determined amounts of time delay. The basic UL requirement on time delay for Class RK-1, RK-15, and J fuses which are marked "time delay" is that the fuse must carry a current equal to five times its continuous rating for a period not less than ten seconds. UL has not developed time-delay tests for all fuse classes. Fuses are available for use where time delay is needed along with current limitation on high-level short circuits. In all cases, manufacturers' literature should be consulted to determine the degree of time delay in relation to the operating characteristics of the circuit being protected.

Types Of Fuses

Plug Fuses: Plug fuses have a screw-shell base and are sometimes used in small commercial buildings for circuits that supply lighting and 120-volt power outlets. Plug fuses are supplied with standard screw bases (Edison-base) or Type S bases. An Edison-base fuse consists of a strip of fusible (capable of being melted) metal in a small porcelain or glass case, with the fuse strip, or link, visible through a "window" in the top of the fuse. The screw base corresponds to the base of a standard medium-base incandescent lamp. Edison-base fuses are permitted only as replacements in existing installations; all new work must use the S-base fuses (*NEC* Section 240-51(b).

- "Plug fuses of the Edison-base type shall be used only for replacements in existing installations where there is no evidence of overfusing or tampering."

The chief disadvantage of the Edison-base plug fuse is that it is made in several ratings from 0 to 30 A, all with the same size base — permitting unsafe replacement of one rating by a higher rating. Type S fuses were developed to reduce the possibility of over-fusing a circuit (inserting a fuse with a rating greater than that required by the circuit). There are 15 classifications of Type S fuses based on current rating: 0-30 amperes. Each Type S fuse has a base of a different size and a matching adapter. Once an adapter is screwed into a standard Edison-base fuseholder, it locks into place and is not readily removed without destroying the fuseholder. As a result, only a Type S fuse with a size same as that of the adapter may be inserted. Two types of plug fuses are shown in Figure 7-14; a Type S adapter is also shown.

Plug fuses also are made in time-delay types that permit a longer period of overload flow before operation, such as on motor inrush current and other higher-than-normal rated currents. They are available in ratings up to 30 amperes, both in Edison base and Type S. Their principle use is in motor circuits, where the starting inrush current to the motor is much higher than the running, or

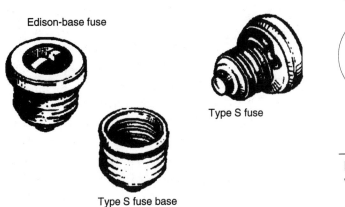

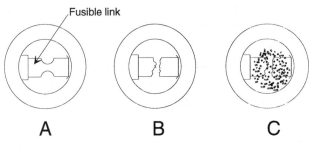

Fusible link

A   B   C

Figure 7-16: Window view of plug fuse showing various conditions of the element

Figure 7-14: Types of common plug fuses

Edison-base fuse

Type S fuse

Type S fuse base

continuous, current. The time-delay fuse will not open on the inrush of high-starting current. If, however, the high current persists, the fuse will open the circuit just as if a short circuit or heavy overload current had developed. All Type S fuses are time-delay fuses. Figure 7-15 shows the internal construction of a time-delay fuse.

Plug fuses are permitted to be used in circuits of no more than 125 volts between phases, but they may be used where the voltage between any ungrounded conductor and ground is not more than 150 volts. The screwshell of the fuseholder for plug fuses must be connected to the load side

circuit conductor; the base contact is connected to the line side or conductor supply. A disconnecting means (switch) is not required on the supply side of a plug fuse.

The plug fuse is a *nonrenewable* fuse; that is, once it has opened the circuit because of a fault or overload, it cannot be used again or renewed. It must be replaced by a new fuse of the same rating and characteristics for safe and effective restoration of circuit operation.

The "window" of a plug fuse can tell much about the condition of the fuse, and also the probable cause if the fuse blows. For example, Figure 7-16 shows window views of a plug fuse in three conditions. If the fuse is good, then the element or link will be whole and clearly visible as shown at A. If the current through the fuse only slightly exceeded the current rating over a period of time, then a melt-through would occur as shown at B. However, if very high currents occurred — as in a short-circuit, the fuse will melt more violently and will spray the zinc or alloy over the inside viewing window as shown in C. The window will also be smudged with black.

CARTRIDGE FUSES

In most industrial and commercial applications, cartridge fuses are used because they have a wider range of types, sizes, and ratings than do plug

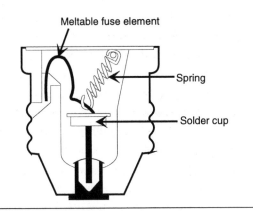

Meltable fuse element

Spring

Solder cup

Figure 7-15: Time-delay plug fuse

fuses. Many cartridge fuses are also provided with a means to renew the fuse by unscrewing the end caps and replacing the links. In large industrial installations where thousands of cartridge fuses are in use to protect motors and machinery, this replacement feature can be a significant savings over a period of a year. There are several different types of cartridge fuses, and although all operate in a similar fashion, all have slightly different characteristics. Each is described in the paragraphs to follow.

Single-Element Cartridge Fuses: The basic component of a fuse is the link. Depending upon the ampere rating of the fuse, the single-element fuse may have one or more links. They are electrically connected to the end blades (or ferrules) and enclosed in a tube or cartridge surrounded by an arc-quenching filler material.

Under normal operation, when the fuse is operating at or near its ampere rating, it simply functions as a conductor. However, if an overload current occurs and persists for more than a short interval of time, the temperature of the link eventually reaches a level that causes a restricted segment of the link to melt; as a result, a gap is formed and an electric arc established. See Figure 7-17. However, as the arc causes the link metal to burn back, the gap becomes progressively larger. Electrical resistance of the arc eventually reaches such a high level that the arc cannot be sustained and is extinguished; the fuse will have then completely cut off all current flow in the circuit. Suppression or quenching of the arc is accelerated by the filler material.

Single-element fuses have a very high speed of response to overcurrents. They provide excellent short-circuit component protection. However, temporary harmless overloads or surge currents may cause nuisance openings unless these fuses are oversized. They are best used, therefore, in circuits not subject to heavy transient surge currents and the temporary overload of circuits with inductive loads such as motors, transformers, and solenoids. Because single-element fuses have a high speed-of-response to short-circuit currents, they are particularly suited for the protection of circuit breakers with low interrupting ratings.

Dual-Element Cartridge Fuses: Unlike single-element fuses, the dual-element fuse can be applied in circuits subject to temporary motor overload and surge currents to provide both high performance short-circuit and overload protection. Oversizing in order to prevent nuisance openings is not necessary. The dual-element fuse contains two distinctly separate types of elements. Electrically, the two elements are series connected. The fuse links similar to those used in the single-element fuse perform the short-circuit protection function; the overload element provides protection against low-level overcurrents or overloads and will hold an overload that is five times greater than the ampere rating of the fuse for a minimum time of ten seconds.

As shown in Figure 7-18 on page 154, the overload section consists of a copper heat absorber and a spring-operated trigger assembly. The heat-absorber strip is permanently connected to the short-circuit link and to the short-circuit link on the opposite end of the fuse by the S-shaped connector of the trigger assembly. The connector electronically joins the one short-circuit link to the heat absorber in the overload section of the fuse. These elements are joined by a "calibrated" fusing alloy. An overload current causes heating of the short-circuit link connected to the trigger assembly. Transfer of heat from the short-circuit link to the heat absorbing strip in the midsection of the fuse begins to raise the temperature of the heat absorber. If the overload is sustained, the temperature of the heat absorber eventually reaches a level that permits the trigger spring to "fracture" the calibrated fusing alloy and pull the connector free. The short-circuit link is electrically disconnected from the heat absorber, the conducting path through the fuse is opened, and overload current is interrupted. A critical aspect of the fusing alloy is that it retains its original characteristic after repeated temporary overloads without degradation. The main purposes of dual-element fuses are as follows:

- Provide motor overload, ground-fault and short-circuit protection.

- Permit the use of smaller and less costly switches.

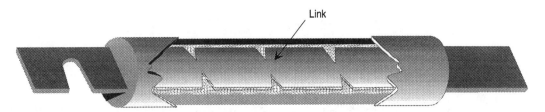

Link

Cut-away view of single-element fuse.

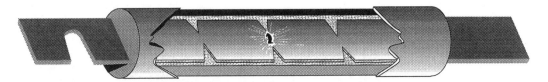

Under sustained overload a section of the link
melts and an arc is established.

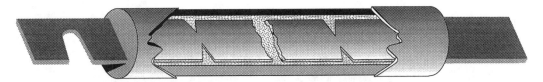

The "open" single-element fuse after opening a
circuit overload.

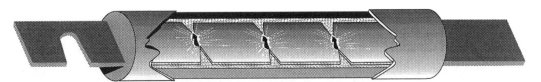

When subjected to a short-circuit, several sections
of the fuse link melt almost instantly.

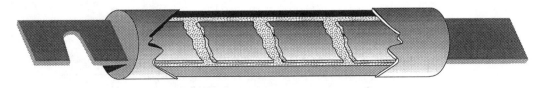

The appearance of an "open" single-element fuse
after opening a short-circuit.

Figure 7-17: Characteristics of single-element fuses

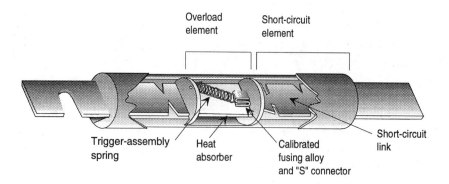

The true dual-element fuse has distinct and separate overload
and short-circuit elements.

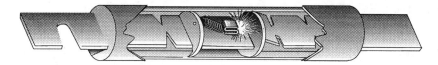

Under sustained overload conditions, the trigger spring fractures
the calibrated fusing alloy and releases the "connector."

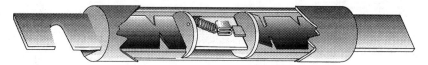

The "open" dual-element fuse after opening under an overload.

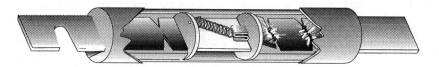

Like the single-element fuse, a short-circuit current causes the
restricted portions of the short-circuit elements to melt and arcing
to burn back the resulting gaps until the arcs are suppressed by
the arc-quenching material and increased arc resistance.

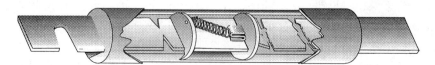

The "open" dual-element fuse after opening under a short-circuit
condition.

Figure 7-18: Characteristics of dual-element time-delay fuses

- Give a higher degree of short-circut protection (greater current limitation) in circuits in which surge currents or temporary overloads occur.

- Simplify and improve blackout prevention (selective coordination).

Dual-element fuses may also be used in circuits other than motor branch circuits and feeders, such as lighting circuits and those feeding mixed lighting and power loads. The low-resistance construction of the fuses offers cooler operation of the equipment, which permits higher loading of fuses in switches and panel enclosures without heat damage and without nuisance openings from accumulated ambient heat.

Fuse Markings

It is a requirement of the *NEC* that cartridge fuses used for branch-circuit or feeder protection must be plainly marked, either by printing on the fuse barrel or by a label attached to the barrel, showing the following:

- Ampere rating

- Voltage rating

- Interrupting rating (if other than 10,000A)

- "Current limiting," where applicable

- The name or trademark of the manufacturer

Underwriters' Laboratories' Fuse Classes

Fuses are tested and listed by Underwriters' Laboratories Inc. in accordance with established standards of construction and performance. There are many varieties of miscellaneous fuses used for special purposes or for supplementary protection of individual types of electrical equipment. However, here we will chiefly be concerned with those fuses used for protection of branch circuits and

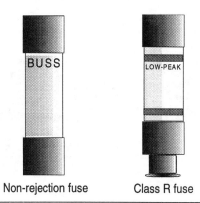

Non-rejection fuse   Class R fuse

Figure 7-19: Comparison of Class H and Class R fuses

feeders on systems operating at 600 volts or below. Cartridge fuses in this category include UL Class RK1, RK5, H, K, G, J, R, T, CC, and L.

Class R Fuses: UL Class R (rejection) fuses are high performance $\frac{1}{10}$ to 600 ampere units, 250 volt and 600 volt, having a high degree of current limitation and a short-circuit interrupting rating of up to 200,000 amperes (rms symmetrical). This type of fuse is designed to be mounted in rejection type fuseclips to prevent older type Class H fuses from being installed. Since Class H fuses are not current limiting and are recognized by UL as having only a 10,000 ampere interrupting rating, serious damage could result if a Class H fuse were inserted in a system designed for Class R fuses. Consequently, *NEC* Section 240-60(b) requires fuseholders for current-limiting fuses to reject non-current limiting type fuses.

Figure 7-19 shows a standard Class H fuse (left) and a Class R fuse (right). A grooved ring in one ferrule of the Class R fuse provides the rejection feature of the Class R fuse in contrast to the lower interrupting capacity, non-rejection type. Figure 7-20 on page 156 shows Class R type fuse rejection clips that accept only the Class R rejection type fuses.

Class CC Fuses: 600-volt, 200,000-ampere interrupting rating, branch circuit fuses with overall dimensions of $\frac{15}{32}$ inch $\times 1\frac{1}{2}$ inches. Their design incorporates rejection features that allow them to be inserted into rejection fuse holders and fuse

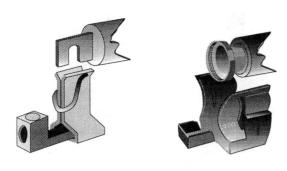

Figure 7-20: Class R fuse rejection clips that accept only Class R fuses

blocks that reject all lower voltage, lower interrupting rated fuses. They are available from $\frac{1}{10}$ ampere through 30 amperes.

Class G Fuses: 300-volt, 100,000 ampere interrupting rating branch-circuit fuses that are size rejecting to eliminate overfusing. The fuse diameter is $\frac{13}{32}$ inch while the length varies from $1\frac{5}{16}$ inches to $2\frac{1}{4}$ inches. These are available in ratings from 1 amp through 60 amps.

Class H Fuses: 250-volt and 600-volt, 10,000 ampere interrupting rating branch circuit fuses that may be renewable or non-renewable. These are available in ampere ratings of 1 amp through 600 amps.

Class J Fuses: These fuses are rated to interrupt 200,000 amperes ac. They are UL labeled as "current limiting," are rated for 600 volts ac, and are not interchangeable with other classes.

Class K Fuses: These are fuses listed by UL as K-1, K-5, or K-9 fuses. Each subclass has designated I^2t and Ip maximums. These are dimensionally the same as Class H fuses (*NEC* dimensions) and they can have interrupting ratings of 50,000, 100,000, or 200,000 amps. These fuses are current limiting, however, they are not marked "current limiting" on their label since they do not have a rejection feature.

Class L Fuses: These fuses are rated for 601 through 6000 amperes, and are rated to interrupt 200,000 amperes ac. They are labeled "current

limiting" and are rated for 600 volts ac. They are intended to be bolted into their mountings and are not normally used in clips. Some Class L fuses have designed-in time-delay features for all-purpose use.

Class T Fuses: A UL classification of fuses in 300 volt and 600 volt ratings from 1 amp through 1200 amps. They are physically very small and can be applied where space is at a premium. They are fast acting fuses, with an interrupting rating of 200,000 amps RMS.

Branch-Circuit Listed Fuses

Branch-circuit listed fuses are designed to prevent the installation of fuses that cannot provide a comparable level of protection to equipment. The characteristics of branch-circuit fuses are as follows:

- They must have a minimum interrupting rating of 10,000 amperes.
- They must have a minimum voltage rating of 125 volts.
- They must be size rejecting such that a fuse of a lower voltage rating cannot be installed in the circuit.
- They must be size rejecting such that a fuse with a current rating higher than the fuseholder rating cannot be installed.

Medium-Voltage Fuses

Fuses above 600 volts are classified under one of three classifications as defined in ANSI/IEEE 40-1981:

- General purpose current-limiting fuse
- Back-up current-limiting fuse
- Expulsion fuse

General Purpose Current-Limiting Fuse: A fuse capable of interrupting all currents from the rated interrupting current down to the current that causes melting of the fusible element in one hour.

Back-Up Current-Limiting Fuse: A fuse capable of interrupting all currents from the maximum rated interrupting current down to the rated minimum interrupting current.

Expulsion Fuse: A vented fuse in which the expulsion effect of gasses produced by the arc and lining of the fuseholder, either alone or aided by a spring, extinguishes the arc.

One should note that in the definitions just given, the fuses are defined as either expulsion or current limiting. A current-limiting fuse is a sealed, non-venting fuse that, when melted by a current within its interrupting rating, produces arc voltages exceeding the system voltage which in turn forces the current to zero. The arc voltages are produced by introducing a series of high resistance arcs within the fuse. The result is a fuse that typically interrupts high fault currents with the first $\frac{1}{2}$ cycle of the fault.

In contrast an expulsion fuse depends on one arc to initiate the interruption process. The arc acts as a catalyst causing the generation of de-ionizing gas from its housing. The arc is then elongated either by the force of the gasses created or a spring. At some point the arc elongates far enough to prevent a restrike after passing through a current zero. Therefore, it is not atypical for an expulsion fuse to take many cycles to clear.

Application Of Medium-Voltage Fuses

Many of the rules for applying expulsion fuses and current-limiting fuses are the same, but because the current-limiting fuse operates much faster on high-fault, some additional rules must be applied.

Three basic factors must be considered when applying any fuse:

- Voltage
- Continuous current-carrying capacity
- Interrupting rating

Voltage: The fuse must have a voltage rating equal to or greater than the normal frequency recovery voltage which will be seen across the fuse under all conditions. On three-phase systems it is a good rule-of-thumb that the voltage rating of the fuse be greater than or equal to the line-to-line voltage of the system.

Continuous Current-Carrying Capacity: Continuous current values that are shown on the fuse represent the level of current the fuse can carry continuously without exceeding the temperature rises as specified in ANSI C37.46. An application that exposes the fuse to a current slightly above its continuous rating but below its minimum interrupting rating may damage the fuse due to excessive heat. This is the main reason overload relays are used in series with back-up current-limiting fuses for motor protection.

Interrupting Rating: All fuses are given a maximum interrupting rating. This rating is the maximum level of fault current that the fuse can safely interrupt. Back-up current-limiting fuses are also given a minimum interrupting rating. When using back-up current-limiting fuses, it is important that other protective devices are used to interrupt currents below this level.

When choosing a fuse, it is important that the fuse be properly coordinated with other protective devices located upstream and downstream. To accomplish this, one must consider the melting and clearing characteristics of the devices. Two curves, the minimum melting curve and the total clearing curve, provide this information. To insure proper coordination, the following rules should be used:

- The total clearing curve of any downstream protective device must be below a curve representing 75 percent of the minimum melting curve of the fuse being applied.

- The total clearing curve of the fuse being applied must lie below a curve representing 75 percent of the minimum melting curve for any upstream protective device.

Current-Limiting Fuses

To insure proper application of a current-limiting fuse, it is important that the following additional rules be applied.

1. Current-limiting fuses produce arc voltages that exceed the system voltage. Care must be taken to make sure that the peak voltages do not exceed the insulation level of the system. If the fuse voltage rating is not permitted to exceed 140 percent of the system voltage, there should not be a problem. This does not mean that a higher rated fuse cannot be used, but points out that one must be assured that the system insulation level (BIL) will handle the peak arc voltage produced. BIL stands for basic impulse level which is the reference impulse insulation strength of an electrical system.

2. As with the expulsion fuse, current-limiting fuses must be properly coordinated with other protective devices on the system. For this to happen, the rules for applying an expulsion fuse must be used at all currents that cause the fuse to interrupt in 0.01 seconds or greater.

When other current-limiting protective devices are on the system, it becomes necessary to use I^2t values for coordination at currents causing the fuse to interrupt in less than 0.01 seconds. These values may be supplied as minimum and maximum values or minimum melting and total clearing I^2t curves. In either case, the following rules should be followed:

1. The minimum melting I^2t of the fuse should be greater than the total clearing I^2t of the downstream current-limiting device.

2. The total clearing I^2t of the fuse should be less than the minimum melting I^2t of the upstream current-limiting device.

The fuse-selection chart in Figure 7-21 should serve as a guide for selecting fuses on circuits of 600 volts or less. The dimensions of various Buss fuses are shown in Figure 7-22, beginning on page 160. Both of these charts will prove invaluable on all types of projects — from residential to heavy industrial applications. Other valuable information may be found in catalogs furnished by manufacturers of overcurrent protective devices. These are usually obtainable from electrical supply houses or from manufacturers' reps. You may also write the various manufacturers for a complete list (and price, if any) for all reference materials offered by them.

NEC REGULATIONS

The "critical" *NEC* sections shown in Figure 7-23 (beginning on page 163) are ones with which every electrician should should become familiar. These are not the only *NEC* requirements of importance when dealing with overcurrent protective devices, but these are considered by many to be the most important.

Summary

Overcurrent protection offers one of the greatest means of protection against electrical faults to both equipment and personnel. However, there are certain precautionary measures that must be observed at all times.

- Make certain the switch or circuit breaker is open before making any inspections or changes to either the overcurrent device or the circuit it is protecting.

- Make certain that all components are compatible with each other and with the system on which they are used; that is, manufacturer, type, normal current ratings, interrupting capacity, voltage (especially voltage), etc.

- Use the proper tools for the job; that is, fuse pullers, fuse wrenches, circuit-breaker screwdrivers, and the like.

Circuit	Load	Ampere Rating	Fuse Type	Symbol	Voltage Rating (ac)	UL Class	Interrupting Rating (K)	Remarks
Main, Feeder and Branch (Conventional dimensions)	All type load (optimum overcurrent protection)	0-600A	LOW-PEAK® (dual-element, time-delay)	LPN-RK	250V	RK1	200	All-purpose fuses. Unequaled for combined short-circuit and overload protection.
				LPS-RK	600V			
		601 to 6000A	LOW-PEAK® time delay	KRP-C	600V	L	200	
	Motors, welders, transformers, capacitor banks (circuits with heavy inrush currents)	0 to 600A	FUSETRON® (dual-element, time-delay)	FRN-R	250V	RK5	200	Moderate degree of current limitation. Time-delay passes surge currents.
				FRS-R	600V			
		601 to 4000A	LIMITRON® (time delay)	KLU	600V	L	200	All-purpose fuse. Time-delay passes surge-currents.
	Non-motor loads (circuits with no heavy inrush currents)	0 to 600A	LIMITRON® (fast-acting)	KTN-R	250V	RK1	200	Same short-circuit protection as LOW-PEAK fuses but must be sized larger for circuits with surge-currents; i.e., up to 300%.
				KTS-R	600V			
	LIMITRON fuses particularly suited for circuit breaker protection	601 to 6000A	LIMITRON® (fast-acting)	KTU	600V	L	200	A fast-acting, high performance fuse.
Note: All shaded areas represent fuses with reduced dimensions for installation in restricted space	All type loads (optimum overcurrent protection)	0 to 600A	LOW-PEAK® (dual-element time-delay)	LPJ	600V	J	200	All-purpose fuses. Unequaled for combined short-circuit and overload protection.
	Non-motor loads (circuits with no heavy inrush currents)	0 to 600A	LIMITRON® (quick-acting)	JKS	600V	J	200	Very similar to KTS-R LIMITRON, but smaller.
		0 to 1200A	T-TRON™	JJN	300V	T	200	The space saver (1/3 the size of KTN-R/KTS-R).
				JJS	600V			

Figure 7-21: Fuse selection chart (600 volts or less)

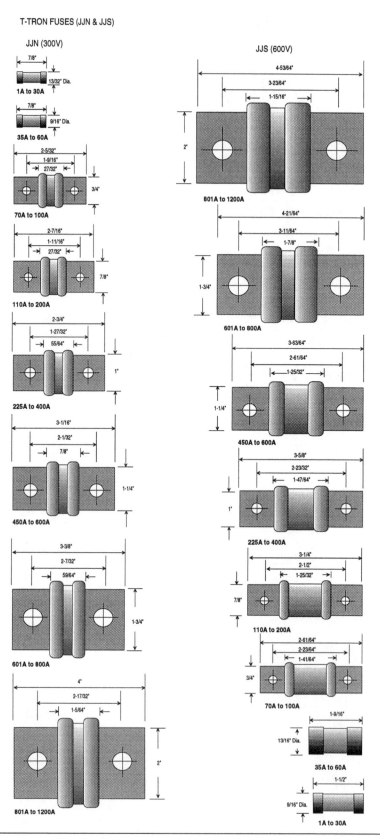

Figure 7-22: Buss fuse dimensional data

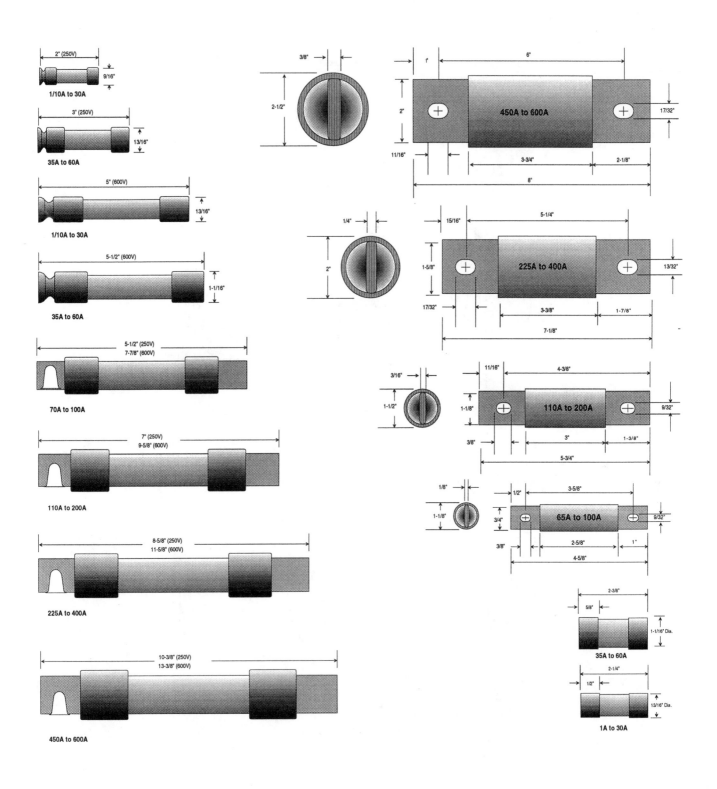

Figure 7-22: Buss fuse dimensional data (*cont.*)

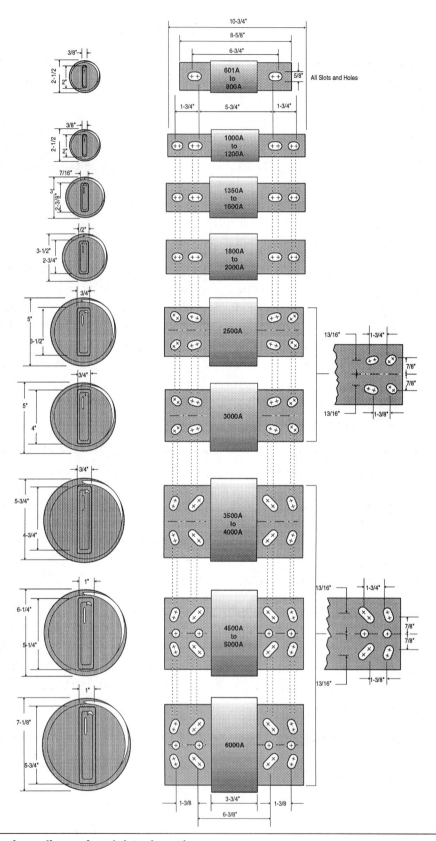

Figure 7-22: Buss fuse dimensional data (*cont.*)

NEC Section	*NEC* Requirements
110-3(b)	**Installation and Use.** Listed or labeled equipment shall be used or installed in accordance with any instructions included in the listing or labeling.
110-9	**Interrupting Rating.** Equipment intended to break current at fault levels shall have an interrupting rating sufficient for the nominal circuit voltage and the current that is available at the terminals of the equipment. Equipment intended to break current at other than fault levels shall have an interrupting rating at nominal circuit voltage sufficient for the current that must be interrupted.
110-10	**Circuit Impedance and Other Characteristics.** The overcurrent protective devices, the total impedance, the component short-circuit withstand ratings, and other characteristics of the circuit to be protected shall be so selected and coordinated as to permit the circuit protective devices that are used to clear a fault without the occurrence of extensive damage to the electrical components of the circuit. The fault shall be assumed to be either between two or more of the circuit conductors, or between any circuit conductor and the grounding conductor or enclosing metal raceway.
110-10	**Available Short-Circuit Current.** Service equipment shall be suitable for the short-circuit current available at its supply terminals.
240-1	**Scope (FPN).** Overcurrent protection for conductors and equipment is provided to open the circuit if the current reaches a value that will cause an excessive or dangerous temperature in conductors or conductor insulation. See also Sections 110-9 and 110-10 for requirements for interrupting capacity and protection against fault currents.
240-11	**Definition of Current-Limiting Overcurrent Protective Devices.** A current-limiting overcurrent protective device is a device that, when interrupting currents in its current-limiting range, will reduce the current flowing in the faulted circuit to a magnitude substantially less than that obtainable in the same circuit if the device were replaced with a solid conductor having comparable impedance.
240-83(c)	**Interrupting Rating.** Every circuit breaker having an interrupting rating other than 5000 amperes shall have its interrupting rating shown on the circuit breaker.

Figure 7-23: *NEC* installation requirements for overcurrent protection

NEC Section	*NEC* Requirements
240-83(e)	**Voltage Marking.** Circuit breakers shall be marked with a voltage rating no less than the nominal system voltage that is indicative of their capability to interrupt fault currents between phases or phase to ground.
250-1	**(FPN No. 1):** Systems and circuit conductors are grounded to limit voltages due to lightning, line surges, or unintentional contact with higher voltage lines, and to stabilize the voltage to ground during normal operation.

Figure 7-23: *NEC* **installation requirements for overcurrent protection (***cont.***)**

Chapter 8
Transformers

The electric power produced by alternators in a generating station is transmitted to locations where it is utilized and distributed to users. Many different types of transformers play an important role in the distribution of electricity. Power transformers are located at generating stations to step up the voltage for more economical transmission. Substations with additional power transformers and distribution equipment are installed along the transmission line. Finally, distribution transformers are used to step down the voltage to a level suitable for utilization.

Transformers are also used quite extensively in all types of control work, to raise and lower ac voltage on control circuits. They are also used in 480Y/277-volt electrical systems to reduce the voltage for operating 208Y/120-volt lighting and other electrically-operated equipment. Buck-and-boost transformers are used for maintaining appropriate voltage levels in certain electrical systems in all types of commercial applications.

It is important for anyone working with electricity to become familiar with transformer operation; that is, how they work, how they are connected into circuits, their practical applications and precautions to take during the installation or while working on them. This chapter is designed to cover these items as well as *National Electrical Code (NEC)* requirements for overcurrent protection and grounding. Specialty transformers suitable for use in commercial wiring systems are also covered.

TRANSFORMER BASICS

A very basic transformer consists of two coils or windings formed on a single magnetic core as shown in Figure 8-1. Such an arrangement will allow transforming a large alternating current at low voltage into a small alternating current at high voltage, or vice versa. But let's start at the beginning. What makes a transformer work?

Mutual Inductance

The term *mutual induction* refers to the condition in which two circuits are sharing the energy

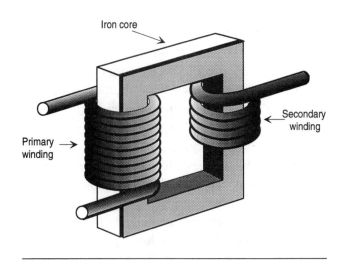

Figure 8-1: Basic parts of a transformer

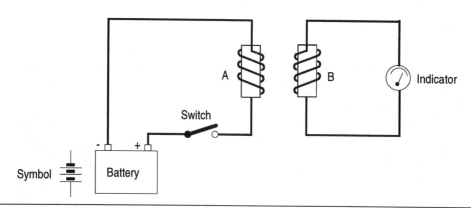

Figure 8-2: Mutual induction circuits

of one of the circuits. It means that energy is being transferred from one circuit to another.

Consider the diagram in Figure 8-2. Coil A is the primary circuit which obtains energy from the battery. When the switch is closed, the current starts to flow and a magnetic field expands out of coil A. Coil A then changes electrical energy of the battery into the magnetic energy of a magnetic field. When the field of coil A is expanding, it cuts across coil B, the secondary circuit, inducing a voltage in coil B. The indicator (a galvanometer) in the secondary circuit is deflected, and shows that a current, developed by the induced voltage, is flowing in the circuit.

The induced voltage may be generated by moving coil B through the flux of coil A. However, this voltage is induced without moving coil B. When the switch in the primary circuit is open, coil A has no current and no field. As soon as the switch is closed, current passes through the coil and the magnetic field is generated. This expanding field moves or "cuts" across the wires of coil B, thus inducing a voltage without the movement of coil B.

The magnetic field expands to its maximum strength and remains constant as long as full current flows. Flux lines stop their cutting action across the turns of coil B because expansion of the field has ceased. At this point the indicator needle on the meter reads zero because no induced voltage exists anymore. If the switch is opened, the field collapses back to the wires of coil A. As it does so, the changing flux cuts across the wires of coil B,

but in the opposite direction. The current present in the coil causes the indicator needle to deflect, showing this new direction. The indicator, then, shows current flow only when the field is changing, either building up or collapsing. In effect, the changing field produces an induced voltage exactly as does a magnetic field moving across a conductor. This principle of inducing voltage by holding the coils steady and forcing the field to change is used in innumerable applications. The transformer is particularly suitable for operation by mutual induction. Transformers are perfect components for transferring and changing ac voltages as needed.

Transformers are generally composed of two coils placed close to each other but not connected together. Refer again to Figure 8-1. The coil that receives energy from the line voltage source, etc., is called the "primary" and the coil that delivers energy to a load is called the "secondary." Even though the coils are not physically connected together they manage to convert and transfer energy as required by a process known as mutual induction.

Transformers, therefore, enable changing or converting power from one voltage to another. For example, generators that produce moderately large alternating currents at moderately high voltages utilize transformers to convert the power to very high voltage and proportionately small current in transmission lines, permitting the use of smaller cable and providing less power loss.

When alternating current (ac) flows through a coil, an alternating magnetic field is generated around the coil. This alternating magnetic field expands outward from the center of the coil and collapses into the coil as the ac through the coil varies from zero to a maximum and back to zero again. Since the alternating magnetic field must cut through the turns of the coil, a self-inducing voltage occurs in the coil which opposes the change in current flow.

If the alternating magnetic field generated by one coil cuts through the turns of a second coil, voltage will be generated in this second coil just as voltage is induced in a coil which is cut by its own magnetic field. The induced voltage in the second coil is called the "voltage of mutual induction," and the action of generating this voltage is called "transformer action." In transformer action, electrical energy is transferred from one coil (called the primary) to another (the secondary) by means of a varying magnetic field.

Induction In Transformers

A simple transformer consists of two coils located very close together and electrically insulated from each other. Alternating current is applied to the *primary*. In doing so, a magnetic field is generated which cuts through the turns of the other coil, and generates a voltage in the secondary. The coils are not physically connected to each other. They are, however, magnetically coupled to each other. Consequently, a transformer transfers electrical power from one coil to another by means of an alternating magnetic field.

Assuming that all the magnetic lines of force from the primary cut through all the turns of the secondary, the voltage induced in the secondary will depend on the ratio of the number of turns in the primary to the number of turns in the secondary. For example, if there are 100 turns in the primary and only 10 turns in the secondary, the voltage in the primary will be 10 times the voltage in the secondary. Since there are more turns in the primary than there are in the secondary, the transformer is called a "step-down transformer." Fig-

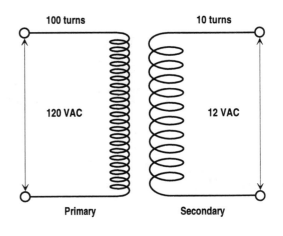

Figure 8-3: Step-down transformer with 10:1 ratio

ure 8-3 shows a diagram of a step-down transformer with a ratio of 100:10 or 10:1. Therefore, if the primary has a potential of 120 volts, then the secondary may be calculated as follows:

$$\frac{10 \text{ turns}}{100 \text{ turns}} = 0.10 = .10 \times 120 = 12 \text{ volts}$$

With fewer turns on the secondary than on the primary, the secondary voltage will not only be proportionately lower than that in the primary, but the secondary current will be that much larger, again in proportion to the current on the primary.

If there are more turns on the secondary than on the primary winding, the secondary voltage will be higher than that in the primary and by the same proportion as the number of turns in the winding. This configuration results in a "step-up" transformer. In this type of transformer, the secondary current, in turn, will be proportionately smaller than the primary current.

Since alternating current continually increases and decreases in value, every change in the primary winding of the transformer produces a similar change of flux in the core. Every change of flux in the core along with every corresponding

167

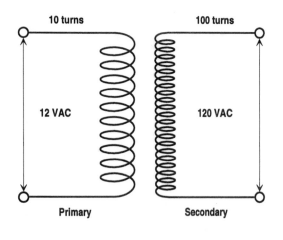

Figure 8-4: Step-up transformer with 1:10 ratio

movement of magnetic field around the core produce a similarly changing voltage in the secondary winding, causing an alternating current to flow in the circuit that is connected to the secondary.

For example, if there are, say, 100 turns in the secondary and only 10 turns in the primary, the voltage induced in the secondary will be 10 times the voltage applied to the primary. See Figure 8-4. Since there are more turns in the secondary than in the primary, the transformer is called a "step-up transformer."

$$\frac{100}{10} = 10 \times 12 = +120 \; volts$$

Note: A transformer does not generate electrical power. It simply transfers electric power from one coil to another by magnetic induction. Transformers are rated in either volt-amperes (VA) or kilo-volt-amperes (kVA).

TRANSFORMER TAPS

If the exact rated voltage could be delivered at every transformer location, transformer taps would be unnecessary. However, this is not possible, so taps are provided on the secondary windings to provide a means of either increasing or decreasing the secondary voltage.

Generally, if a load is very close to a substation or power plant, the voltage will consistently be above normal. Near the end of the line the voltage may be below normal.

In large transformers, it would naturally be very inconvenient to move the thick, well-insulated primary leads to different tap positions when changes in source-voltage levels make this necessary. Therefore, taps are used, such as shown in the wiring diagram in Figure 8-5. In this transformer, the permanent high-voltage leads would be connected to H_1 and H_2, and the secondary leads, in their normal fashion, to X_1 and X_2, X_3, and X_4. Note, however, the tap arrangements available at taps 2 through 7. Until a pair of these taps is interconnected with a jumper wire, the primary circuit is not completed. If this were, say, a typical 7200-volt primary, the transformer would have a normal 1620 turns. Assume 810 of these turns are between H_1 and H_6 and another 810 between H_3 and H_2. Then, if taps 6 and 3 were connected together with a flexible jumper on which lugs have already been installed, the primary circuit is completed, and we have a normal ratio transformer that could deliver 120/240 volts from the secondary.

Between taps 6 and either 5 or 7, 40 turns of wire exist. Similarly, between taps 3 and either 2 or 4, 40 turns are present. Changing the jumper from 3 to 6 to 3 to 7 removes 40 turns from the left half of the primary. The same condition would apply on the right half of the winding if the jumper were between taps 6 and 2. Either connection would boost secondary voltage by $2\frac{1}{2}$ percent. Had taps 2 and 7 been connected, 80 turns would have been omitted and a 5 percent boost would result. Placing the jumper between taps 6 and 4 or 3 and 5 would reduce the output voltage by 5 percent.

Caution: *Before changing any transformer taps, make sure that the primary is de-energized and that the circuit has been "tagged out."*

Commercial Electrical Wiring

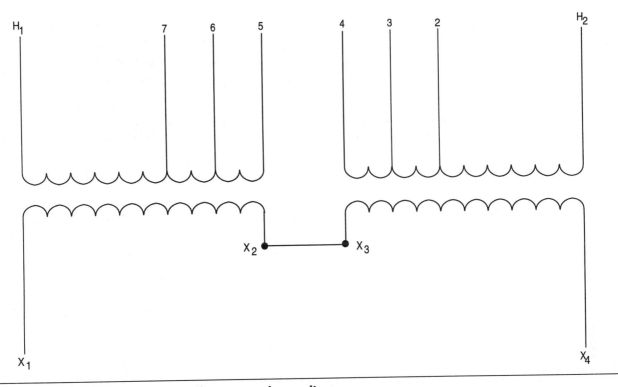

Figure 8-5: Transformer taps to adjust secondary voltage

TRANSFORMER CONNECTIONS — BASIC

Transformer connections are many, and space does not permit the description of all of them here. However, an understanding of a few will give the basic requirements and make it possible to use manufacturer's data for others should the need arise on any commercial electrical installation.

former is mounted on a concrete pad (called *pad-mount*) so that the electric service may be routed underground to the point of utilization. Underground services, as discussed in Chapter 5, is called *service lateral*.

In either case, the transformer connections are identical, and both provide 120/240-volt, single-phase, three-wire power.

Single-Phase For Light And Power

The diagram in Figure 8-6 is a transformer connection used quite extensively for residential and small commercial applications. It is the most common single-phase distribution system in use today. It is known as the 240/120-volt, single-phase, three-wire system and is used where 120 and 240 volts are used simultaneously.

In application, a single transformer is mounted on a power pole close to the premises where the power is to be used. In some instances, the trans-

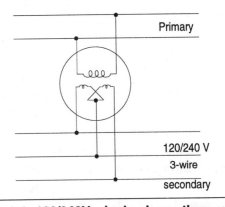

Figure 8-6: 120/240V, single-phase, three-wire

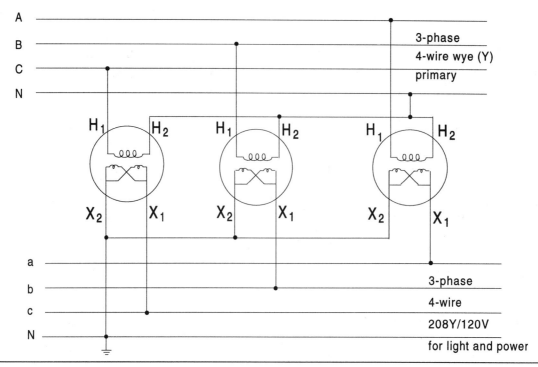

Figure 8-7: Three-phase, four-wire, Y-Y connected transformer system

Y-Y For Light and Power

The primaries of the transformer connection in Figure 8-7 are connected in wye — sometimes called *star* connection. When the primary system is 2400/4160Y volts, a 4160-volt transformer is required when the system is connected in delta-Y. However, with a Y-Y system, a 2400-volt transformer can be used, offering a saving in transformer cost. It is necessary that a primary neutral be available when this connection is used, and the neutrals of the primary system and the transformer bank are tied together as shown in the diagram. If the three-phase load is unbalanced, part of the load current flows in the primary neutral. For these reasons, it is essential that the neutrals be tied together as shown. If this tie were omitted, the line-to-neutral voltages on the secondary would be very unstable. That is, if the load on one phase were heavier than on the other two, the voltage on this phase would drop excessively and the voltage on the other two phases would rise. Also, varying voltages would appear between lines and neutral, both in the transformers and in the secondary system,

in addition to the 60-hertz component of voltage. This means that for a given value of rms voltage, the peak voltage would be much higher than for a pure 60-hertz voltage. This overstresses the insulation both in the transformers and in all apparatus connected to the secondaries.

Delta-Connected Transformers

The delta-connected system in Figure 8-8 operates a little differently from the previously described Y-Y system. While the wye-connected system is formed by connecting one terminal from three equal voltage transformer windings together to make a common terminal, the delta-connected system has its windings connected in series, forming a triangle or the Greek delta symbol Δ. Note in Figure 8-8 that a center-tap terminal is used on one winding to ground the system. On a 240/120-volt system, there are 120 volts between the center-tap terminal and each ungrounded terminal on either side; that is, phases A and C. There are 240 volts across the full winding of each phase.

170

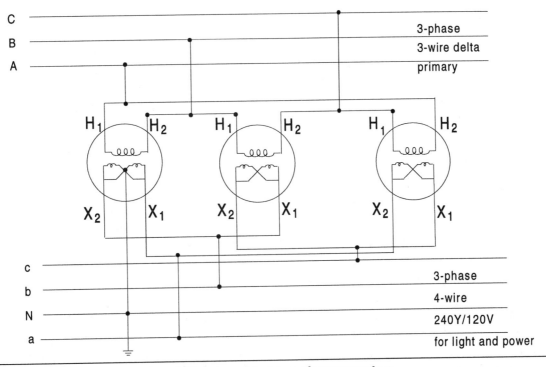

Figure 8-8: Three-phase, four-wire, delta-connected transformer system

Refer to Figure 8-9 and note that a high leg results at point "B." This is known in the trade as the "high leg," "red leg," or "wild leg." This high leg has a higher voltage to ground than the other two phases. The voltage of the high leg can be determined by multiplying the voltage to ground of either of the other two legs by the square root of 3. Therefore, if the voltage between phase A to ground is 120 volts, the voltage between phase B to ground may be determined as follows:

$$120 \times \sqrt{3} \quad = \quad 207.84 = 208 \text{ volts}$$

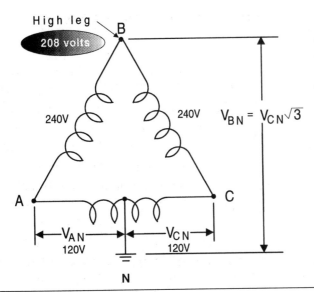

On a 3-phase, 4-wire, 120/240V delta-connected system, the midpoint of one phase winding is grounded to provide 120V between phase A and ground; also between phase C and ground. Between phase B and ground, however, the voltage is higher and may be calculated by multiplying the voltage between C and ground (120V) by the square root of 3 or 1.73. Consequently, the voltage between phase B and ground is approximately 208 volts. Thus, the name "high leg."

The *NEC* requires that conductors connected to the high leg of a 4-wire delta system be color-coded with orange insulation or tape.

Figure 8-9: Characteristics of a center-tap, delta-connected system

---

WARNING!   Always use caution when working on a center-tapped, four-wire, delta-connected system. Phase B has a higher voltage to ground than phases A and C. Never connect 120-volt circuits to the high leg. Doing so will result in damage to the circuits and equipment.

From this, it should be obvious that no single-pole breakers should be connected to the high leg of a center-tapped, four-wire delta-connected system. In fact, *NEC* Section 215-8 requires that the phase busbar or conductor having the higher voltage to ground to be permanently marked by an outer finish that is orange in color. By doing so, this will prevent future workers from connecting 120-volt single-phase loads to this high leg which will probably result in damaging any equipment connected to the circuit. Remember the color *orange*; no 120-volt loads are to be connected to this phase.

transformers becomes damaged. The damaged transformer is disconnected from the circuit and the remaining two transformers carry the load. In doing so, the three-phase load carried by the open delta bank is only 86.6 percent of the combined rating of the remaining two equal sized units. It is only 57.7 percent of the normal full-load capability of a full bank of transformers. In an emergency, however, this capability permits single- and three-phase power at a location where one unit has burned out and a replacement was not readily available. The total load must be curtailed to avoid another burnout.

Open Delta

Three-phase, delta-connected systems may be connected so that only two transformers are used; this arrangement is known as *open delta* as shown in Figure 8-10. This arrangement is frequently used on a delta system when one of the three

Tee-Connected Transformers

When a delta-wye transformer is used, we would usually expect to find three primary and three secondary coils. However, in a tee-connected three-phase transformer, only two primary and two secondary windings are used as shown in Figure 8-11. If an equilateral triangle is drawn as indicated by the dotted lines in Figure 8-11 so that the distance between H_1 and H_3 is 4.8 inches, you would find that the distance between H_2 to the midpoint of H_1 - H_3 measures 4.16 inches. Therefore, if the voltage between outside phases is 480 volts, the voltage between H_2 to the midpoint of H_1 - H_3 will equal 480 volts $\times$ .866 = 415.68 or 416 volts. Also, if you were to place an imaginary dot exactly in the center of this triangle it would lay on the horizontal winding — the one containing 416 volts. If you measured the distance from this dot to H_2, you would find it to be twice as long as the distance between the dot and the midpoint of H_1 to H_3. The measured distances would be 2.77 inches and 1.385 inches or the equivalent of 277 volts and 138½ volts respectively.

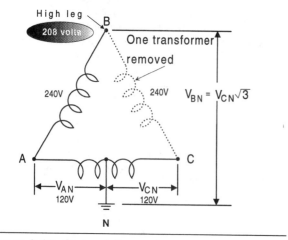

Figure 8-10: Open delta system

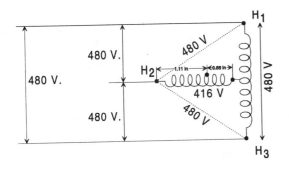

Figure 8-11: Typical tee-connected transformer

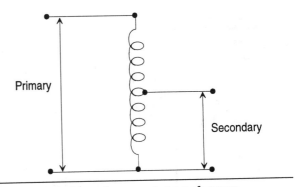

Figure 8-13: Step-down autotransformer

Now, let's look at the secondary winding in Figure 8-12. By placing a neutral tap X_0 so that $\frac{1}{3}$ the number of turns exist between it and the midpoint of X_1 and X_3, as exist between it and X_2, we then can establish X_0 as a neutral point which may be grounded. This provides 120 volts between X_0 and any of the three secondary terminals and the three-phase voltage between X_1, X_2, and X_3, will be 208 volts.

AUTOTRANSFORMERS

An autotransformer is a transformer whose primary and secondary circuits have part of a winding in common and therefore the two circuits are not isolated from each other. See Figure 8-13. The

application of an autotransformer is a good choice for some users where a 480Y/277- or 208Y/120-volt, three-phase, four-wire distribution system is utilized. Some of the advantages are as follows:

- Lower purchase price
- Lower operating cost due to lower losses
- Smaller size; easier to install
- Better voltage regulation
- Lower sound levels

For example, when the ratio of transformation from the primary to secondary voltage is small, the most economical way of stepping down the voltage is by using autotransformers as shown in Figure 8-14 on page 174. For this application, it is necessary that the neutral of the autotransformer bank be connected to the system neutral.

An autotransformer, however, cannot be used on a 480- or 240-volt, three-phase, three-wire delta system. A grounded neutral phase conductor must be available in accordance with *NEC* Article 210-9, which states:

NEC Section 210-9: Circuits Derived from Autotransformers. Branch circuits shall not be derived from autotransformers.

Exception: Where the system supplied has a grounded conductor that is electrically connected to a grounded conductor of the system supplying the autotransformer.

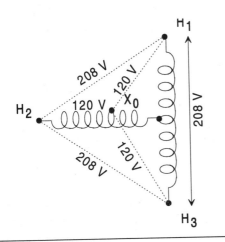

Figure 8-12: Secondary voltage on tee-connected system

Primary: 3-phase, 4-wire

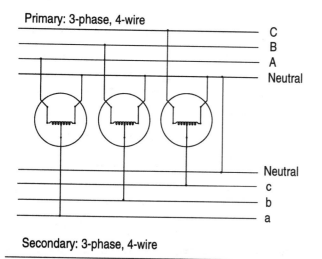

Secondary: 3-phase, 4-wire

Figure 8-14: Autotransformer supplying power from a three-phase, four-wire system

Another Exception: An autotransformer used to extend or add an individual branch circuit in an existing installation for an equipment load without the connection to a similar grounded conductor when transforming from a nominal 208 volts to a nominal 240 volt supply or similarly from 240 volts to 208 volts.

The *NEC*, in general, requires that separately derived alternating-current systems be grounded. The secondary of a two-winding insulated trans-

former is a separately derived system. Therefore, it must be grounded in accordance with *NEC* Section 250-26.

A typical drawing for this *NEC* requirement is shown in Figure 8-15. In the case of an autotransformer, the grounded conductor of the supply is brought into the transformer to a common terminal, and the ground is established to satisfy the *NEC*.

TRANSFORMER CONNECTIONS — DRY TYPE

Electricians performing work on commercial installations will more often be concerned with the installation and connection of dry-type transformers as opposed to oil-filled ones. Dry-type transformers are available in both single- and three-phase with a wide range of sizes from the small control transformers to those rated at 500 kVA or more. Such transformers have wide application in electrical systems of all types.

NEC Section 450-11 requires that each transformer must be provided with a nameplate giving the manufacturer; rated kVA; frequency; primary and secondary voltage; impedance of transformers 25 kVA and larger; required clearances for transformers with ventilating openings; and the amount

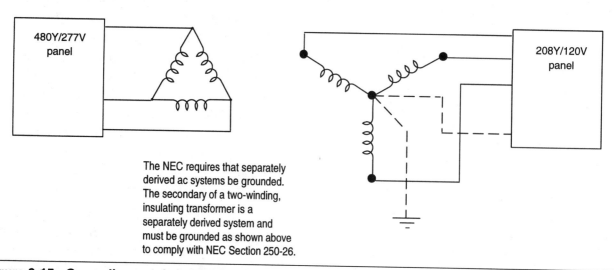

The NEC requires that separately derived ac systems be grounded. The secondary of a two-winding, insulating transformer is a separately derived system and must be grounded as shown above to comply with NEC Section 250-26.

Figure 8-15: Grounding requirements for autotransformers

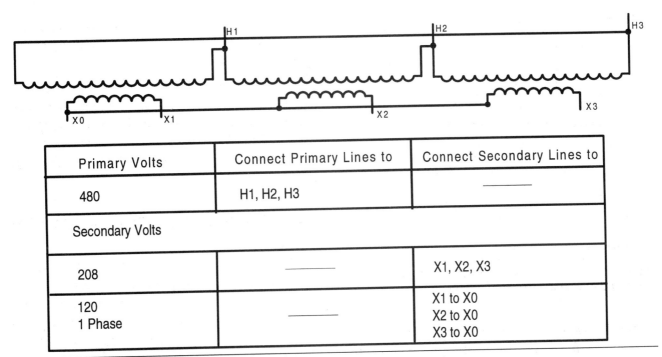

Primary Volts	Connect Primary Lines to	Connect Secondary Lines to
480	H1, H2, H3	———
Secondary Volts		
208	———	X1, X2, X3
120 1 Phase	———	X1 to X0 X2 to X0 X3 to X0

Figure 8-16: Typical transformer manufacturer's wiring diagram — delta-wye

and kind of insulating liquid where used. In addition, the nameplate of each dry-type transformer must include the temperature class for the insulation system.

In addition, most manufacturers include a wiring diagram and a connection chart as shown in Figure 8-16 for a 480-volt delta primary to 208Y/120-volt secondary. It is recommended that all transformers be connected as shown on the manufacturer's nameplate.

In general, this wiring diagram and accompanying table indicate that the 480-volt, three-phase, three-wire primary conductors are connected to terminals H_1, H_2, and H_3, respectively — regardless of the desired voltage on the primary. A neutral conductor, if required, is carried from the primary through the transformer to the secondary. Two variations are possible on the secondary side of this transformer: 208-volt, three-phase, three- or four-wire or 120-volt, single-phase, two-wire. To connect the secondary side of the transformer as a 208-volt, three-phase, three-wire system, the secondary conductors are connected to terminals X_1, X_2, and X_3; the neutral is carried through with

conductors usually terminating at a solid-neutral bus in the transformer.

Zig-Zag Connections

There are many occasions where it is desirable to upgrade a building's lighting system from 120-volt fixtures to 277-volt fluorescent lighting fixtures. Oftentimes these buildings have a 480/240-volt, three-phase, four-wire delta system. One way to obtain 277 volts from a 480/240-volt system is to connect 480/240-volt transformers in a zig-zag fashion as shown in Figure 8-17 on page 176. In doing so, the secondary of one phase is connected in series with the primary of another phase, thus changing the phase angle.

The zig-zag connection may also be used as a grounding transformer where its function is to obtain a neutral point from an ungrounded system. With a neutral being available, the system may then be grounded. When the system is grounded through the zig-zag transformer, its sole function is to pass ground current. A zig-zag transformer is

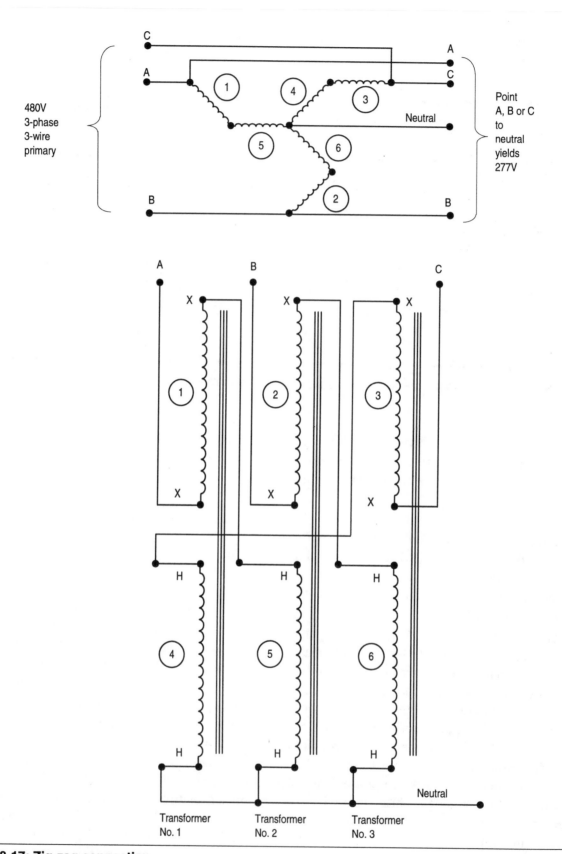

Figure 8-17: Zig-zag connection

essentially six impedances connected in a zig-zag configuration.

The operation of a zig-zag transformer is slightly different from that of the conventional transformer. We will consider current rather than voltage. While a voltage rating is necessary for the connection to function, this is actually line voltage and is not transformed. It provides only exciting current for the core. The dynamic portion of the zig-zag grounding system is the fault current. To understand its function, the system must also be viewed backward; that is, the fault current will

flow into the transformer through the neutral as shown in Figure 8-18.

The zero sequence currents are all in phase in each line; that is, they all hit the peak at the same time. In reviewing Figure 8-18, we see that the current leaves the motor, goes to ground, flows up the neutral, and splits three ways. It then flows back down the line to the motor through the fuses which then open — shutting down the motor.

The neutral conductor will carry full fault current and must be sized accordingly. It is also time rated (0-60 seconds) and can therefore be reduced

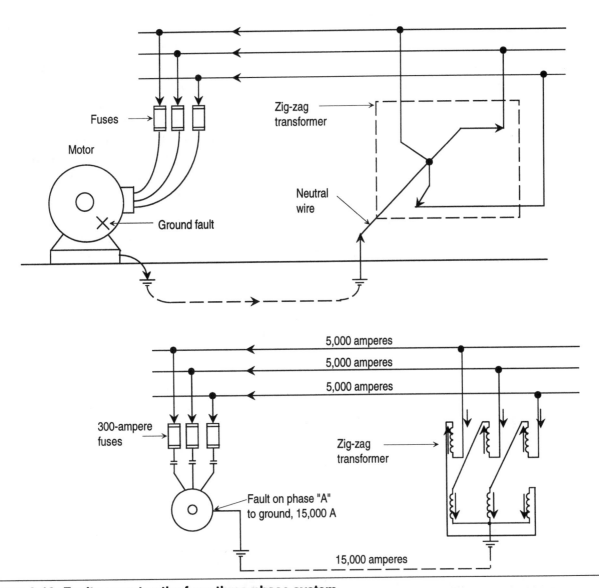

Figure 8-18: Fault-current paths for a three-phase system

in size. This should be coordinated with the manufacturer's time/current curves for the fuse. See Chapter 6 — Overcurrent Protection.

To determine the size of a zig-zag grounding transformer, proceed as follows:

Step 1. Calculate the system line-to-ground asymmetrical fault current.

Step 2. If relaying is present, consider reducing the fault current by installing a resistor in the neutral.

Step 3. If fuses or circuit breakers are the protective device, you may need all the fault current to quickly open the overcurrent protective devices.

Step 4. Obtain time/current curves of relay, fuses, or circuit breakers.

Step 5. Select zig-zag transformer for:
   a. *Fault current — the line-to-ground*
   b. *Line-to-line voltage*
   c. *Duration of fault (determined from time/current curves)*
   d. *Impedance per phase at 100 percent; for any other, contact manufacturer*

BUCK-AND-BOOST TRANSFORMERS

The buck-and-boost transformer is a very versatile unit for which a multitude of applications exist. Buck-and-boost transformers, as the name implies, is designed to raise (boost) or lower (buck) the voltage in an electrical system or circuit. In their simplest form, these insulated units will deliver 12 or 24 volts when the primaries are energized at 120 or 240 volts respectively. Their prime use and value, however, lies in the fact that the primaries and the secondaries can be interconnected — permitting their use as an autotransformer.

Let's assume that an installation is supplied with 208Y/120V service, but one piece of equipment in the installation is rated for 230 volts. A buck-and-boost transformer may be used on the 230-volt circuit to increase the voltage from 208 volts to 230 volts. See Figure 8-19. With this connection, the transformer is in the "boost" mode and delivers 228.8 volts at the load. This is close enough to 230 volts that the load equipment will function properly.

If the connections were reversed, this would also reverse the polarity of the secondary with the result that a voltage would be 208 volts minus 20.8 volts = 187.2 volts. The transformer is now operating in the "buck" mode.

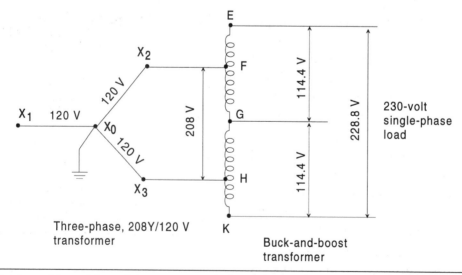

Figure 8-19: Buck-and boost transformer connected to a 208-volt system to obtain 230 volts

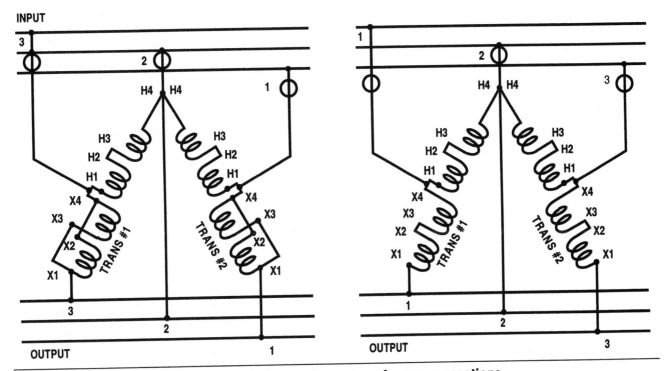

Figure 8-20: Open delta, three-phase, buck-and-boost transformer connections

Transformer connections for typical three-phase buck-and-boost open-delta transformers are shown in Figure 8-20. The connections shown are in the "boost" mode; to convert to "buck" mode, reverse the input and output.

Another three-phase buck-and-boost transformer connection is shown in Figure 8-21; this time wye-connected. While the open-delta transformers (Figure 8-20) can be converted from buck to boost or vice versa by reversing the input/output connections, this is not the case with the three-phase, wye-connected transformer. The connection shown (Figure 8-21) is for the boost mode only.

Several typical single-phase buck-and-boost transformer connections are shown in Figure 8-22 on page 180. Other diagrams may be found on the transformer's nameplate or with packing instructions that come with each new transformer.

Manufacturers of buck-and-boost transformers normally offer "easy-selector" charts to quickly select a buck-and-boost transformer for practically any application. These charts may be obtained

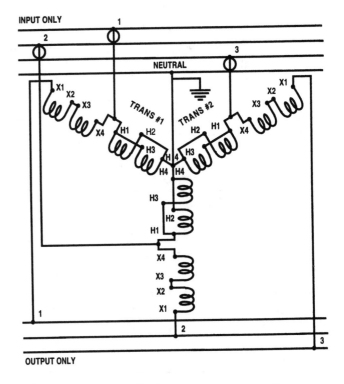

Figure 8-21: Three-phase, wye-connected buck-and-boost transformer in the "boost" mode

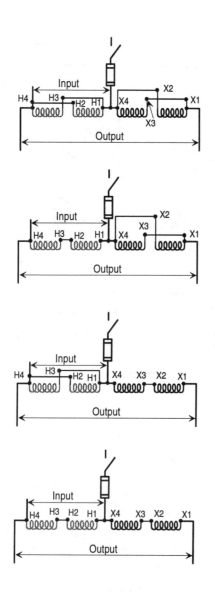

Figure 8-22: Typical single-phase buck-and-boost transformer connections

from electrical equipment suppliers or ordered directly from the manufacturer often at no charge. Instructions accompanying these charts will enable anyone familiar with transformers and electrical circuits to use them. However, a brief review of principles involved in using buck-and-boost transformers is in order.

When reviewing the easy-selector charts mentioned in the above paragraph, it may astonish many people; that is, how can these transformers handle a load so much greater that its nameplate rating? For example, a typical 1 kVA buck-and-boost transformer can easily handle a 10 kVA load when the voltage boost is only 10 percent! Let's see how this is possible.

Assume that we have a 1 kVA (1000 VA) insulating transformer designed to transform 120/240 volts to 12/24 volts. This results in a transformer winding with a ratio of 10:1. The primary current may be found by the following equation:

$$Primary\ current = \frac{1000\ VA}{240\ Volts}$$
$$= 4.166\ amperes$$

Therefore, 4.166 amperes rounds off to 4.17 amperes. Because the transformation ratio is 10 to 1, the secondary amperes will be 41.7 amperes; that is, 4.17 multiplied by 10 = 41.7, or the amperage may be determined by the following equation:

$$Secondary\ current = \frac{1000\ VA}{24\ Volts}$$
$$= 41.66\ amperes$$

Figure 8-23 shows a wiring diagram of the transformer under consideration. Note that a 240-volt source delivers 10 kVA, but the secondary winding of the 1 kVA buck-and-boost transformer has been placed in series with the line to the load.

$$P(kVA) = EI,\ kVA = kV\ times\ I$$
$$or$$
$$I = \frac{kVA}{kV}$$

Therefore, by substituting I = 10 over 0.24, the secondary current equals 41.7 amperes. Thus, the current from the source is 41.7 amperes. The 1 kVA buck-and-boost transformer at full load has a secondary current that also equals 41.7 amperes.

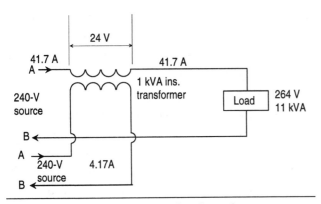

Figure 8-23: Transformer circuit under consideration

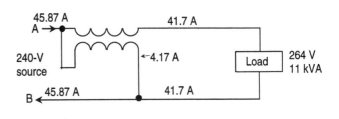

Figure 8-24: Simplified diagram of Figure 8-23

Consequently, there is no harm in connecting it in the line because its secondary current rating is adequate to handle the load current. Because we started with 240 volts at the source and now add 24 volts to it, the load actually gets 264 volts. To find the kVA ratings of this system at the load, multiply volts divided by 1000 times amperes:

$$kVA = \frac{Volts}{1000} \times amperes$$
$$= \frac{264}{1000} \times 41.7 = 11kVA$$

Ten kVA comes from the source and the extra 1 kVA from the booster secondary. The total current consumed by the circuit is 41.7 amperes plus 4.17 or 45.87 amperes.

In actual practice, four leads would not be brought out and connected to the source as shown in Figure 8-23. Rather, the circuit would probably be simplified as shown in Figure 8-24. Since the two source lines marked A in Figure 8-23 are the same point and the two Bs are identical, they may be connected together as shown in Figure 8-24. The connections are identical to Figure 8-23 except now there are only two lines running to the power source and the combined current may be shown as 45.87 amperes.

In actual practice, drawing diagrams such as the ones shown in Figures 8-23 and 8-24 are usually simplified even more — in kind of a "ladder" or schematic diagram as shown in Figure 8-25. Actually, all three of these wiring diagrams (Figures 8-23, 8-24, and 8-25) indicate the same thing, and following the connections on any of these drawings will produce the same results at the load.

It should now be evident how a little 1 kVA buck-and-boost transformer, when connected in the circuit as described previously, can actually carry 11 kVA in its secondary winding.

CONTROL TRANSFORMERS

Control transformers are available in numerous types, but most control transformers are dry-type step-down units with the secondary control circuit

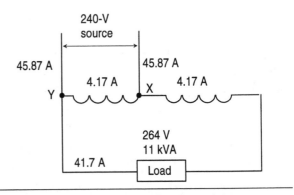

Figure 8-25: Further simplification of the transformer circuit under consideration

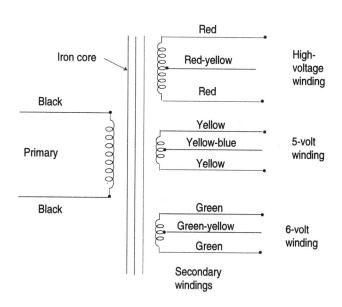

Figure 8-26: Typical control-transformer wiring diagram

isolated from the primary line circuit to assure maximum safety. See Figure 8-26. These transformers and other components are usually mounted within an enclosed control box or control panel, which has a pushbutton station or stations independently grounded as recommended by the *NEC*. Industrial control transformers are especially designed to accommodate the momentary current inrush caused when electromagnetic components are energized, without sacrificing secondary voltage stability beyond practical limits.

Other types of control transformers, sometimes referred to as control and signal transformers, normally do not have the required industrial control transformer regulation characteristics. Rather, they are constant-potential, self-air-cooled transformers used for the purpose of supplying the proper reduced voltage for control circuits of electrically operated switches or other equipment and, of course, for signal circuits. Some are of the open type with no protective casing over the winding, while others are enclosed with a metal casing over the winding.

In seeking control transformers for any application, the loads must be calculated and completely analyzed before the proper transformer selection can be made. This analysis involves every electrically energized component in the control circuit. To select an appropriate control transformer, first determine the voltage and frequency of the supply circuit. Then determine the total inrush volt-amperes (watts) of the control circuit. In doing so, do not neglect the current requirements of indicating lights and timing devices that do not have inrush volt-amperes, but are energized at the same time as the other components in the circuit. Their total volt-amperes should be added to the total inrush volt-amperes.

POTENTIAL AND CURRENT TRANSFORMERS

In general, a potential transformer is used to supply voltage to instruments such as voltmeters, frequency meters, power-factor meters, and watt-hour meters. The voltage is proportional to the primary voltage, but it is small enough to be safe for the test instrument. The secondary of a potential transformer may be designed for several different voltages, but most are designed for 120 volts. The potential transformer is primarily a distribution transformer especially designed for good voltage regulation so that the secondary voltage under all conditions will be as nearly as possible a definite percentage of the primary voltage.

Current Transformers

A current transformer (Figure 8-27) is used to supply current to an instrument connected to its secondary, the current being proportional to the primary current, but small enough to be safe for the instrument. The secondary of a current transformer is usually designed for a rated current of five amperes.

A current transformer operates in the same way as any other transformer in that the same relation

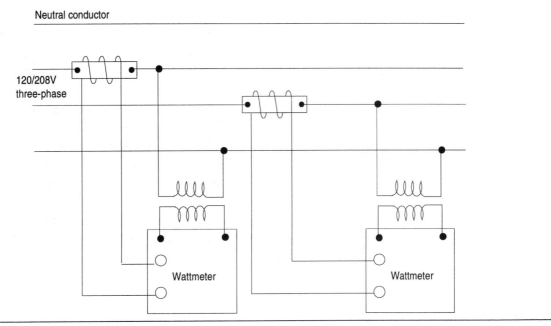

Neutral conductor

120/208V three-phase

Wattmeter

Wattmeter

Figure 8-27: Current and potential transformers used in conjunction with watt-hour meter

exists between the primary and the secondary current and voltage. A current transformer is connected in series with the power lines to which it is applied so that line current flows in its primary winding. The secondary of the current transformer is connected to current devices such as ammeters, wattmeters, watt-hour meters, power-factor meters, some forms of relays, and the trip coils of some types of circuit breakers.

When no instruments or other devices are connected to the secondary of the current transformer, a short-circuit device or connection is placed across the secondary to prevent the secondary circuit from being opened while the primary winding is carrying current. There will be no secondary ampere turns to balance the primary ampere turns, so the total primary current becomes exciting current and magnetizes the core to a high flux density. This produces a high voltage across both primary and secondary windings and endangers the

life of anyone coming in contact with the meters or leads.

NEC REQUIREMENTS

Transformers must normally be accessible for inspection except for dry-type transformers under certain specified conditions. Certain types of transformers with a high voltage or kVA rating are required to be enclosed in transformer rooms or vaults when installed indoors. The construction of these vaults is covered in *NEC* Sections 450-41 through 450-48.

In general, the *NEC* specifies that the walls and roofs of vaults must be constructed of materials that have adequate structural strength for the conditions with a minimum fire resistance of three hours. However, where transformers are protected with an automatic sprinkler system, water spray,

---

WARNING!   The secondary circuit of a current transformer should never be opened while the primary is carrying current.

---

carbon dioxide, or halon, the fire resistance construction may be lowered to only one hour. The floors of vaults in contact with the earth must be of concrete and not less than 4 inches thick. If the vault is built with a vacant space or other floors (stories) below it, the floor must have adequate structural strength for the load imposed thereon and a minimum fire resistance of three hours. Again, if the fire extinguishing facilities are provided, as outlined above, the fire resistance construction need only be one hour. The *NEC* does not permit the use of studs and wall board construction for transformer vaults.

Overcurrent Protection For Transformers

The overcurrent protection for transformers is based on their rated current, not on the load to be served. The primary circuit may be protected by a device rated or set at not more than 125 percent of the rated primary current of the transformer for transformers with a rated primary current of nine amperes or more.

Instead of individual protection on the primary side, the transformer may be protected only on the secondary side if all the following conditions are met.

- The overcurrent device on the secondary side is rated or set at not more than 125 percent of the rated secondary current.
- The primary feeder overcurrent device is rated or set at not more than 250 percent of the rated primary current.

For example, if a 12 kVA transformer has a primary current rating of:

12,000 watts/480 volts = 25 amperes

and a secondary current rated at:

12,000 watts/120 volts = 100 amperes

the individual primary protection must be set at:

1.25 x 25 amperes = 31.25 amperes

In this case, a standard 30-ampere cartridge fuse rated at 600 volts could be used, as could a circuit breaker approved for use on 480 volts. However, if certain conditions are met, individual primary protection for the transformer is not necessary in this case if the feeder overcurrent-protective device is rated at not more than:

2.5 x 25 amperes = 62.5 amperes

and the protection of the secondary side is set at not more than:

1.25 x 100 amperes = 125 amperes

A 125-ampere circuit breaker could be used.

Note: The example cited above is for the transformer only; not the secondary conductors. The secondary conductors must be provided with overcurrent protection as outlined in NEC Section 210-20.

The requirements of *NEC* Section 450-3 cover only transformer protection; in practice, other components must be considered in applying circuit overcurrent protection. For circuits with transformers, requirements for conductor protection per *NEC* Articles 240 and 310 for panelboards per *NEC* Article 384 must be observed. Refer to *NEC* Sections 240-3(f); 240-21, and *NEC* Section 384-16(d).

Primary Fuse Protection Only (NEC Table 450-3(b)1): If secondary fuse protection is not provided, then the primary fuses must not be sized larger than 125 percent of the transformer primary full-load amperes except if the transformer primary full-load amperes (F.L.A.) is that shown in *NEC* Table 450-3(b)1. (See Figure 8-28).

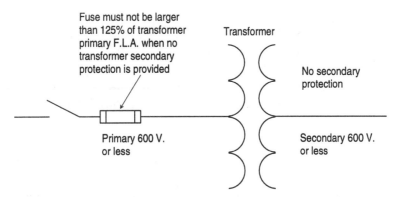

Primary Current	Primary Fuse Rating
9 amperes or more	125% or next higher standard rating if 125% does not correspond to a standard fuse size
2 amperes or more	167% maximum
Less than 2 amperes	300% maximum

Figure 8-28: Transformer circuit with primary fuse only

Individual transformer primary fuses are not necessary where the primary circuit fuse provides this protection.

Primary and Secondary Protection: In unsupervised locations, with primary over 600 volts, the primary fuse can be sized at a maximum of 300 percent. If the secondary is also over 600 volts, the secondary fuses can be sized at a maximum of 250 percent for transformers with impedances not greater than 6 percent; 225 percent for transformers with impedances greater than 6 percent and not more than 10 percent. If the secondary is 600 volts or below, the secondary fuses can be sized at a maximum of 125 percent. Where these settings do not correspond to a standard fuse size, the next higher standard size is permitted.

In supervised locations, the maximum settings are as shown in Figure 8-29 on page 186 except for secondary voltages of 600 volts or below, where the secondary fuses can be sized at a maximum of 250 percent.

Primary Protection Only: In supervised locations, the primary fuses can be sized at a maximum of 250 percent, or the next larger standard size if 250 percent does not correspond to a standard fuse size.

Note: The use of "Primary Protection Only" does not remove the requirements for compliance with NEC Articles 240 and 384. See (FPN) in NEC Section 450-3 which references NEC Sections 240-3 and 240-100 for proper protection for secondary conductors.

Protection For Small Power Transformers

Low amperage, E-rated medium voltage fuses are general purpose current limiting fuses. The E rating defines the melting-time-current characteristic of the fuse and permits electrical interchangeability of fuses with the same E rating. For a general purpose fuse to have an E rating the condition must be met.

- The current responsive element shall melt in 300 seconds at an rms current within the range of 200 percent to 240 percent of the continuous current rating of the fuse, fuse refill, or link. (ANSI C37.46).

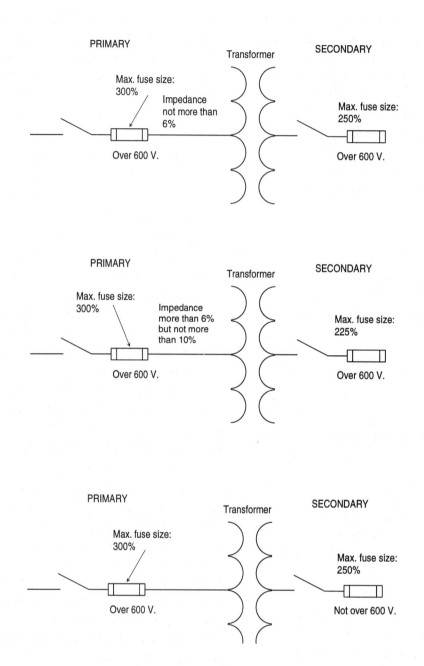

Figure 8-29: Minimum overcurrent protection for transformers in supervised locations

Low amperage, E-rated fuses are designed to provide primary protection for potential, small service, and control transformers. These fuses offer a high level of fault current interruption in a self-contained non-venting package which can be mounted indoors or in an enclosure.

As for all current-limiting fuses, the basic application rules found in the *NEC* and manufacturer's literature should be adhered to. In addition, potential transformer fuses must have sufficient inrush capacity to successfully pass through the magnetizing inrush current of the transformer. If the fuse is not sized properly, it will open before the load is energized. The maximum magnetizing inrush currents to the transformer at system voltage and the duration of this inrush current varies with the transformer design. Magnetizing inrush currents are usually denoted as a percentage of the transformer full load current, i.e., 10X, 12X, 15X, etc. The inrush current duration is usually given in seconds. Where this information is available, an easy check can be made on the appropriate minimum-melting curve to verify proper fuse selection. In lieu of transformer inrush data, the rule-of-thumb is to select a fuse size rated at 300 percent of the primary full load current or the next larger standard size.

Example

The transformer manufacturer states that an 800 VA 240-volt, single-phase potential transformer has a magnetizing inrush current of 12X lasting for 0.1 second.

A. $I_{FL} = 800VA/2400V = 0.333$ ampere

Inrush Current $= 12 \times 0.333 = 4$ amperes

Since the voltage is 2400 volts we can use either a JCW or a JCD fuse. The proper fuse would be a JCW-1E, or JCD-1E.

B. Using the rule-of-thumb — 300 percent of .333 ampere is .999 ampere.

Therefore we would choose a JCW-1E or JCD-1E.

Typical Potential Transformer Connections: The typical potential transformer connections encountered in industry can be grouped into two categories:

- Those connections that require the fuse to pass only the magnetizing inrush of one potential transformer.

- Those connections that must pass the magnetizing inrush of more than one potential transformer.

Fuses for Medium Voltage Transformers and Feeders: E-rated, medium-voltage fuses are general purpose current limiting fuses. The fuses carry either an "E" or an "X" rating which defines the melting-time-current characteristic of the fuse. The ratings are used to allow electrical interchangeability among different manufacturers' fuses.

For a general purpose fuse to have an E rating, the following conditions must be met:

- The current responsive element with ratings 100 amperes or below shall melt in 300 seconds at an rms current within the range of 200 percent to 240 percent of the continuous current rating of the fuse unit. (ANSI C37.46).

- The current responsive element with ratings above 100 amperes shall melt in 600 seconds at an rms current within the range of 220 percent to 264 percent of the continuous current rating of the fuse unit. (ANSI C37.46).

A fuse with an "X" rating does not meet the electrical interchangeability for an E-rated fuse but offers the user other ratings that may provide better protection for his particular application.

Transformer protection is the most popular application of E-rated fuses. The fuse is applied to the primary of the transformer and is solely used to prevent rupture of the transformer due to short circuits. It is important, therefore, to size the fuse so that it does not clear on system inrush or permissible overload currents. Magnetizing inrush must also be considered with sizing a fuse. In general, power transformers have a magnetizing inrush current of 12X the full load rating for a duration of $1/10$ second.

TRANSFORMER GROUNDING

Grounding is necessary to remove static electricity and also as a precautionary measure in case the transformer windings accidentally come in contact with the core or enclosure. All should be grounded and bonded to meet *NEC* requirements and also local codes, where applicable.

The tank of every power transformer should be grounded to eliminate the possibility of obtaining static shocks from it or being injured by accidental grounding of the winding to the case. A grounding lug is provided on the base of most transformers for the purpose of grounding the case and fittings.

The *NEC* specifically states the requirements of grounding and should be followed in every respect. Furthermore, certain advisory rules recommended by manufacturers provide additional protection beyond that of the *NEC*. In general, the *NEC* requires that separately derived alternating current systems be grounded as stated in Article 250-26.

Figure 8-30 summarizes *NEC* regulations governing the grounding of transformers to provide for fault current to trip overcurrent protective devices.

The noncurrent-carrying metal parts of transformers must be effectively bonded together
NEC Section 250-71(a)

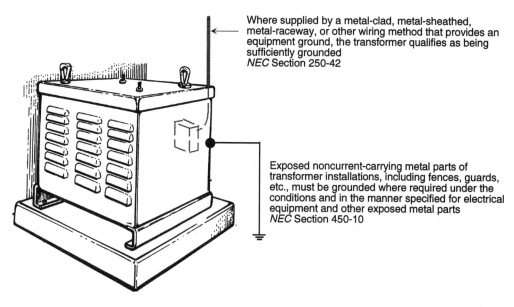

Where supplied by a metal-clad, metal-sheathed, metal-raceway, or other wiring method that provides an equipment ground, the transformer qualifies as being sufficiently grounded
NEC Section 250-42

Exposed noncurrent-carrying metal parts of transformer installations, including fences, guards, etc., must be grounded where required under the conditions and in the manner specified for electrical equipment and other exposed metal parts
NEC Section 450-10

The path to ground must be permanent and electrically continuous; have capacity to donduct safely any fault current likely to be imposed on it; and have sufficiently low imedance to limit the voltage to ground and to facilitiate the operation of the overcurrent protective devices
NEC Section 250-51

The main and equipment bonding jumpers must be of copper or other corrosion-resistant material
NEC Section 250-79(a)

Figure 8-30: Summary of *NEC* requirements for transformer grounding

Rectifier Transformers

Transformers used in connection with rectifiers have to be specially designed to be able to provide single-phase, as well as three-phase, alternating current to the rectifier. The transformation of primary voltage to the secondary voltage makes the rectified current output smoother and more efficient. Because of complicated connections and special requirements, rectifier transformers are considerably more expensive than power transformers of equivalent kVA rating.

Reactors

Reactors are really transformers with only one winding. They are designed so that, for a given current through the reactor, a definite voltage drop exists across the winding. The current flowing through a reactor is all exciting current. The magnetic circuit, therefore, is designed to give the required exciting current at the desired voltage drop.

There are two kinds of reactors, namely, iron-core reactors and air-core reactors. Since the same characteristics can be obtained with either the air-core or the iron-core reactor for a desired range of current and voltage, the decision as to which of the two types to use for a given application is generally made on the basis of comparative costs. Iron-core reactors are usually cheaper wherever the short-time overloads of the reactor are less than three times its continuous-current capacity. For heavier short-time overloads, air-core reactors are ordinarily used.

Typical applications of iron-core reactors are:

● To increase the reactance of one branch of parallel circuits so the load will divide properly among the branches.

● To compensate for the leading charging current of a long transmission line, as when used for shunt reactors operating at nearly constant voltage.

● To limit the current of an arc furnace transformer.

The principal application of air-core reactors is as current-limiting reactors, where the short-circuit current of the system may be many times the normal current. Values from 10 to 20 times the normal current are typical. Air-core reactors are also used for large-capacity shunt reactors. Reactors used to limit the current required for the starting of motors may be of either type, depending on the time required by the motor to come up to speed.

Step-Voltage Regulators

Regulators of the step-voltage type are small transformers (not above 2500-kVA three-phase or 833-kVA single-phase) provided with load tap changers. They are used to raise or to lower the voltage of a circuit in response to a voltage regulating relay or other voltage-control device. Regulators are usually designed to provide secondary voltages ranging from 10 percent below the supply voltage to 10 percent above it, or a total change of 20 percent in 32 steps of $\frac{5}{8}$ percent each.

Specialty Transformers

Specialty transformers make up a large class of transformers and autotransformers used for changing line voltage to some particular value best adapted to the load device. The primary voltage is generally 600 volts, or less. Examples of specialty transformers are sign-lighting transformers, where 120 volts is stepped down to 25 volts for low-voltage tungsten sign lamps; arc-lamp autotransformers, where 240 volts is stepped down to the voltage required for best operation of the arc; and transformers used to change 240 volts power to 120 volts for operating portable tools, fans, welders, and other devices. Also included in this specialty class are neon-sign transformers that step 120 volts up to between 2000 and 15,000 volts for the operation of neon signs.

Summary

When the ac voltage needed for an application is lower or higher than the voltage available from the source, a transformer is used. The essential parts of a transformer are the primary winding, which is connected to the source, and the secondary winding, which is connected to the load, both wound on an iron core. The two windings are not physically connected. The alternating voltage in the primary winding induces an alternating voltage in the secondary winding. The ratio of the primary and secondary voltages is equal to the ratio of the number of turns in the primary and secondary windings. Transformers may step up the voltage applied to the primary winding and have a higher voltage at the secondary terminals, or they may step down the voltage applied to the primary winding and have a lower voltage available at the secondary terminals. Transformers are applied in ac systems only, single-phase and polyphase, and would not work in dc systems since the induction of voltage depends on the rate of change of current.

A transformer is constructed as a single-phase or a three-phase apparatus. A three-phase transformer has three primary and three secondary windings which may be connected in delta (Δ) or wye (Y). Combinations such as Δ-Δ, Δ-Y, Y-Δ, and Y-Y are possible connections of the primary and secondary windings. The first symbol indicates the connection of the primary winding, and the second, that of the secondary winding. A bank of three single-phase transformers can serve the same purpose as one three-phase transformer.

Chapter 9
Wiring Methods

Several types of wiring methods are used for commercial electrical installations. The methods used on a given project are determined by several factors:

- The installation requirements set forth in the *National Electrical Code (NEC)*

- Local codes and ordinances

- Type of building construction

- Location of the wiring in the building

- Importance of the wiring system's appearance

- Costs and budget

In general, two types of basic wiring methods are used in the majority of electrical systems for commercial buildings:

- Open wiring

- Concealed wiring

In open-wiring systems, the outlets and cable or raceway systems are installed on the surfaces of the walls, ceilings, columns, and the like where they are in view and readily accessible. Such wiring is often used in areas where appearance is not important and where it may be desirable to make changes in the electrical system at a later date. You will frequently find open-wiring systems in mechanical rooms and in interior parking areas of commercial buildings.

Concealed wiring systems have all cable and raceway runs concealed inside of walls, partitions, ceilings, columns, and behind baseboards or molding where they are out of view and not readily accessible. This type of wiring system is generally used in all new construction with finished interior walls, ceilings, floors and is the preferred type where good appearance is important.

CABLE SYSTEMS

Several types of cable systems are used to construct commercial electrical systems, and include the following:

Type NM Cable: is manufactured in two- or three-wire assemblies, and with varying sizes of conductors. In both two- and three-wire cables, conductors are color-coded: one conductor is black while the other is white in two-wire cable; in three-wire cable, the additional conductor is red. Both types will also have a grounding conductor which is usually bare, but is sometimes covered with a green plastic insulation — depending upon the manufacturer. The jacket or covering consists of rubber, plastic, or fiber. Most will also have markings on this jacket giving the manufacturer's name or trademark, the wire size, and the number of conductors. For example, "NM 12-2 W/GRD" indicates that the jacket contains two No. 12 AWG conductors along with a grounding wire; "NM 12-3 W/GRD" indicates three conductors plus a grounding wire. This type of cable may be concealed in the framework of buildings, or in some instances, may be run exposed on the building surfaces. It may not be used in any building exceeding three floors above grade; as a service-entrance cable; in commercial garages having hazardous locations; in theaters and similar locations; places of assembly; in motion picture studios; in storage battery rooms; in hoistways; embedded in poured concrete, or aggregate; or in any hazardous location except as otherwise permitted by the *NEC*. Nonmetallic sheathed cable is frequently referred to as *Romex* on the job. See Figure 9-1.

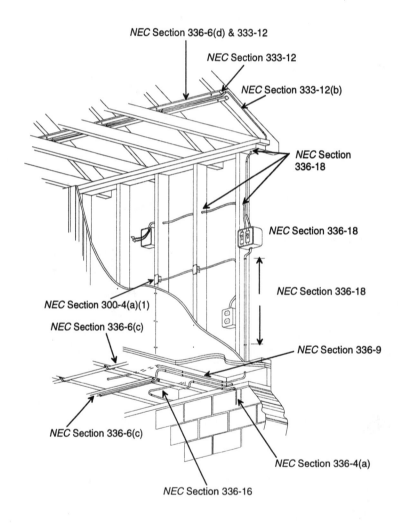

Figure 9-1: *NEC* installation requirements for Type NM cable; also Type SE cable with all conductors insulated

Type AC (Armored) Cable: Type AC cable — commonly called "BX" — is manufactured in two-, three-, and four-wire assemblies, with varying sizes of conductors, and is used in locations similar to those where Type NM cable is allowed. The metallic spiral covering on BX cable offers a greater degree of mechanical protection than with NM cable, and the metal jacket also provides a continuous grounding bond without the need for additional grounding conductors.

BX cable may be used for under-plaster extensions, as provided in the *NEC*, and embedded in plaster finish, brick, or other masonry, except in damp or wet locations. It may also be run or "fished" in the air voids of masonry block or tile walls, except where such walls are exposed or subject to excessive moisture or dampness or are below grade. This type of cable is a favorite for connecting 2×4 troffer-type lighting fixtures in commercial installations. See Figure 9-2.

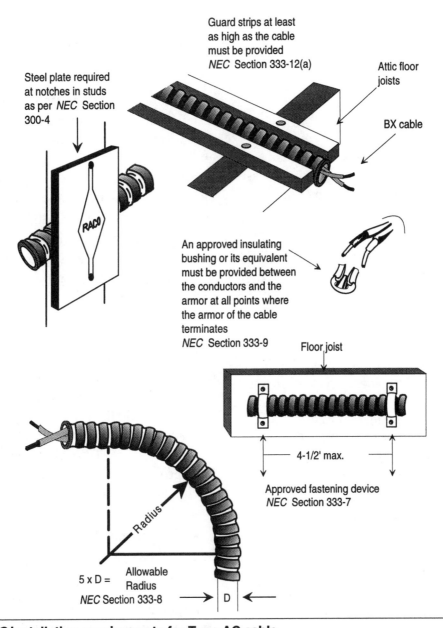

Figure 9-2: *NEC* installation requirements for Type AC cable

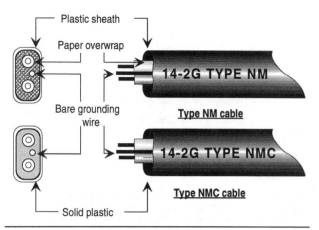

Figure 9-3: Type NM and NMC cable

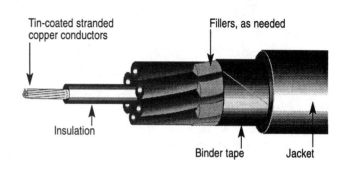

Figure 9-4: Typical Type TC cable

Type NMC Cable: This type of cable is similar in appearance and used as conventional Type NM cable except that Type NM cable is suitable for dry locations only while Type NMC cable is permitted for dry, moist, damp, or corrosive locations. See Figure 9-3 for the difference in Type NM and NMC cable.

Underground Feeder Cable: Type UF cable may be used underground, including direct burial in the earth, as a feeder or branch-circuit cable when provided with overcurrent protection at the rated ampacity as required by the *NEC*. When Type UF cable is used above grade where it will come in direct contact with the rays of the sun, its outer covering must be sun resistant. Furthermore, where Type UF cable emerges from the ground, some means of mechanical protection must be provided. This protection may be in the form of conduit or guard strips. Type UF cable resembles Type NM cable in appearance. The jacket, however, is constructed of weather resistant material to provide the required protection for direct-burial wiring installations.

Power and Control Tray Cable: Type TC is a factory assembly of two or more insulated conductors, with or without associated bare or covered grounding conductors under a nonmetallic sheath. This is a rugged assembly, as indicated in the description in *NEC* Article 340. This cable is intended for use in cable trays, in raceways, or where supported by messenger wire. It may be installed in wet locations as permitted by Article 340-3(a) and in hazardous locations as permitted by Article 340-4(3). See Figure 9-4.

Service Entrance Cable: Type SE cable, when used for electrical services, must be installed as specified in *NEC* Article 230. This cable is available with the grounded conductor bare for outside service conductors, and also with an insulated grounded conductor (Type SER) for interior wiring systems.

Type SE and SER cable are permitted for use on branch circuits or feeders provided all current-carrying conductors are insulated; this includes the grounded or neutral conductor. When Type SE cable is used for interior wiring, all *NEC* regulations governing the installation of Type NM cable also apply to Type SE cable. There are, however, some exceptions. Type SE cable with an uninsulated grounded conductor may be used on the following appliances:

● Electric range

● Wall-mounted oven

● Counter-mounted cooking unit

● Clothes dryer

Figure 9-5 summarizes the installation rules for Type SE cable — for both exterior and interior wiring.

194

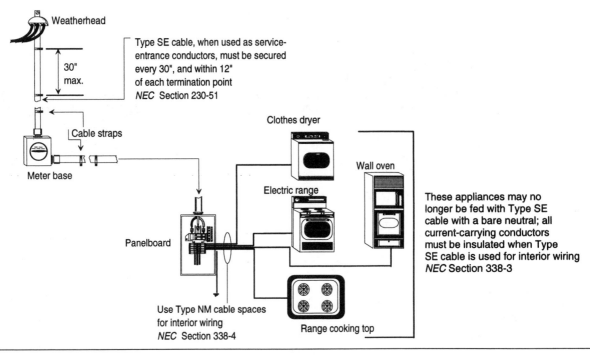

Figure 9-5: *NEC* installation requirements for Type SE cable

Underground Service-Entrance Cable: Type USE cable is similar in appearance to Type SE cable except that it is approved for underground use and must be manufactured with a moisture-resistant covering. If a flame-retardant covering is not provided, it is not approved for indoor use.

Flat Conductor Cable: Type FCC cable consists of three or more flat copper conductors placed edge-to-edge and separated and enclosed within an insulating assembly. FCC systems consist of cable and associated shielding, connectors, terminators, adapters, boxes and receptacles. These systems are designed for installation under carpet squares on hard, sound, smooth, continuous floor surfaces made of concrete, ceramic, composition floor, wood, and similar materials. If used on heated floors with temperatures in excess of 86°F, the cable must be identified as suitable for use at these temperatures.

FCC systems must not be used outdoors or in corrosive locations; where subject to corrosive vapors; in any hazardous location; or in residential, school, or hospital buildings.

Flat-Cable Assemblies: This is Type FC cable assembly and should not be confused with Type FCC cable; there is a big difference. A Type FC wiring system is an assembly of parallel, special-stranded copper conductors formed integrally with an insulating material web specifically designed for field installation in surface metal raceway. The assembly is made up of three- or four-conductor cable, cable supports, splicers, circuit taps, fixture hangers, insulating end caps and other fittings as shown in Figure 9-6 on page 196. Guidelines for the use of this system are given in *NEC* Article 363. In general, the assembly is installed in an approved U-channel surface-metal raceway with one side open. Tap devices can be inserted anywhere along the channel. Connections from the tap devices to the flat-cable assembly are made by pin-type contacts when the tap devices are secured in place. The pin-type contacts penetrate the insulation of the cable assembly and contact the multistranded conductors in a matched phase sequence. These taps can then be connected to either lighting fixtures or power outlets. See Figure 9-7 on page 196.

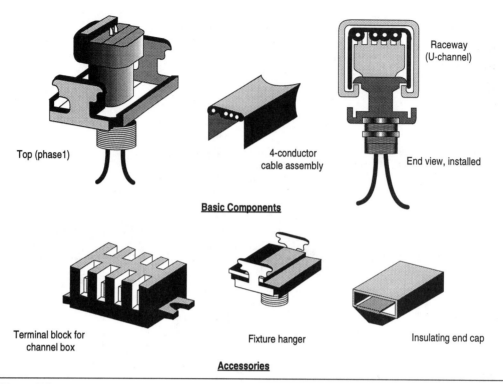

Basic Components

Top (phase1)

4-conductor cable assembly

Raceway (U-channel)

End view, installed

Terminal block for channel box

Fixture hanger

Insulating end cap

Accessories

Figure 9-6: Type FC assembly accessories

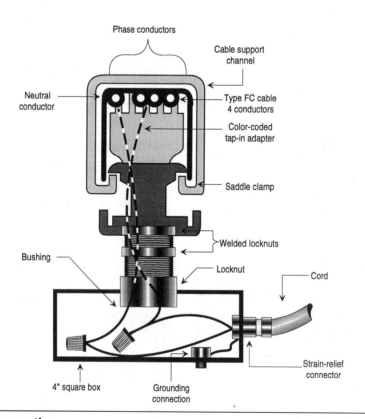

Phase conductors

Cable support channel

Neutral conductor

Type FC cable 4 conductors

Color-coded tap-in adapter

Saddle clamp

Welded locknuts

Bushing

Locknut

Cord

4" square box

Grounding connection

Strain-relief connector

Figure 9-7: Type FC connection

Flat-cable assemblies must be installed for exposed work only and must not be installed in locations where they will be subjected to severe physical damage.

Mineral-Insulated Metal-Sheathed Cable: Type MI cable is a factory assembly of one or more conductors insulated with a highly compressed refractory mineral insulation and enclosed in a liquid-tight and gas-tight continuous copper sheath. It may be used for electric services, feeders, and branch circuits in dry, wet, or continuously moist loctions. Furthermore, it may be used indoors or outdoors, embedded in plaster, concrete, fill, or other masonry, whether above or below grade. This type of cable may also be used in hazardous locations, where exposed to oil or gasoline, where exposed to corrosive conditions not deteriorating to the cable's sheath, and in underground runs where suitably protected against physical damage and corrosive conditions. In other words, MI cable may be used in practically any electrical installation.

Power and Control Tray Cable: Type TC power and control tray cable is a factory assembly of two or more insulated conductors, with or without associated bare or covered grounding conductors, under a nonmetallic sheath, approved for installation in cable trays, in raceways, or where supported by a messenger wire. The use of this cable is limited to commercial and industrial applications where the conditions of maintenance and supervision assure that only qualified persons will service the installation.

Metal-Clad Cable: Type MC cable is a factory assembly of one or more conductors, each individually insulated and enclosed in a metallic sheath of interlocking tape or a smooth or corrugated tube. This type of cable may be used for services, feeders, and branch circuits; power, lighting, control, and signal circuits; indoors or outdoors; where exposed or concealed; direct buried; in cable tray; in any approved raceway; as open runs of cable; as aerial cable on a messenger; in hazardous locations as permitted in *NEC* Articles 501, 502, and 503; in dry locations; and in wet locations under certain conditions as specified in the *NEC*.

High-Voltage Cable

Electrical circuits above 600 volts may be encountered in some commercial wiring installations such as circuits feeding sports lighting in stadiums, sign lighting, electric services for the larger highrise buildings, and similar installations. When these higher voltages are encountered, the appropriate wiring system must be used; that is, cable devices, cable splices and terminations must be designed for the installation. Figure 9-8 gives the basic characteristics of high-voltage cable and a brief description follows:

Medium-Voltage Cable: Type MV cable is covered in *NEC* Article 326 and consists of one or more insulated conductors encased in an outer jacket. The cable is suitable for use with voltages

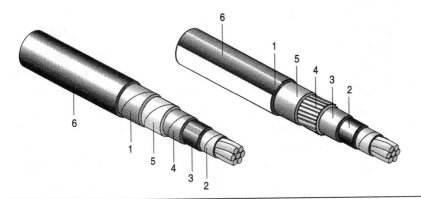

1. Strand shielding

2. Primary insulation

3. Shield bedding

4. Cable shielding

5. Cable bedding

6. Jacket

Figure 9-8: Characteristics of high-voltage cable

ranging from 2001 to 35,000 volts. It may be installed in wet and dry locations and may be buried directly in the earth.

High-Voltage Shielded Cable: Shielding of high-voltage cable protects the conductor assembly against surface discharge or burning (due to corona discharge in ionized air), which can be destructive to the insulation and jacketing.

Electrostatic shielding of cables makes use of both nonmetallic and metallic materials.

INSTRUMENTATION CONTROL WIRING

Instrumentation control wiring links the field-sensing, controlling, printout, and operating devices that form an electronic instrumentation control system. The style and size of instrumentation control wiring must be matched to a specific job.

Instrumentation control wiring usually has two or more insulated conductors as shown in Figure 9-9. An outer layer called the jacket protects the pairs of conductors inside. The number of pairs in a multi-conductor cable depends on the size of the wire used. A multi-pair cable may consist of as many as 60 or more pairs of conductors.

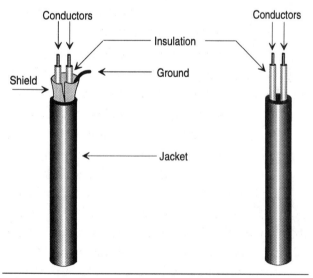

Figure 9-9: Two-conductor instrumentation control wiring

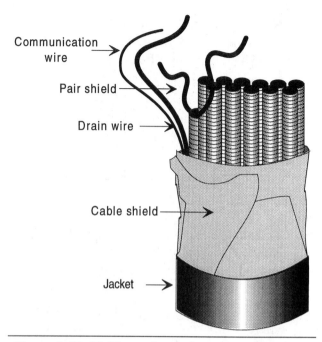

Figure 9-10: Multi-conductor instrumentation control cable with shield

Shields: Shields are provided on instrumentation control wiring to protect the electrical signals traveling through the conductors from electrical interference or noise. Shields are usually constructed of aluminum foil bonded to a plastic film as shown in Figure 9-10. If the wiring is not properly shielded, electrical distrubances may cause erratic or erroneous control signals, false indications, and improper operation of control devices.

Grounding: A ground wire is a bare copper wire used to provide continuous contact with a specified grounding terminal. A ground wire allows connections of all the instruments within a loop to a common grounding system. In some electronic systems the grounding wire is called a *drain wire*.

In most cases, instruments connected to the system are not grounded at both ends of the circuit. This is to prevent unwanted ground loops in the system. If the ground is not to be connected at the end of the wire, do not remove the ground wire. Rather, fold it back and tape it to the cable. This is called *floating the ground*. This is done in case the ground at the opposite end of the conductors develops a problem.

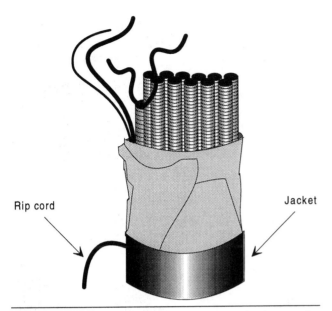

Figure 9-11: Control cable with outer jacket and rip cord

Jackets: A synthetic plastic jacket covers and protects the components within the cable. Polyethylene and PVC jackets are the most commonly used. See Figure 9-11. Some cable jackets have a nylon rip cord that allows the jacket to be peeled back without the use of a knife or cable cutter. This eliminates the possible nicking of the conductor insulation when preparing for terminations of the conductors.

RACEWAY SYSTEMS

A raceway wiring system consists of an electrical wiring system in which one or more individual conductors are pulled into a conduit or similar housing, usually after the raceway system has been completely installed. The basic raceways are rigid steel conduit, electrical metallic tubing (EMT), and PVC (polyvinyl chloride) plastic. Other raceways include surface metal moldings and flexible metallic conduit.

These raceways are available in standardized sizes and serve primarily to provide mechanical protection for the wires run inside and, in the case of metallic raceways, to provide a continuously grounded system. Metallic raceways, properly installed, provide

the greatest degree of mechanical and grounding protection and provide maximum protection against fire hazards for the electrical system. However, they are more expensive to install.

Most electricians prefer to use a hacksaw with a blade having 18 teeth per inch for cutting rigid conduit and 32 teeth per inch for cutting the smaller sizes of conduit. For cutting larger sizes of conduit ($1\frac{1}{2}$ inches and above), a special conduit cutter should be used to save time. While quicker to use, the conduit cutter almost always leaves a hump inside the conduit and the burr is somewhat larger than made by a standard hacksaw. If a power band saw is available on the job, it is preferred for cutting the larger sizes of conduit. Abrasive cutters are also popular for the larger sizes of conduit.

Conduit cuts should be made square and the inside edge of the cut must be reamed to remove any burr or sharp edge that might damage wire insulation when the conductors are pulled inside the conduit. After reaming, most experienced electricians feel the inside of the cut with their finger to be sure that no burrs or sharp edges are present.

Lengths of conduit to be cut should be accurately measured for the size needed and an additional $\frac{3}{8}$ inch should be allowed on the smaller sizes of conduit for terminations; the larger sizes of conduit will require approximately $\frac{1}{2}$ inch for locknuts, bushings, and the like at terminations.

A good lubricant (cutting oil) is then used liberally during the thread-cutting process. If sufficient lubricant is used, cuts may be made cleaner and sharper, and the cutting dies will last much longer.

Full threads must be cut to allow the conduit ends to come close together in the coupling or to firmly seat in the shoulders of threaded hubs of conduit bodies. To obtain a full thread, run the die up on the conduit until the conduit barely comes through the die. This will give a good thread length adequate for all purposes. Anything longer will not fit into the coupling and will later corrode because threading removes the zinc or other protective coating from the conduit.

Clean, sharply cut threads also make a better continuous ground and save much trouble once the system is in operation.

Some *NEC* regulations governing the installation of rigid metal conduit are shown in Figure 9-12.

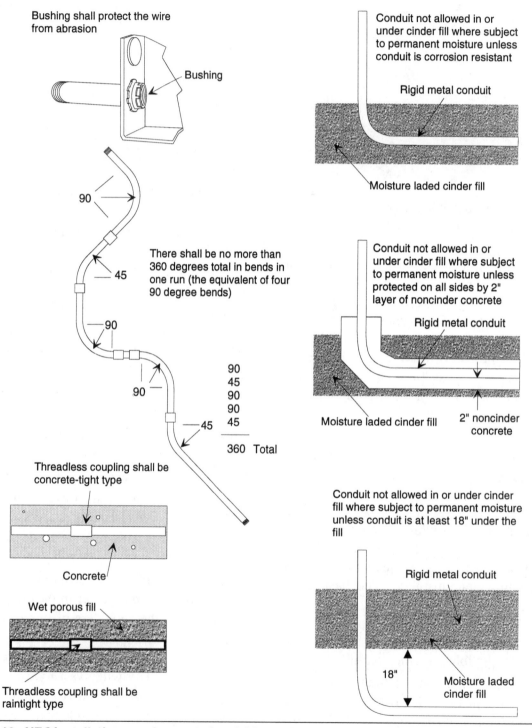

Figure 9-12: *NEC* installation requirements for rigid metal conduit

Electrical Metallic Tubing

Electrical metallic tubing (EMT) may be used for both exposed and concealed work except where it will be subjected to severe damage during use, in cinder concrete, or in fill where subjected to permanent moisture unless some means to protect it is provided; the tubing may be installed a minimum of 18 inches under the fill.

Threadless couplings and connectors are used for EMT installation and these should be installed so that the tubing will be made up tight. Where buried in masonry or installed in wet locations, couplings and connectors, as well as supports, bolts, straps, and screws, should be of a type approved for the conditions.

Bends in the tubing should be made with a tubing bender so that no injury will occur and so the internal diameter of the tubing will not be effectively reduced. The bends between outlets or termination points should contain no more than the equivalent of four quarter-bends (360° total), including those bends located immediately at the outlet or fitting (offsets).

All cuts in EMT are made with either a hacksaw, power hacksaw, tubing cutter, or other approved device. Once out, the tubing ends should be reamed with a screwdriver handle or pipe reamer to remove all burrs and sharp edges that might damage conductor insulation.

Flexible Metal Conduit

Flexible metal conduit generally is manufactured in two types, a standard metal-clad type and a liquid-tight type. The former type cannot be used in wet locations unless the conductors pulled in are of a type specially approved for such conditions. Neither type may be used where they will be subjected to physical damage or where any combination of ambient and/or conductor temperature will produce an operating temperature in excess of that for which the material is approved. Other uses are fully described in Articles 350 and 351 of the *NEC*.

When this type of conduit is installed, it should be secured by an approved means at intervals not exceeding 4½ feet and within 12 inches of every outlet box, fitting, or other termination points. In some cases, however, exceptions exist. For example, when flexible metal conduit must be finished in walls, ceilings, and the like, securing the conduit at these intervals would not be practical. Also, where more flexibility is required, lengths of not more than 3 feet may be utilized at termination points.

Flexible metal conduit may be used as a grounding means where both the conduit and the fittings are approved for the purpose. In lengths of more than 6 feet, it is best to install an extra grounding conductor within the conduit for added insurance.

Surface Metal Molding

When it is impractical to install the wiring in concealed areas, surface metal molding is a good compromise, even though it is visible, since proper painting to match the color of the ceiling and walls makes it very inconspicuous. Surface metal molding is made from sheet metal strips drawn into shape and comes in various shapes and sizes with factory fittings to meet nearly every application found in finished areas of commercial buildings. A complete list of fittings can be obtained at your local electrical equipment supplier.

The running of straight lines of surface molding is simple. A length of molding with the coupling is slipped in the end, out enough so that the screw hole is exposed, and then the coupling is screwed to the surface to which the molding is to be attached. Then another length of molding is slipped on the coupling.

Factory fittings are used for corners and turns or the molding may be bent (to a certain extent) with a special bender. Matching outlet boxes for surface mounting are also available, and bushings are necessary at such boxes to prevent the sharp edges of the molding from injuring the insulation on the wire.

Clips are used to fasten the molding in place. The clip is secured by a screw and then the molding is slipped into the clip, wherever extra support of the molding is needed, and fastened by screws. When parallel runs of molding are installed, they may be secured in place by means of a multiple strap. The joints in runs of molding are covered by slipping a connection cover over the joints. Such runs of molding should be grounded the same as any other metal raceway, and this is done by use of grounding clips. The current-carrying wires are normally pulled in after the molding is in place.

The installation of surface metal molding requires no special tools unless bending the molding is necessary. The molding is fastened in place with screws, toggle bolts, and the like, depending on the materials to which it is fastened. All molding should be run straight and parallel with the room or building lines, that is, baseboards, trims, and other room moldings. The decor of the room should be considered first and the molding made as inconspicuous as possible.

It is often desirable to install surface molding not used for wires in order to complete a pattern set by other surface molding containing current-carrying wires, or to continue a run to make it appear to be part of the room's decoration.

Wireways

Wireways are sheet-metal troughs with hinged or removable covers for housing and protecting wires and cables and in which conductors are held in place after the wireway has been installed as a complete system. They may be used only for exposed work and shouldn't be installed where they will be subject to severe physical damage or corrosive vapor nor in any hazardous location except Class II, Division 2 of the *NEC*.

The wireway structure must be designed to safely handle the sizes of conductors used in the system. Furthermore, the system should not contain more than 30 current-carrying conductors at any cross section. The sum of the cross-sectional areas of all contained conductors at any cross section of a wireway shall not exceed 20 percent of the interior cross-sectioned area of the wireway.

Splices and taps, made and insulated by approved methods, may be located within the wireway provided they are accessible. The conductors, including splices and taps, shall not fill the wireway to more than 75 percent of its area at that point.

Wireways must be securely supported at intervals not exceeding 5 feet, unless specially approved for supports at greater intervals, but in no case shall the distance between supports exceed 10 feet.

Busways

There are several types of busways or duct systems for electrical transmission and feeder purposes. Lighting duct, trolley duct, and distribution bus duct are just a few. All are designed for a specific purpose, and the electrician or electrical designer should become familiar with all types before an installation is laid out.

Lighting duct, for example, permits the installation of an unlimited amount of footage from a single working platform. As each section and the lighting fixtures are secured in place, the complete assembly is then simply transported to the area of installation and installed in one piece.

Trolley duct is widely used for industrial applications, and where the installation requires a continuous polarization to prevent accidental reversal, a polarizing bar is used. This system provides polarization for all trolley, permitting standard and detachable trolleys to be used on the same run.

Plug-in bus duct is also widely used for industrial applications, and the system consists of interconnected prefabricated sections of bus duct so formed that the complete assembly will be rigid in construction and neat and symmetrical in appearance.

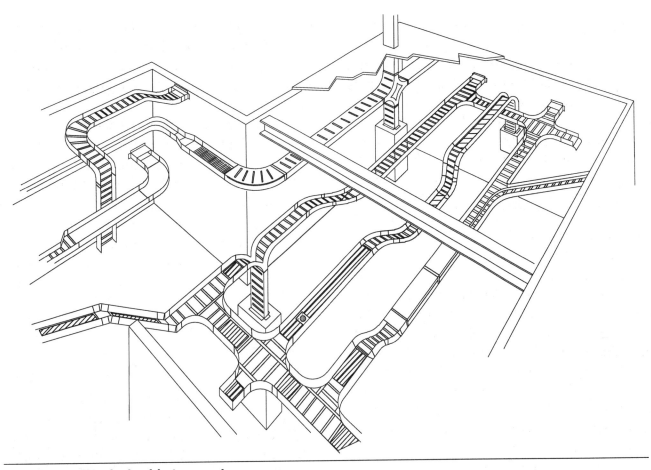

Figure 9-13: Typical cable-tray system

Cable Trays

Cable trays are used to support electrical conductors used mainly in industrial applications, but are sometimes used for communication and data processing conductors in large commercial establishments. The trays themselves are usually made up into a system of assembled, interconnected sections and associated fittings, all of which are made of metal or other noncombustible material. The finished system forms into a rigid structural run to contain and support single, multiconductor, or other wiring cables. Several styles of cable trays are available, including ladder, trough, channel, solid-bottom trays, and similar structures. See Figure 9-13.

Other types of raceway systems, including the complete installation procedures for each, are covered in Chapter 10, *Raceways, Boxes, and Fittings.*

CONDUCTORS FOR RACEWAY SYSTEMS

A variety of materials is used to transmit electrical energy, but copper, due to its excellent cost-to-conductivity ratio, still remains the basic and most ideal conductor. Electrolytic copper, the type used in most electrical conductors, can have three general characteristics:

- Method of stranding
- Degree of hardness (temper)
- Bare, tinned, or coated

Method of Stranding: Stranding refers to the relative flexibility of the conductor and may consist of only one strand or many thousands, depending on the rigidity or flexibility required for a specific need. For example, a small-gauge wire that is to be used in a fixed installation is normally solid (one strand), whereas a wire that will be

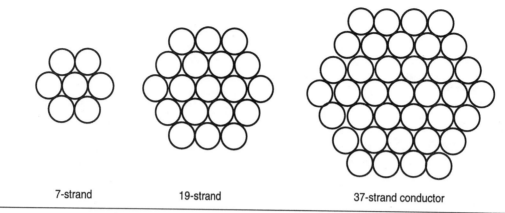

7-strand   19-strand   37-strand conductor

Figure 9-14: Common number of strands used in conductors

constantly flexed requires a high degree of flexibility and would contain many strands.

- Solid wire is the least flexible form of a conductor and is merely one strand of copper.

- Stranded refers to more than one strand in a given conductor and may vary from 3 to 37 depending on size. See Figure 9-14.

- Flexible simply indicates that there are a greater number of strands than are found in normal stranded construction.

Degree of Hardness (Temper): Temper refers to the relative hardness of the conductor and is noted as soft drawn-annealed (SD), medium hard drawn (MHD), and hard drawn (HD). Again, the specific need of an installation will determine the required temper. Where greater tensile strength is indicated, MHD would be used over SD, and so on.

Bare, Tinned, or Coated: Untinned copper is plain bare copper that is available in either solid, stranded, or flexible and in the various tempers just described. In this form it is often referred to as *red* copper.

Bare copper is also available with a coating of tin, silver, or nickel to facilitate soldering, to impede corrosion, and to prevent adhesion of the copper conductor to rubber or other types of conductor insulation. The various coatings will also affect the electrical characteristics of copper.

Conductor Size: The American Wire Gage (AWG) is used in the United States to identify the sizes of wire and cable up to and including No. 4/0 (0000), which is commonly pronounced in the electrical trade as "four-aught" or "four-naught." These numbers run in reverse order as to size; that is, No. 14 AWG is smaller than No. 12 AWG and so on up to size No. 1 AWG. To this size (No. 1 AWG), the larger the gauge number, the smaller the size of the conductor. However, the next larger size after No. 1 AWG is No. 1/0 AWG, then 2/0 AWG, 3/0 AWG, and 4/0 AWG. At this point, the AWG designations end and the larger sizes of conductors are identified by circular mils (CM or cmil). From this point, the larger the size of wire, the larger the number of circular mils. For example, 300,000 cmil is larger than 250,000 cmil. In writing these sizes in circular mils, the "thousand" decimal is replaced by the letter k, and instead of writing, say, 500,000 cmil, it is usually written 500 kcmil — pronounced *five-hundred kay-cee-mil.* See Figure 9-15 for a comparison of the different wire sizes.

Compressed Conductors

Compressed aluminum conductors are those which have been compressed so as to reduce the air space between the strands. Figure 9-16 shows a cross-section of a 37-strand compressed conductor.

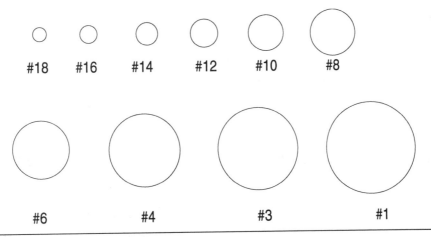

Figure 9-15: Comparison of some wire sizes

Notice that the once round strands are compressed into a sort of rectangular shape that reduces the air space between strands.

The purpose of compressed conductors is to reduce the overall diameter of the cable so that it may be installed in a conduit that is smaller than that required for standard conductors of the same wire size. Compressed conductors are especially useful when increasing the ampacity of an existing service or feeder circuit.

For example, let's say an existing service is rated at 250 amperes and is fed with four 350 kcmil THW conductors in 3-inch conduit. Should it be-

come necessary to increase the ampacity of the service to, say, 300 amperes, 500 kcmil THW compressed conductors may replace the 350 kcmil conductors without increasing the size of conduit.

Both standard and compressed conductors are covered in this chapter, including conductor insulation and practical applications of a few selected types, such as THW and THHN. *NEC* requirements are followed at all times.

PROPERTIES OF CONDUCTORS

Various *NEC* tables define the physical and electrical properties of conductors. Electricians use these tables to select the type of conductor and the size of conduit, or other raceway, to enclose the conductors in specific applications.

NEC tables tabulate properties of conductors as follows:

- Name
- Operating temperature
- Application
- Insulation
- Physical properties
- Electrical resistance
- AC resistance and reactance

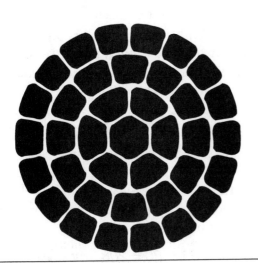

Figure 9-16: Cross-section of compressed conductors

NEC Table 310-13 gives the name, operating temperature, application, and insulation of various types of conductors, while tables in *NEC* Chapter 9 (5, 6, 7, 8, and 9) give the physical properties and electrical resistance.

To gain an understanding of these tables and how they are used in practical applications, let's take a 4/0 THHN copper conductor and see what properties may be determined from *NEC* tables.

Step 1. Turn to *NEC* Table 310-13 and scan down the second column from the left (Type Letter) until "THHN" is found. Scan to the left in this row to see that the trade name of this conductor is Heat-Resistant Thermosplastic.

Step 2. Scanning to the right in this row, note that the maximum operating temperature for this wire type is 90°C (194°F). Continuing to the right in this row, under the column headed "Application Provisions," we find that this wire type is suitable for use in dry and damp locations. The next column reveals that the insulation is flame-retardant, heat-resistant thermoplastic.

Step 3. Continuing to the right in this row, the next column lists insulation thickness for various AWG or kcmil wire sizes. The insulation thicknesses for Type THHN wire, from No. 14 AWG to 1000 kcmil, are as follows:

AWG or kcmil size	Thickness in Mils
14 - 12	15 mils
10	20 mils
8 - 6	30 mils
4 - 2	40 mils
1 - 4/0	50 mils
250 - 500	0 mils
501 - 1000	70 mils

Consequently, the insulation thickness for 4/0 THHN is 50 mils.

Step 4. Looking in the very right-hand column ("Outer Covering"), this wire type has a nylon jacket or equivalent.

If it is desired to find the maximum current-carrying capacity of this size and type conductor, when used in a raceway, turn to *NEC* Table 310-16 and proceed as follows:

Step 1. Scan down the left-hand column until the wire size is found.

Step 2. Scan to the right in this row until the 90°C column is found. This column covers the insulation types that are rated for 90°C maximum operating temperature, and include type THHN conductor insulation.

Step 3. Note that the current-carrying rating for this size and type of conductor is 260 amperes. This rating, however, is for not more than three conductors in a raceway. If more than three conductors, such as four conductors in a three-phase, four-wire feeder to a subpanel, this figure (260 amperes) must be derated as follows:

Step 4. Turn to "Notes to Ampacity Tables of 0 to 2000 volts" which precedes the *NEC* Ampacity Tables 310-16 through 310-19. Look under 310-15(b)(2), "Adjustment Factors" which state:

Where the number of current-carrying conductors in a raceway or cable exceeds three, the allowable ampacites shall be reduced as shown in the following table (in the NEC book):

Step 5. Since we want to know the allowable maximum current-carrying capacity of four 4/0 THHN conductors in one raceway, it is necessary to multiply the previous amperage (260 amperes) by 80 percent or 0.80.

260 x .80 = 208 amperes

These amperage tables are also based on the conductors being installed in areas where the ambient air temperature is 30°C (86°F). If the conductors are installed in areas with different ambient temperatures, a further deduction is required. For example, if this same set of four 4/0 THHN conductors were installed in an industrial area where the ambient temperature averaged, say, 35°C, look in the correction-factor tables at the bottom of *NEC* Table 310-16 through 310-19. In doing so, we find that a correction or derating factor for our situation is 0.96. Consequently, our present current-carrying capacity of 208 amperes must be multiplied by 0.96 to obtain the actual current-carrying capacity of the four conductors.

$$208 \times .96 = 199.68 \text{ amperes}$$

Other sizes and types of conductors are handled in a similar manner; that is, find the appropriate table, determine the listed ampacity and then multiply this ampacity by the appropriate factors in the correction-factor tables.

It sometimes becomes necessary to know additional properties of conductors for some conductor calculations, especially for voltage-drop calculations. There are many useful tables in *NEC* Chapter 9.

Identifying Conductors

The *NEC* specifies certain methods of identifying conductors used in wiring systems of all types. For example, the high leg of a 120/240-volt grounded three-phase delta system must be marked with an orange color for identification; a grounded conductor must be identified either by the color of its insulation, by markings at the terminals, or by other suitable means. Unless allowed by *NEC* exceptions, a grounded conductor must have a white or natural gray finish. When this is not practical for conductors larger than No. 6 AWG, marking the terminals with white color is an acceptable method of identifying the conductors.

Color Coding

Conductors contained in cables are color-coded so that identification may be easily made at each access point. The following lists the color-coding for cables up through five-wire cable:

- Two-conductor cable: one white wire, one black wire, and a grounding conductor (usually bare)
- Three-conductor cable: one white, one black, one red, and a grounding conductor
- Four-conductor cable: fourth wire blue
- Five-conductor cable: fifth wire yellow
- The grounding conductor may be either green or green with yellow stripes

Although some control-wiring and communication cables contain 60, 80, or more pairs of conductors — using a combination of colors — the ones listed are the most common and will be encountered the most on electrical installations.

When conductors are installed in raceway systems, any color insulation is permitted for the ungrounded phase conductors except the following:

White or gray	Reserved for use as the grounded circuit conductor
Green	Reserved for use as a grounding conductor only

Changing Colors

Should it become necessary to change the actual color of a conductor to meet *NEC* requirements or to facilitate maintenance on circuits and equipment, the conductors may be reidentified with colored tape or paint.

For example, assume that a two-wire cable containing a black and white conductor is used to feed a 240-volt, two-wire single-phase motor. Since the white colored conductor is supposed to be reserved for the grounded conductor, and none is required in this circuit, the white conductor may be marked with a piece of red tape at each end of the circuit so that everyone will know that this wire is not a grounded conductor.

CONDUCTOR PROTECTION

All conductors are to be protected against over-currents in accordance with their ampacities as set forth in *NEC* Section 240-3. They must also be protected against short-circuit current damage as required by *NEC* Sections 240-1 and 110-10.

Ampere ratings of overcurrent-protective devices must not be greater than the ampacity of the conductor. There is, however, an exception. *NEC* Section 240-3(b) states that if such conductor rating does not correspond to a standard size overcurrent-protective device, the next larger size overcurrent-protective device may be used provided its rating does not exceed 800 amperes and when the conductor is not part of a multi-outlet branch circuit supplying receptacles for cord-and-plug connected portable loads. When the ampacity of busway or cablebus does not correspond to a standard overcurrent-protective device, the next larger standard rating may be used even through this rating may be greater than 800 amperes (*NEC* Sections 364-10 and 365-5).

Standard fuse sizes stipulated in *NEC* Section 240-6 are: 1, 3, 6, 10, 15, 20, 25, 30, 35, 40, 45, 50, 60, 70, 80, 90, 100, 110, 125, 150, 175, 200, 225, 250, 300, 350, 400, 450, 500, 600, 700, 800, 1000, 1200, 1600, 2000, 2500, 3000, 4000, 5000, and 6000 amperes.

Note: The small fuse ampere ratings of 1, 3, 6, and 10 have recently been added to the NEC to provide more effective short-circuit and ground-fault protection for motor circuits in accordance with Sections 430-40 and 430-52 and UL requirements for protecting the overload relays in controllers for very small motors.

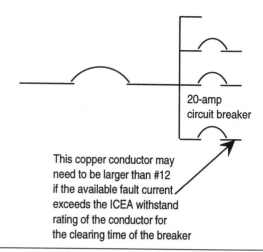

This copper conductor may need to be larger than #12 if the available fault current exceeds the ICEA withstand rating of the conductor for the clearing time of the breaker

20-amp circuit breaker

Figure 9-17: Conductor protection with non-current-limiting device

Protection of conductors under short-circuit conditions is accomplished by obtaining the maximum short-circuit current available at the supply end of the conductor, the short-circuit withstand rating of the conductor, and the short-circuit let-through characteristics of the overcurrent device.

When a non-current-limiting device is used for short-circuit protection, the conductor's short-circuit withstand rating must be properly selected based on the overcurrent protective device's ability to protect. See Figure 9-17.

It is necessary to check the energy let-through of the overcurrent device under short-circuit conditions and select a wire size of sufficient short-circuit withstand ability.

In contrast, the use of a current-limiting fuse permits a fuse to be selected which limits short-circuit current to a level less than that of the conductors short-circuit withstand rating — doing away with the need of oversized ampacity conductors. See Figure 9-18.

In many applications, it is desirable to use the convenience of a circuit breaker for a disconnecting means and general overcurrent protection, supplemented by current-limiting fuses at strategic points in the circuits.

Flexible cord, including tinsel cord and extension cords, must be protected against overcurrent in accordance with their ampacities.

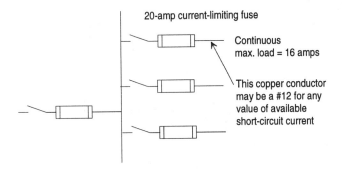

Figure 9-18: Circuits protected by current-limiting devices

Location Of Fuses In Circuits

In general, fuses must be installed at points where the conductors receive their supply; that is, at the beginning or lineside of a branch circuit or feeder. Exceptions to this rule are given in *NEC* Section 240-21.

Exception No. 1: Fuses are not required at the conductor supply if the fuses protecting one conductor are small enough to protect a small conductor connected thereto.

Exception No. 2: Fuses are not required at the conductor supply if a feed tap conductor is not over 10 feet long; is enclosed in raceway, does not extend beyond the switchboard, panelboard or control device which it supplies, and has an ampacity not less than the combined computed loads supplied and not less than the rating of the device supplied unless the tap conductors are terminated in a fuse not exceeding the tap conductors' ampacities. For field installed taps, the ampacity of the tap conductor must be at least 10 percent of the overcurrent device rating.

Exception No. 3: Fuses are not required at the conductor supply if a feeder tap conductor is not over 25 feet long; is suitably protected from physical damage; has an ampacity not less than $\frac{1}{3}$ that of the feeder conductors or fuses from which the tap conductors receive their supply; and terminates in a single set of fuses sized not more than the tap conductor ampacity.

Exception No. 4: Fuses are not required at the conductor supply if a transformer feeder tap has primary conductors at least $\frac{1}{3}$ ampacity and/or secondary conductors at least $\frac{1}{3}$ ampacity when multiplied by the approximate transformer turns ratio of the fuse or conductors from which they are tapped; the total length of one primary plus one secondary conductor (excluding any portion of the primary conductor that is protected at its ampacity) is not over 25 feet in length; the secondary conductors terminate in a set of fuses rated at the ampacity of the tap conductors; and if the primary and secondary conductors are suitably protected from physical damage.

Exception No. 5: Fuses are not required at the conductor supply if:

- A feeder tap is not over 25 feet long horizontally

- Not over 100 feet long total length in high bay manufacturing buildings when only qualified persons will service such a system

provided that, in all of the above cases, that all of the following are provided:

- The ampacity of the tap conductors is not less than $\frac{1}{3}$ of the fuse rating from which they are supplied

- The tap conductors are at least No. 6 AWG copper or No. 4 AWG aluminum

- The tap conductors do not penetrate walls, floors, or ceilings, and are tapped no less than 30 feet from the floor.

WARNING!   Smaller conductors tapped to larger conductors can be a serious hazard. If not protected against short-circuit conditions, these unprotected conductors can vaporize or incur severe insulation damage.

Exception No. 6: Transformer secondary conductors of separately derived systems do not require fuses at the transformer terminals when all of the following conditions are met:

- Must be an industrial location

- Secondary conductors must be less than 25 feet long

- Secondary conductor ampacity must be at least equal to secondary full-load current of transformer and sum of terminating, grouped, overcurrent devices

- Secondary conductors must be protected from physical damage

Note: *Switchboard and panelboard protection (NEC Section 384-16) and transformer protection (NEC Section 450-3) must still be observed.*

Lighting/Appliance Loads

The branch-circuit rating must be classified in accordance with the rating of the overcurrent protective device. Classifications for those branch circuits other than individual loads must be: 15, 20, 30, 40, and 50 amperes as specified in *NEC* Section 210-3.

Branch-circuit conductors must have an ampacity of the rating of the branch circuit and not less than the load to be served (*NEC* Section 210-19). The minimum size branch-circuit conductor that can be used is No. 14 (*NEC* Section 210-19). However, there are some exceptions as specified in *NEC* Section 210-19.

Branch-circuit conductors and equipment must be protected by a fuse whose ampere rating conforms to *NEC* Section 210-20. Basically, the branch-circuit conductor and fuse must be sized for the actual non-continuous load and 125 percent for all continuous loads. The fuse size must not be greater than the conductor ampacity. Branch circuits rated 15 through 50 amperes with two or more outlets (other than receptacle circuits) must be fused at their rating and the branch-circuit conductor sized according to *NEC* Table 210-24.

Feeder Circuits With No Motor Load

The feeder fuse ampere rating and feeder conductor ampacity must be at least 100 percent of the non-continuous load plus 125 percent of the continuous load as calculated per *NEC* Article 220. The feeder conductor must be protected by a fuse not greater than the conductor ampacity. Motor loads shall be computed in accordance with Article 430.

Service Equipment

Each ungrounded service-entrance conductor must have a fuse in series with a rating not higher than the ampacity of the conductor. These service fuses shall be part of the service disconnecting means or be located immediately adjacent thereto. (*NEC* Section 230-91).

Service disconnecting means can consist of one to six switches or circuit breakers for each service or for each set of service-entrance conductors permitted in *NEC* Section 230-2. When more than one switch is used, the switches must be grouped together (*NEC* Section 230-71).

Transformer Secondary Conductors

Field installations indicate nearly 50 percent of transformers installed do not have secondary protection. The *NEC* recommends that secondary conductors be protected from damage by the proper overcurrent-protective device. For example, the primary overcurrent device protecting a three-wire transformer cannot offer protection of the secondary conductors. Also see *NEC* exception in Section 240-3 for two-wire primary and secondary circuits.

Motor Circuit Protection

Motors and motor circuits have unique operating characteristics and circuit components. Therefore, these circuits must be dealt with differently from other types of loads. Generally, two levels of overcurrent protection are required for motor branch circuits:

- Overload protection — Motor running overload protection is intended to protect the system components and motor from damaging overload currents.

- Short-circuit protection (includes ground-fault protection) — Short-circuit protection is intended to protect the motor circuit components such as the conductors, switches, controllers, overload relays, motor, etc. against short-circuit currents or grounds. This level of protection is commonly referred to as motor branch-circuit protection applications. Dual-element fuses are designed to provide this protection provided they are sized correctly.

There are a variety of ways to protect a motor circuit — depending upon the user's objective. The ampere rating of a fuse selected for motor protection depends on whether the fuse is of the dual-element time-delay type or the non-time-delay type. If circuit breakers are used, the type and time-delay rating must also be considered.

In general, non-time-delay fuses can be sized at 300 percent of the motor full-load current for ordinary motors so that the normal motor starting current does not affect the fuse. Dual-element, time-delay fuses are able to withstand normal motor-starting current and can be sized closer to the actual motor rating than can non-time-delay fuses.

A summary of *NEC* regulations governing overcurrent protection is covered in the table in Figure 9-19 on page 212, while the table in Figure 9-20 on page 214 gives generalized fuse application guidelines for motor branch circuits.

Summary

Reliable overcurrent-protective devices prevent or minimize costly damage to transformers, conductors, motors, and the other many components and electrical loads that make up the complete electrical distribution system. Consequently, reliable circuit protection is essential to avoid the severe monetary losses which can result from power blackouts and prolonged downtime of various types of facilities. Knowing these facts, the NFPA — via the *NEC* — has set forth various minimum requirements dealing with overcurrent-protective devices, and how they should be installed in various types of electrical systems.

Knowing how to select the type of overcurrent devices for specific applications is one of the basic requirements of every electrician. It is one of the best ways to ensure a safe, fault-free electrical installation that will give years of maintenance-free service.

Always remember that the conductor size and type must match the load, and then overcurrent protection must be provided for both; that is, the conductor and the load. These two items dictate the size of overcurrent protection.

Application	Rule	*NEC* Section
Scope (FPN)	Overcurrent protection for conductors and equipment is provided to open the circuit if the current reaches a value that will cause an excessive or dangerous temperature in conductors or conductor insulation. See also Sections 110-9 and 110-10 for requirements for interrupting capacity and protection against fault currents.	Section 240-1
Protection Required	Each ungrounded service-entrance conductor must have overcurrent protection. Device must be in series with each ungrounded conductor.	Section 230-90(a)
Number of Devices	Up to six circuit breakers or sets of fuses may be considered as the overcurrent device.	Section 230-90(a)
Location in Building	The overcurrent device must be part of the service disconnecting means or be located immediately adjacent to it.	Section 230-91
Accessibility	In a property comprising more than one building under single management, the ungrounded conductors supplying each building served shall be protected by overcurrent devices, which may be located in the building served or in another building on the same property, provided they are accessible to the occupants of the building served. In a multiple-occupancy building each occupant shall have access to the overcurrent protective devices.	Section 230-72(c)
Location in Circuit	The overcurrent device must protect all circuits and devices, except equipment which may be connected on the supply side including: 1) Service switch, 2) Special equipment, such as lightning arresters, 3) Circuits for emergency supply and load management (where separately protected), 4) Circuits for fire alarms or fire-pump equipment (where separately protected), 5) Meters, with all metal housing grounded, (600 volts or less), 6) Control circuits for automatic service equipment if suitable overcurrent protection and disconnecting means are provided.	Section 230-94

Figure 9-19: *NEC* **regulations concerning overcurrent protection**

Application	Rule	NEC Section
Installation and Use	Listed or labeled equipment shall be used or installed in accordance with any instructions included in the listing or labeling.	Section 110-3(b)
Interrupting Rating	Equipment intended to break current at fault levels shall have an interrupting rating sufficient for the system voltage and the current which is available at the line terminals of the equipment.	Section 110-9
Circuit Impedance And Other Characteristics	The overcurrent protective devices, the total impedance, the component short-circuit withstand ratings, and other characteristics of the circuit to be protected shall be so selected and coordinated as to permit the circuit protective devices used to clear a fault without the occurrence of extensive damage to the electrical components of the circuit.	Section 110-10
Effective Grounding Path	The path to ground from circuits, equipment, and conductor enclosures shall have capacity to conduct safely any fault current likely to be imposed on it.	Section 250-68(b)
General	Bonding shall be provided where necessary to assure electrical continuity and the capacity to conduct safely any fault current likely to be imposed.	Section 250-98(a)

Figure 9-19: *NEC* **regulations concerning overcurrent protection (***cont.***)**

Type of Motor	Dual-Element, Time-Delay Fuses		Non-Time-Delay Fuses	
	Desired Level of Protection			
	Motor Overload and Short-Circuit	**Backup Overload and Short-Circuit**	**Short-Circuit Only**	**Short-Circuit Only**
Service Factor 1.15 or Greater or 40°C Temp. Rise or Less	100 to 115%	115% or next standard size	150% to 175%	300%
Service Factor Less Than 1.15 or Greater Than 40°C Temp Rise	100 to 115%	115% or next standard size	150% to 175%	300%

Fuses give overload and short-circuit protection

Overload relay gives overload protection and fuses provide backup overload protection

Overload relay provides overload protection and fuses provide only short-circuit protection

Overload relay provides overload protection and fuses provide only short-circuit protection

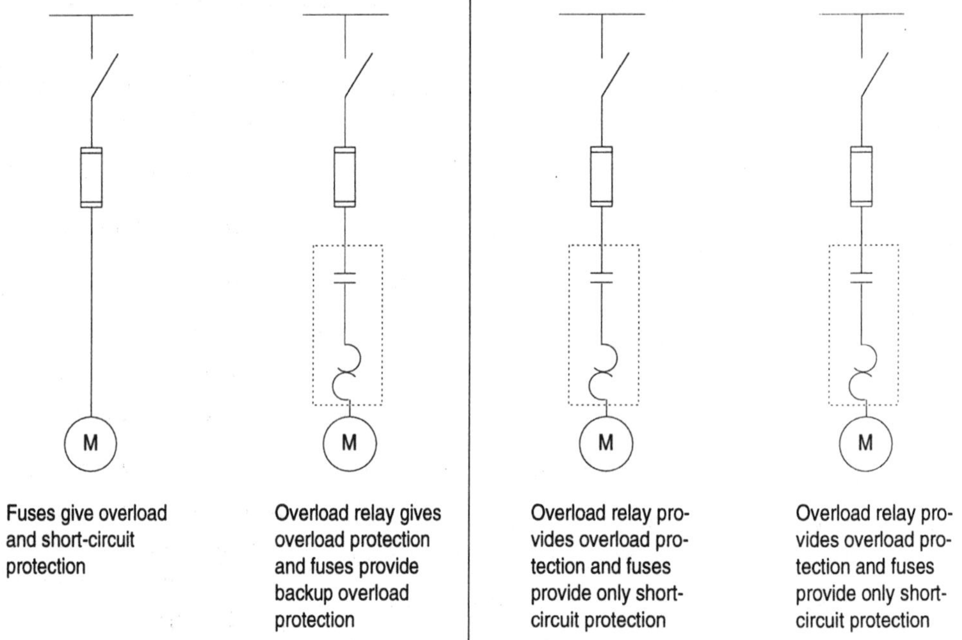

Figure 9-20: Sizing of fuses as a percentage of motor full-load current

Chapter 10
Raceways, Boxes and Fittings

A raceway is any channel used for holding wires, cables, or busbars, which is designed and used solely for this purpose. Types of raceways include rigid metal conduit, intermediate metal conduit (IMC), rigid nonmetallic conduit, flexible metal conduit, liquid-tight flexible metal conduit, electrical metallic tubing (EMT), underfloor raceways, cellular metal floor raceways, cellular concrete floor raceways, surface metal raceways, wireways, and auxiliary gutters. Raceways are constructed of either metal or insulating material.

Raceways provide mechanical protection for the conductors that run in them and also prevent accidental damage to insulation and the conducting metal. They also protect conductors from the harmful chemical attack of corrosive atmospheres and prevent fire hazards to life and property by confining arcs and flame due to faults in the wiring system.

One of the most important functions of metal raceways is to provide a path for the flow of fault current to ground, thereby preventing voltage build-up on conductor and equipment enclosures. This feature, of course, helps to minimize shock hazards to personnel and damage to electrical equipment. To maintain this feature, it is extremely important that all metal raceway systems be securely bonded together into a continuous conductive path and properly connected to a grounding electrode such as a water pipe or a ground rod.

A box or fitting must be installed at:

- Each conductor splice point
- Each outlet, switch point, or junction point
- Each pull point for the connection of conduit and other raceways

Furthermore, boxes or other fittings are required when a change is made from conduit to open wiring. Electrical workers also install pull boxes in raceway systems to facilitate the pulling of conductors.

In each case — raceways, outlet boxes, pull and junction boxes — the *NEC* specifies specific maximum fill requirements; that is, the area of conductors in relation to the box, fitting, or raceway system. This chapter is designed to cover these *NEC* requirements and apply these rules to practical applications.

CONDUIT FILL REQUIREMENTS

The *NEC* provides rules on the maximum number of conductors permitted in raceways. In conduits, for either new work or rewiring of existing raceways, the maximum fill must not exceed 40 percent of the conduit cross-sectional area. In all such cases, fill is based on using the actual cross-sectional areas of the particular types of conductors used. Other derating rules specified by the *NEC* may be found in Article 310. For example, if more than three conductors are used in a single conduit, a reduction in currrent-carrying capacity is required. Ambient temperature is another consideration that may call for derating of wires below the values given in *NEC* tables.

The allowable number of conductors in a raceway system is calculated as percentage of fill as specified in Table 1 of *NEC* Chapter 9 (Figure 10-1). When using this table, remember that equipment grounding or bonding conductors, where installed, must be included when calculating conduit or tubing fill. The actual dimensions of the equipment grounding or bonding conductor (insulated or bare) must be used in the calculation.

Conduit fill may be determined in one of two different ways:

- Calculating the fill as a percentage of the conduit's inside diameter (ID)
- Using tables in *NEC* Appendix C

When determining the number of conductors (all the same size) for use in trade sizes of conduit or tubing ½ inch through 6 inches, refer to Tables C1 through C12 in *NEC* Appendix C. For example, let's assume that four 500 kcmil THHN conductors must be installed in a rigid conduit run. What size of conduit is required for four 500 kcmil THHN conductors?

Turn to Table C8 in *NEC* Appendix C and scan down the left-hand column until the insulation type (THHN) is found. Once the insulation type has been located, move to the second column (Conductor Size AWG/kcmil) and scan down this column until 500 kcmil is found. Now, scan across this row until the desired number of conductors (four in this case) is found. When the number "4" has been located, move up this column to see the size of conduit required at the top

Table 1. Percent of Cross Section of Conduit and Tubing for Conductors

Number of Conductors	All Conductors Types
1	53
2	31
Over 2	40

FPN No. 1: Table 1 is based on common conditions of proper cabling and alignment of conductors where the length of the pull and the number of bends are within reasonable limits. It should be recognized that, for certain conditions, larger size conduit or a lesser conduit fill should be considered.

FPN No. 2: When pulling three conductors or cables into a raceway, if the ratio of the raceway (inside diameter) to the conductor or cable (outside diameter) is between 2.8 and 3.2, jamming can occur. While jamming can occur when pulling four or more conductors or cables into a raceway, the probability is very low.

Figure 10-1: Percent of cross-section of conduit for conductors

of the page. In doing so, we can see that 3-inch conduit is the size to use.

If compact conductors (all the same size) are used, refer to Table C8(A) in *NEC* Appendix C for trade sizes of conduit or tubing ½ inch through 4 inches. For example, let's assume that four 500 kcmil compact conductors are to be installed in a raceway system. What size of conduit is required?

Refer to Table C8(A) (*NEC* Appendix C). Again, the conduit size is 3 inches — the same as in our previous example. Therefore, no savings in conduit size will be realized in this case if compact conductors are used. Furthermore, the compact conductors will be more costly and it will be to everyone's advantage to stick with conventional conductors in this situation.

When working with conductors larger than 750 kcmil or combinations of conductors of different sizes, Tables 4 through 8 of *NEC* Chapter 9 should be used to obtain the dimensions of con-

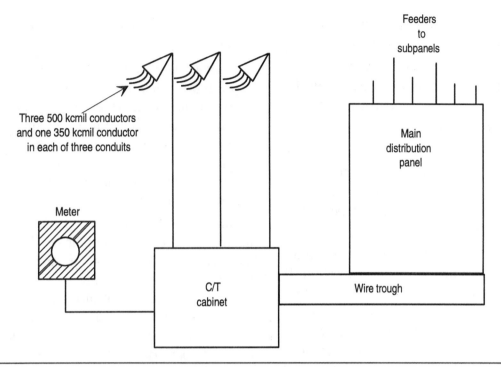

Figure 10-2: Power-riser diagram of a 1200-ampere service-entrance

ductors, conduit, and tubing. These tables give the nominal size of conductors or tubing for use in computing the required size of conduit or tubing for various combinations of conductors. The dimensions represent average conditions only, and variations will be found in dimensions of conductors and conduits of different manufacture.

Note: *Where the calculated conductors, all of the same size (total cross-sectional area including insulation), include a decimal fraction, the next higher whole number must be used where the decimal is 0.8 or larger.*

Let's take a situation where a 1200-ampere service-entrance is to be installed utilizing three parallel conduits, each containing three 500 kcmil THHN ungrounded conductors and one 350 kcmil grounded THHN conductor (neutral) as shown in Figure 10-2. What size rigid conduit is required?

Step 1. Refer to *NEC* Chapter 9, Table 5 to determine the area (in square inches) of 500 kcmil THHN conductor. The area is found to be .7073 sq. in.

Step 2. Since there are three conductors of this area in each conduit, .7073 is multiplied by 3 to obtain the total square inches for all three conductors.

$$.7073 \times 3 = 2.1219$$

Step 3. Refer again to *NEC* Chapter 9, Table 5 to obtain the area of the one 350 kcmil grounded conductor in each conduit. The area is found to be .5242 inch.

Step 4. Add the total area of the three 500 kcmil conductors (2.1219) plus the area of the one 350 kcmil conductor (.5242) to obtain the total area occupied by all four conductors in the conduit.

$$2.1219 + .5242 = 2.6461$$

Step 5. Referring to Table 4 of *NEC* Chapter 9, we look in the "over 2 Wires 40%" column and scan down until we come to

217

a figure that is the closest to 2.6461 inches without going under this figure. In doing so, we find that the closest area is 3.00. Scan to the left from this figure (in the same column) and note that the conduit size is 3 inches.

Therefore, each of the three conduits containing three 500 kcmil THHN conductors and one 350 kcmil THHN conductor must be 3 inches ID to comply with the *NEC*.

Refer again to Figure 10-2 for the conduit routed from the C/T cabinet to the meter location. If this conduit contains six #12 AWG THHN conductors, what size of rigid conduit is required?

Since all conductors are the same size, refer to Table C8 of *NEC* Appendix C, scan down the left-hand column until THHN insulation is found, and then move to the right one column and find #12 AWG conductor. Note that the next column to the right lists ten #12 conductors in ½ inch conduit. Since six conductors is all that is in this conduit run, and since ½ inch is the smallest trade size conduit, this will be the size to use.

INSTALLING CONDUIT

The normal installation of rigid metal conduit, intermediate metal conduit (IMC) and electric metallic tubing (EMT) requires the provision of many changes of direction in the conduit runs — ranging from simple offsets at the point of termination at outlet boxes and cabinets to complicated angular offsets at columns, beams, cornices, and the like.

Unless the contract specifications dictate otherwise, such changes in direction are accomplished, particularly in the case of smaller sizes, by bending the conduit or tubing as required. In the case of 1¼-inch and larger sizes, right-angle changes of direction are sometimes accomplished with the use of "factory" elbows or conduit bodies. In most cases, however, such changes in direction are ac-

complished more economically by making conduit bends in the field.

Another good reason for making on-the-job bends is when multiple runs of the larger conduit sizes are installed. Truer parallel alignment of multiple runs is maintained by using on-the-job conduit bends rather than using factory elbows. Such bends can all be made from the same center, using the bend of the largest conduit in the run as the pattern for all other bends.

Since many raceway systems are run exposed, learning to install neat-looking conduit systems in an efficient and workmanlike manner is the basic trademark of a good electrician. Every electrician working on commercial installations must therefore learn how to calculate and fabricate conduit bends — with both hand and power conduit benders, and take pride in performing the best work possible.

NEC Requirements

In general, the *NEC* requires that metal conduit bends must be made so that the conduit will not be damaged during the operation, and that the internal diameter of the conduit will not be effectively reduced in size. To accomplish this, the *NEC* further specifies the minimum radius of the inner-edge curve of a conduit bend which, in general, requires that the inside radius of an elbow must be at least six times the internal diameter of the conduit when conductors without lead sheath are to be installed. See Figure 10-3. There is good reason for this rule. When the inside radius of an elbow is less than six times the inside diameter, wire pulling becomes extremely difficult and the insulation on the conductors may be damaged.

When conductors without lead sheath are to be installed, the inside radius must be increased according to *NEC* Table 346-10.

The *NEC* further states that no more than four quarter bends (360 degrees total) may be made in any one conduit run between boxes, cabinets, panels, or junction boxes; that is, between pull points.

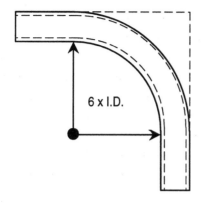

Figure 10-3: Inside radius requirements

Some electricians are prone to consider that offsets, kicks and saddles are not bends — especially in areas where the electrical inspectors are lax. These electricians count only those bends that are actually a quarter circle (90 degrees). The misconception of this is quickly apparent when wires are pulled in. Offsets and saddles add just as much resistance to pulling in conductors as any 90-degree elbow. An offset, for example, takes two 45-degree bends which equal one 90-degree bend. A "saddle" should be counted as two quarter bends of 180 degrees.

A 15-degree "kick" in a conduit run may seem insignificant, but after several dozen of these kicks are incorporated into the run, the difficulty of pulling wire becomes apparent. The number of degrees in each kick should be included in the total count, and in no case, should the total number (number of bends × number of degrees in each bend) exceed 360 degrees. This means that a maximum of 24 15-degree kicks may be incorporated into one conduit run; 12 kicks and two 90-degree bends; 12 kicks and two offsets; 12 kicks and one saddle, etc. These are the maximum number allowed. Many electricians prefer to install pull boxes at closer intervals to reduce the number of bends — especially when the larger conductor sizes are being pulled. The additional cost of the pull boxes and the labor to install them is often

offset by the labor saved in pulling the conductors. A lesser number of bends also offers greater insurance against damaging the conductors during the pull.

Types Of Bends

Several types of bends are used in most conduit systems. A brief description of each follows:

Elbow: An elbow or "ell" is a 90-degree bend that is used when a conduit must turn at a 90-degree angle. In single conduit runs, when the larger sizes of conduit are being installed, factory elbows are frequently used to save labor on setting up a power bending machine. However, in multiple conduit runs, a neater job will result if on-the-job "sweep" or concentric bends are properly calculated, and installed as shown in Figure 10-4.

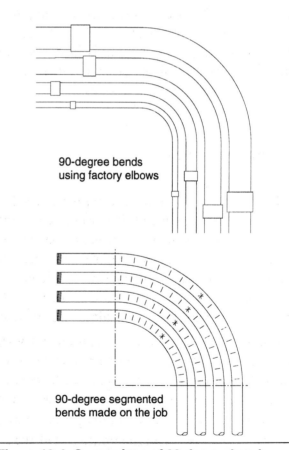

90-degree bends using factory elbows

90-degree segmented bends made on the job

Figure 10-4: Comparison of 90-degree bends — factory and made-on-the-job sweep bends

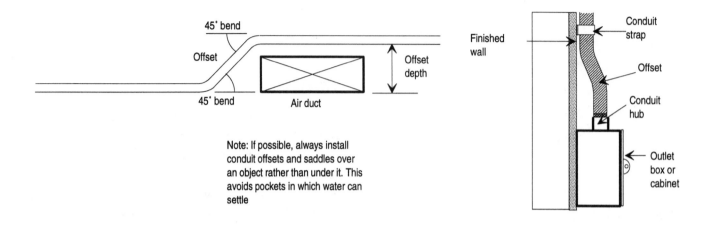

Figure 10-5: Application of conduit offsets

Offset: An offset consists of two 45-degree bends and is used when the conduit run must run over, under, or around an obstacle. An offset is also used at outlet boxes, cabinets, panelboards, and pull boxes as shown in Figure 10-5.

Saddle: A saddle is used to cross a small obstruction or other runs of conduit. A saddle is made by marking the conduit at a point where the saddle is required, and placing a bender a few inches ahead of this point. Bends are made as shown in Figure 10-6 in approximately 30-degree increments. In most cases, the bends should be as close together as the bender will permit. If the obstruction is very large, use two offsets instead.

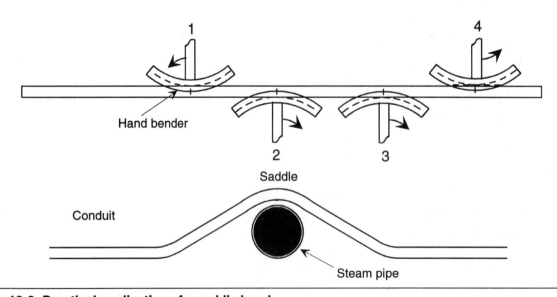

Figure 10-6: Practical application of a saddle bend

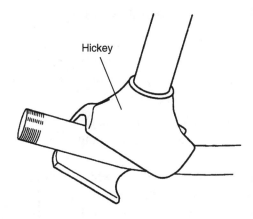

Figure 10-7: A kick in a conduit run is a minor change in direction

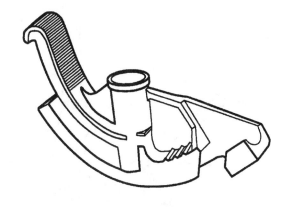

Figure 10-8: Typical hand bender

Kick: A kick is a minor change in direction of a conduit run. It is used mostly where the conduit run will be concealed as in "deck work." The first bend in an offset, for example, is really a "kick" as shown in Figure 10-7; another kick (in the opposite direction), however, transforms the kick into an offset.

Bending Conduit

The art of conduit bending could fill many volumes. Consequently, this chapter sticks to the techniques required to make various bends in the smaller-size of conduits (½ inch through 1¼ inch sizes), using only hand benders.

Figure 10-8 shows a typical hand bender. Hand benders are convenient to use on the job because they are portable and no electrical power is required. Such benders have a shape that supports the walls of the conduit being bent. The hand bender is used for bending EMT, rigid conduit, and IMC. Hand benders provide a bending radius that conforms to the *NEC* for any size of conduit.

Most hand benders are designed to bend rigid conduit and EMT of corresponding sizes. That is, a single hand bender can bend either ¾-inch EMT or ½-inch rigid conduit. This is because the corresponding sizes of conduit have nearly equal outside diameters.

There are two basic ways to use a hand bender. One way is to place the conduit on the floor and apply pressure to the bender with one foot on the step of the bender; then pull the handle toward yourself. A second method is for saddle or offset bends. The handle of the bender is placed on the floor and braced between one foot and the knee. Pressure is then applied to the conduit to form the bend.

A *hickey* (Figure 10-7) should not be confused with a hand bender. The hickey is designed for bending small sizes of rigid conduit only because very little support is given to the walls of the conduit being bent. Consequently, the hickey functions quite differently from the hand bender.

When a hickey is used to bend conduit, both the bend and radius must be formed simultaneously. In doing so, care must be taken so as not to flatten or kink the conduit.

To use a hickey to bend rigid metal conduit, first make a small bend of approximately 10 degrees. Then the hickey is moved to a new position on the conduit and another small bend is made; again, not more than 10 degrees. This process is continued until the bend is completed. In the hands of an experienced electrician, the hickey is an excellent tool. It can be used to effectively stub-up conduits in slabs and decks.

Another bending tool that is frequently used on commercial projects is a special-purpose device

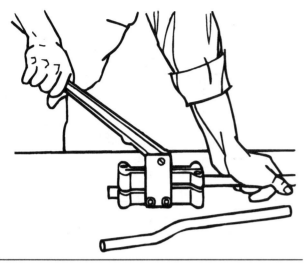

Figure 10-9: Little Kicker being used to make offsets in EMT

called the "Little Kicker." This tool is designed to produce offsets in EMT with one motion of the tool's handle. Two models are available: one for ½-inch EMT and the other for ¾-inch EMT. The Little Kicker is an excellent time-saving device for making large numbers of offsets in the smaller sizes of EMT for terminating into junction or outlet boxes. See Figure 10-9.

Geometry Of Conduit Bending

Bending conduit requires some knowledge of basic geometry. You are probably already familiar with most of the concepts needed, but a brief review should prove helpful.

Right Triangle: A right triangle is defined as any triangle with one 90-degree angle. The side directly opposite the 90-degree angle is called the hypotenuse and the side on which the triangle sits is the base. The vertical side is called the height. On the job, the right triangle is applied when making offset bends; that is, the offset forms the hypotenuse of a right triangle as shown in Figure 10-10.

Circle: A circle is defined as a closed curved line whose points are all the same distance from its center. The distance from the center point to the edge of the circle is called the radius. The length from one edge of the circuit to the other edge is the diameter. The distance around the circle is called the circumference. A circle can be divided into four equal quadrants. Each quadrant accounts for 90 degrees, making a total of 360 degrees. When a 90-degree conduit bend is made, one-fourth of a circle (one quadrant) is used.

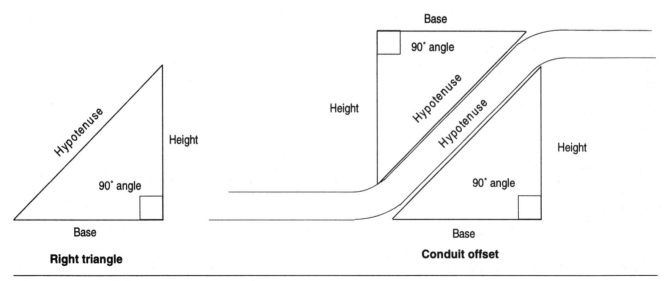

Figure 10-10: Right triangle and its relationship to an offset bend

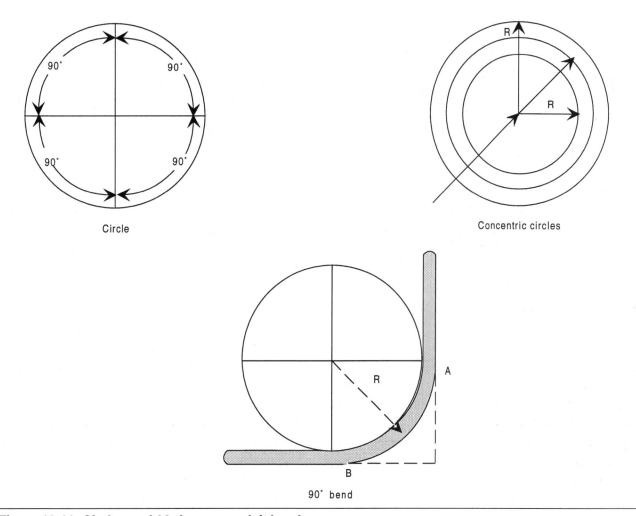

Figure 10-11: Circles and 90-degree conduit bends

Concentric circles are several circles that have a common center but the radius of each circle is different. The concept of concentric circles can be applied to concentric 90-degree bends in conduit. Such bends have the same center point, but the radius of each is different. This relationship is shown in Figure 10-11. Also refer to Figure 10-4 to review the appearance of actual concentric sweep conduit bends.

To calculate the circumference of a circle, use the equation:

$$C = \pi D$$

In this equation, C = circumference, D = diameter, and π = 3.14.

Another way of stating the equation for circumference is:

$$C = \pi 2R$$

In this latter equation, R = the radius or $\frac{1}{2}$ the diameter.

To calculate the arc of a quadrant use the following equation:

$$Length\ of\ arc = (.25)2\pi R = 1.57R$$

In other words, the arc of a quadrant equals $\frac{1}{4}$ the circumference of the circle. Therefore, the arc of a true 90-degree conduit bend also equals $\frac{1}{4}$ the circumference of a circle; four 90-degree conduit bends equals a full circle or 360 degrees.

EMT	Rigid	Take-Up
½″	—	5″
¾″	½″	6″
1″	¾″	8″
1¼″	1″	11″

Figure 10-12: Typical conduit bender take-up distances

Making 90-Degree Bends

The 90-degree stub-up bend is probably the most basic bend of all, especially on deck jobs for stubbing up conduit for outlet boxes prior to the concrete pour, or when the conduit must make a 90-degree change in direction. Before beginning to make the bend, two measurements must be known:

- The desired rise or height of the stub-up

- The take-up distance of the bender

"Take-up" is the amount of conduit the bender will use to form the bend. Take-up distances are usually listed in the bender manufacturer's instruction manual; sometimes these figures are inscribed directly on the side of the bender. A sample table is shown in Figure 10-12.

Once the take-up has been determined, subtract it from the stub-up height. Measure back from the end of the conduit and mark that distance on the conduit with a felt-tip marker; make the mark all the way around the conduit. The mark will indicate the point at which you will begin to bend the conduit. Line up the starting point on the conduit (the mark just made) with the starting point on the bender. The latter is usually in the form of an arrowhead or other mark on the side of the bender.

Once you have lined up the bender with the mark that you made on the conduit, use one foot to hold the conduit steady, keeping the heel of this foot on the floor for balance. Apply pressure on the bender foot pedal with your other foot. Make sure you hold the bender handle level and as far up on the handle as possible to get maximum leverage. Then bend the conduit in one smooth motion, pulling as evenly as possible.

When using the take-up method to bend conduit, the bender is always placed on the conduit and the bend is made facing the end of the conduit from which the measurements were taken. See Figure 10-13 for the measurements and bender placement on ½-inch conduit.

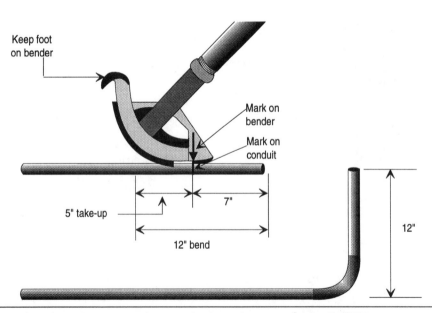

Figure 10-13: Measurements used to make a 12-inch stub-up on ½-inch EMT

After finishing the bend, check to make sure that the correct angle and measurements were made. Use the following steps to check a 90-degree bend:

Step 1. With the conduit on the floor, in the same position that it was while the bend was being made, measure from the floor to the top of the stub-up to make sure it is the correct height.

Step 2. With the back of the bend on the floor, measure to the end of the conduit to make sure it is the right length.

Step 3. Check the 90-degree angle of the bend with a square or at the angle formed by the floor and a wall. You may also use a magnetic torpedo level.

If the conduit is slightly overbent past the desired angle, use the bender to bend the conduit back to the correct angle.

The method just described will produce a 90-degree "one-shot" conduit bend; that is, it took one motion of the bender to form the bend. A segment bend is any bend that is formed by a series of bends of a few degrees each. A "shot" is actually one bend in a segment bend.

Bender Take-Up: Hand benders are available in four sizes:

- $\frac{1}{2}''$
- $\frac{3}{4}''$
- $1''$
- $1\frac{1}{4}''$

Each size has a definite "take-up" or "rise." When a full 90-degree bend is made from the floor, the end of the arc will be 5 inches off the floor with $\frac{1}{2}$-inch EMT, 6 inches off the floor with $\frac{3}{4}$-inch EMT, and 8 inches off the floor with 1-inch EMT.

When working on the floor, a full 90-degree bend is easily made by pulling the bender handle toward the operator until the end of the conduit is pulled from its horizontal position to a vertical position. Maintain constant pressure on the step of the bender with your foot to prevent kinking the conduit. A 45-degree bend is made by pulling only until the bender handle points straight up.

Gain: The gain is the distance saved by the arc of a 90-bend. Knowing the gain can help to pre-cut, ream, and pre-thread both ends of the conduit before the bend is made. This will make the work go quicker because it is easier to work with conduit while it is

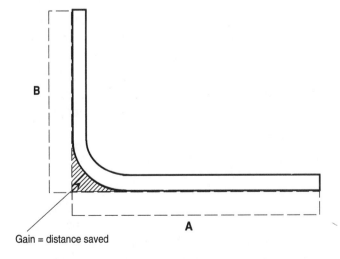

Gain = distance saved

Rigid Conduit Size	*NEC* Radius	90°Gain
$\frac{1}{2}''$	4''	$2\frac{5}{8}''$
$\frac{3}{4}''$	5''	$3\frac{1}{4}''$
1''	6''	4''
$1\frac{1}{4}''$	8''	$5\frac{5}{8}''$

Figure 10-14: Conduit gain specifications

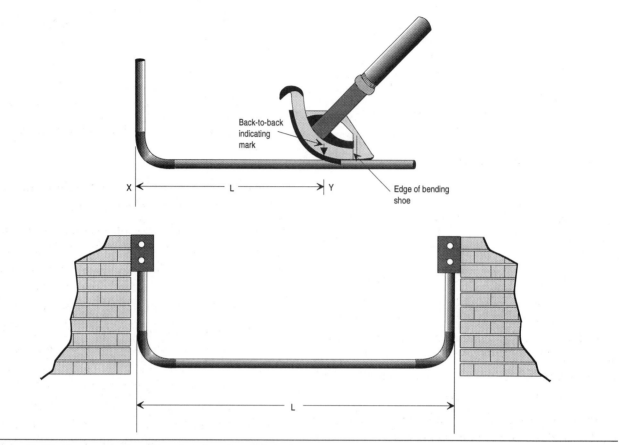

Figure 10-15: Process of making back-to-back bends

straight. Figure 10-14 shows that the overall length of a piece of conduit with a 90-degree bend is less than the sum of the horizontal and vertical distances, when measured square to the corner.

Developed length = (A + B) - Gain

A sample conduit-bender manufacturer's gain table is also shown. These tables are short-cut devices to calculate the gain for any conduit size and will prove useful on all jobs.

Back-To-Back Bends

Back-to-back or "U" bends consist of two 90-degree bends, placed back-to-back as shown in Figure 10-15.

To make back-to-back bends, first bend a conventional 90-degree bend as discussed previously and labeled "X" in Figure 10-15. To make the second bend, measure the required distance between the bends from the back of the first bend. This distance is labeled "L" in Figure 10-15. Reverse the bender on the length of conduit and place the bender's "back-to-back" indicating mark at point "Y" on the conduit. Note that the measurements from point "X" to point "Y" are taken from the outside edges of the conduit. Now, holding the bender in the reverse position and properly aligned, apply foot pressure and complete the second bend.

Offsets

Many situations require that conduit be bent so that it can pass over objects such as beams and other conduits, or for entering meter cabinets, panelboards, and junction boxes. Bends used for this purpose are called "offsets." To produce an

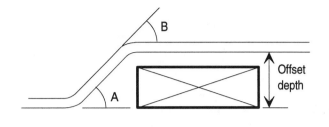

Figure 10-16: Typical offset

Angle	Multiplier
10°	5.76
15°	3.86
22.5°	2.61
30°	2.00
45°	1.41

Figure 10-18: Offset multipliers for various angles

offset, two equal bends of 90-degrees or less are required, a specified distance apart as shown in Figure 10-16.

In making conduit offsets, be aware that the degree of bend dictates the space requirements for the total offset; that is, the smaller the angle of bend, the larger the space needed to complete the change in elevation. If space is not a problem, it is best to keep the angle of the bends small so that pulling the wire will be easier.

Once the angle of the bends and the required height of the offset have been determined, the distance between the bends may be calculated. Bender manufacturers typically provide guides that can help you determine this distance at a glance. A sample chart appears in Figure 10-17.

Charts are also available that provide a multiplier for each angle used in conduit offsets. One such table is shown in Figure 10-18. The distance

between bends is equal to the height of the offset times the multiplier for the angle used.

For example, a bend of 45-degrees with a height of 10 inches requires 14 inches between the first and second bend. So, in this case, you would place indicating marks on the conduit 14 inches apart to indicate where the first and second bends should start.

Once the angle of bend is known along with the height of the offset and the distance between bends, the location of the first bend can be determined. Keep in mind that the conduit will "shrink" a certain amount per inch of offset depth. Using the chart in Figure 10-17, it can be seen that the amount of conduit will shrink $3\frac{3}{4}$ inches for the example cited above. This means that the bender should be placed $3\frac{3}{4}$ inches ahead of the mark for the first bend. Remember that most benders have an arrow or other indicating mark that shows where to place the bender for various types of bends.

To continue, make the bend at the first mark and then make the bend at the second mark, keeping the conduit running in the same direction through the bender's shoe.

It is important to be precise when measuring and bending offsets. A few degrees too much will make the offset too high, and bending less than the required number of degrees will make the offset too low. By the same token, rounding off numbers too much during calculations will result in a length of conduit that won't fit.

Another important consideration in making offset bends is to make certain that the bends are precisely in line with each other. After making the bends, lay the length of conduit on the floor to test for alignment.

Offset Depth in Inches	Distance Between Bends	Angle of Bends	Conduit Length-Loss in Inches
1	6	10°	$\frac{1}{16}$
2	$5\frac{1}{4}$	$22\frac{1}{2}$°	$\frac{3}{8}$
3	6	30°	$\frac{3}{4}$
4	8	30°	1
5	7	45°	$1\frac{7}{8}$
6	$8\frac{1}{2}$	45°	$2\frac{1}{4}$
7	$9\frac{3}{4}$	45°	$2\frac{5}{8}$
8	$11\frac{1}{4}$	45°	3
9	$12\frac{1}{2}$	45°	$3\frac{3}{8}$
10	14	45°	$3\frac{3}{4}$

Figure 10-17: Chart giving measurements for making conduit offsets

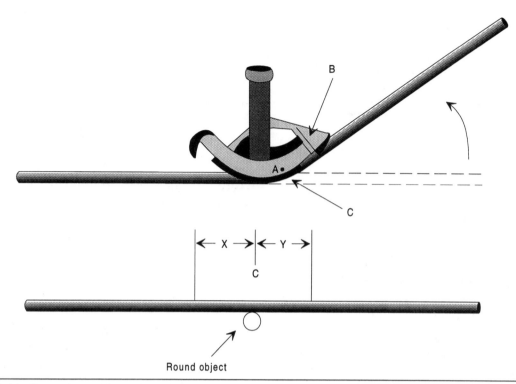

Figure 10-19: Steps in making a saddle bend

Saddle Bends

A saddle bend is a series of bends that are used to change elevation to clear an obstruction and then return to the original elevation. You can think of a saddle as a "double offset." To bend a saddle, place a length of conduit across the object to be saddled as shown in Figure 10-19.

Point "C" will be the center of the finished saddle. Calculate twice the diameter of the object to be saddled. Lay out "X" that distance from "C." "Y" is the same distance in the opposite direction from "C."

Place the conduit in the bender so that "C" on the conduit is at the indicating notch on the bender and make a 45-degree bend. Remember, a 45-degree bend is reached when the bender handle is straight up. Rotate the conduit 180 degrees in the bender, place the "X" on the conduit at "B" on the bender and make a return bend of $22\frac{1}{2}$ degrees. Duplicate this procedure by placing "Y" on the conduit at "B" on the bender.

CUTTING, REAMING, AND THREADING CONDUIT

Rigid conduit, intermediate metal conduit (IMC), and EMT are manufactured in standard 10-foot lengths. When installing conduit, it is cut to fit the job requirements.

In general, there are two methods used to cut metal conduit:

- Hacksaw method

- Pipe-cutter method

To cut conduit with a hacksaw, use the following steps:

Step 1. Inspect the blade of the hacksaw and replace it, if needed. A blade with 18, 24, or 32 teeth per inch is recommended for conduit. Use 24 or 42 teeth per inch for EMT and 18 teeth per inch for rigid con-

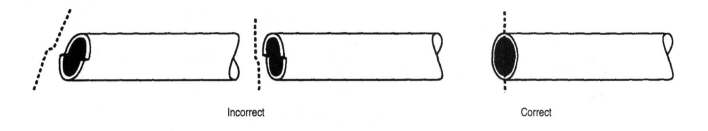

Incorrect   Correct

Figure 10-20: Conduit ends after cutting with a hacksaw

duit and IMC. The teeth of the blade should be pointed toward the front of the saw.

Step 2. Secure the conduit in a pipe vise.

Step 3. Rest the middle of the hacksaw blade on the conduit where the cut is to be made. Position the saw so the end of the blade is pointing slightly down and the handle is slightly up. Push forward gently until the cut is started. Make even strokes until the cut is finished.

Step 4. Check the cut. The end of the conduit should be straight and smooth. Figure 10-20 shows conduit ends with correct and incorrect cuts.

A pipe cutter can also be used to cut conduit. Use the following steps when operating a pipe cutter:

Step 1. Secure the conduit in a pipe vise and mark a place for the cut.

Step 2. Open the cutter and place it over the conduit with the cutter wheel on the mark.

Step 3. Tighten the cutter by rotating the screw handle, but do not overtighten the cutter.

Step 4. Rotate the cutter counterclockwise to start the cut as shown in Figure 10-21.

Step 5. Tighten the cutter handle one-quarter turn for each full turn around the conduit. Again, make sure that you don't overtighten.

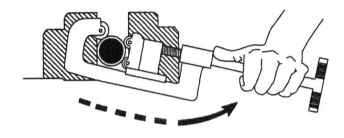

Figure 10-21: Rotating the pipe cutter

Step 6. Add a few drops of cutting oil to the groove and continue cutting until finished.

Step 7. When the cut is almost finished, stop cutting and snap the conduit to finish the cut. This prevents a ridge from forming on the inside of the conduit.

Step 8. Clean the conduit and cutter with a shop towel or rag.

Reaming Conduit: After the conduit is cut, the inside edge is sharp and usually contains burrs. This edge will damage the conductor insulation and must be removed to avoid this damage. A reamer is used for this operation and two types in common use are shown in Figure 10-22 on page 230. One has a ratchet handle and the other has a shank for insertion into a bit brace.

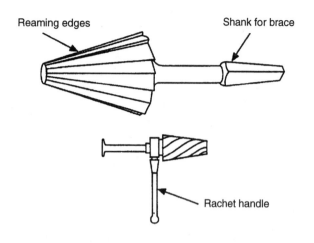

Figure 10-22: Two types of conduit reamers

Use the following steps to ream the inside edge of a length of conduit:

Step 1. Place the conduit in a pipe vise.
Step 2. Insert the reamer tip in the conduit.
Step 3. Apply light forward pressure and start rotating the reamer as shown in Figure 10-23. The reamer can be damaged if it is rotated in the wrong direction. The reamer should "bite" as soon as the proper pressure is applied.
Step 4. Remove the reamer by pulling back on it while continuing to rotate it. Check the progress and then reinsert the reamer if necessary. Rotate the reamer until the inside edge is smooth. You should stop when all burrs have been removed.

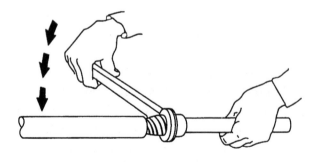

Figure 10-23: Rotation direction for reaming

If a conduit reamer is not available, a half-round metal-cutting file can be used for rigid conduit and IMC. The smaller sizes of EMT may be reamed with the nose of side-cutting pliers, small hand reamers, or even the shank of square-shanked screwdrivers.

JOINING CONDUIT

After conduit is cut and reamed, it must be properly joined and secured in place before conductors are pulled in.

EMT is joined by couplings. Two types are in common use:

- Set-screw couplings
- Compression couplings

Both types are shown in Figure 10-24 and a brief description of each follows.

As its name implies, the set-screw coupling relies on set screws to hold the EMT to the coupling. This type of coupling does not provide a seal and is not permitted to be used in wet locations. However, this type of coupling can be imbedded in concrete.

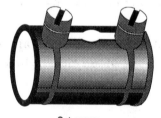

Set-screw
connector

Compression
connector

Figure 10-24: The two types of EMT couplings

Compression couplings provide a tight seal around the conduit, and may be used in some wet locations as stated in the *NEC*.

Galvanized rigid steel conduit (GRC) threadless connectors can also be used with the larger sizes of EMT; that is, 2½ inches and larger, because in these sizes, the outside diameter matches the outside diameter of rigid steel conduit. However, the reverse is not permitted; that is, EMT connectors may not be used to connect rigid steel conduit.

Although threadless couplings and connectors may be used under certain conditions with rigid steel conduit, there are many cases when rigid steel conduit and IMC must be threaded.

The tool used to cut conduit threads is called a "thread-cutting die" or just plain "die." Conduit dies are designed to cut threads with a taper of ¾ inch per foot. The number of threads per inch varies from 8 to 18, depending upon the diameter of the conduit. A thread gauge is used to measure the number of threads per inch.

The threading dies are contained in a die head. The die head can be used with a hand-operated ratchet threader (Figure 10-25) or with a portable power drive.

To thread conduit with a hand-operated threader, perform the following steps:

Step 1. Insert the conduit in a pipe vise. Make sure the vise is fastened to a strong sur-face. Place supports, if necessary, to help secure the conduit.

Step 2. Determine the correct die and head. Inspect the die for damage or broken teeth. Never use a damaged die.

Step 3. Insert the die securely in the head. Make sure the proper die is in the appropriately numbered slot on the head.

Step 4. Determine the correct thread length to cut for the conduit sized used.

Step 5. Lubricate the die with cutting oil at the beginning and throughout the threading operation.

Step 6. Cut threads to the proper length. Make sure that the conduit enters the tapered side of the die. Apply pressure and start turning the head. You should back off the head each quarter-turn to clear away chips.

Step 7. Remove the cutter when the cut is complete. Threads should be cut only to the length of the die. Overcutting will leave the threads exposed to the elements and will corrode.

Step 8. Inspect the threads to make sure they are clean, sharp, and properly made. Use a thread gauge to measure the threads. The finished end should allow for a wrench-tight fit with not more than one thread exposed.

Step 9. Ream the conduit once again to remove any burrs and edges. Cutting oil must be removed from the inside and outside of the conduit.

Die heads can also be used with portable power drives; follow the same steps as listed previously.

Threading machines are often used on larger sizes of conduit and where much threading is to be done. Threading machines hold and rotate the conduit while the die is fed onto the conduit for cutting. When using a conduit threading machine, make sure the threader's legs are secured properly and follow the manufacturer's instructions.

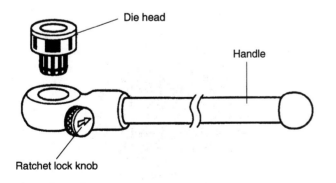

Figure 10-25: Hand-operated ratchet conduit threader

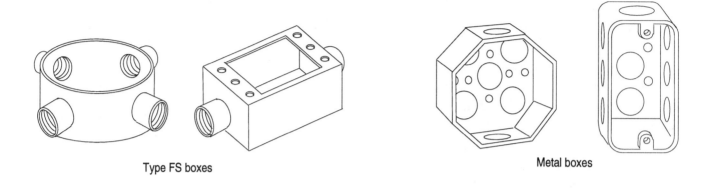

Type FS boxes   Metal boxes

Figure 10-26: Various types of outlet boxes used on commercial electrical installations

CONDUIT-TO-BOX CONNECTIONS

A box is installed at each outlet, switch, or junction point for all wiring installations. Most boxes used in commercial applications are made from sheet steel with the surface of the metal boxes galvanized to resist corrosion and to provide continuous bonding throughout the system. Non-metallic boxes made of PVC or Bakelite are also used to some extent. Various types of outlet boxes are shown in Figure 10-26.

Metal boxes are made with removable circular sections called *knockouts* or *pryouts*. Knockouts are removed to make openings for conduit or cable connections to the boxes. The basic knockout is a

Figure 10-27: Method used to remove a pryout from an outlet box

half-way cut disk that is easily removed when sharply hit by a hammer and punch. Some knockouts, however, are concentric, in which case there are several sections which can be removed to fit the desired conduit or connector.

A pryout is a variation of the knockout. In the former, a slot is cut into the center of a metal tab. To remove a pryout, insert a screwdriver blade into the slot and twist it to break the solid tab as shown in Figure 10-27.

Knockout Punches

Often conduit must enter boxes (or cabinets and panels) that do not have pre-cut knockouts. In these cases, a knockout punch (Figure 10-28) may be used to make a hole for the conduit or connector connection. Knockout punches are available in sizes up to 5 inches and may be used on metal with thicknesses up to 10 gauge.

To use a knockout punch, a hole must first be drilled through the box or cabinet. This hole must be large enough to accept the drive screw in the punch. Then the punch and die are reassembled with the metal between them. The drive nut is tightened, causing the punch to cut through the metal.

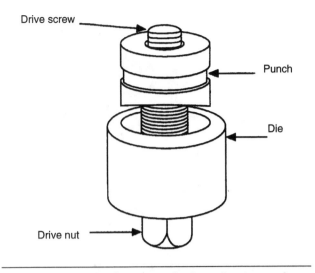

Figure 10-28: Basic parts of a knockout punch

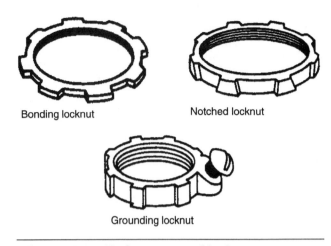

Figure 10-30: Various types of locknuts

Bushings And Locknuts

Conduit is joined to metal boxes by bushings and locknuts. Locknuts secure the conduit or connector to the box while bushings protect the conductors from the sharp edges of the conduit. Bushings are usually made of plastic or metal, and several types are available as shown in Figure 10-29. Note that the grounding bushing has a lug with a set-screw for terminating the bonding wire.

Various types of locknuts appear in Figure 10-30 and are frequently used on both the inside and outside walls of the enclosure to which the conduit terminates. However, if the bushing is metal and fits tightly against the inside wall of the enclosure, only the external locknut is needed on systems 250 volts or less.

A conduit-to-box connection is shown in Figure 10-31 on page 234. To make the connection as shown, use the following steps:

Step 1. Thread the external locknut onto the conduit end. Run the locknut to the bottom of the threads.
Step 2. Insert the conduit into the box opening.
Step 3. If an inside locknut or grounding locknut is required, screw it onto the conduit inside the box opening.
Step 4. Screw the bushing onto the threads projecting into the box opening. Make sure the bushing is tightened as much as possible.
Step 5. Tighten the external locknut to secure the conduit to the box.

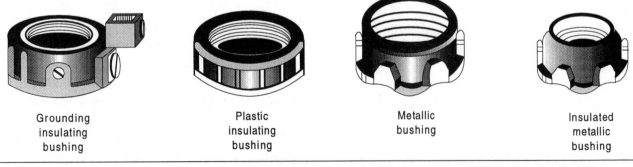

Figure 10-29: Various types of bushings used on commercial electrical installations

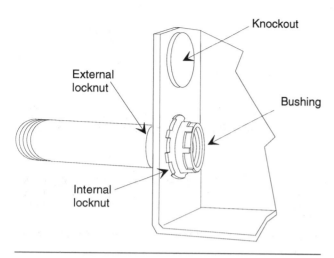

Figure 10-31: Typical conduit-to-box connection

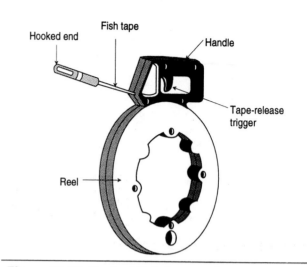

Figure 10-32: Typical fish-tape reel

It is important that the bushings and locknuts fit tightly against the box or cabinet. For this reason, the conduit must enter straight into the box. This may require an offset in the conduit (as discussed previously) to ensure a good fit.

CONDUCTORS IN CONDUIT

In most cases, the installation of conductors in raceway or conduit systems is routine for the experienced electrician. However, there are certain practices that can reduce labor, materials, and help prevent damage to the conductors. The use of modern equipment, such as vacuum fish-tape systems, is one way to reduce labor during this phase of the wiring installation. Furthermore, the proper size and length of the fish tape, as well as the type, should be one of the first considerations. For example, if most of the runs between outlets are only 20 feet or less, a short fish tape of, say, 25 feet will easily handle the job and will not have the weight and bulk of a larger one. When longer runs are encountered, the required length of the fish tape should be enclosed in one of the metal or plastic fish-tape reels as shown in Figure 10-32. This way the fish tape can be rewound on the reel as the pull is being made to avoid having an excessive length of tape lying around on the floor or deck.

When several bends are present in the raceway system, the insertion of the fish tape may be made easier by using flexible fish tape leaders on the end of the fish tape.

The combination blower and vacuum fish-tape systems are ideal for use on long runs and can save much time. Basically, the system consists of a tank and air pump with accessories. An electrician can vacuum or blow a line or tape in any size of conduit from $\frac{1}{2}$ inch through 4 inches, or even up to 6-inch conduit with optional accessories.

After the fish tape is inserted in the raceway system, the conductors must be firmly attached by some approved means. On short runs, where only a few conductors are involved, all that is required is to strip the insulation from the ends of the wires, bend these ends around the hook in the fish tape, and securely tape them in place. Where several wires are to be pulled together, the wires should be staggered and the fish tape securely taped at the point of attachment so that the overall diameter is not increased any more than is absolutely necessary. Staggering is done by attaching one wire to the fish tape and then attaching the second wire a short distance behind this to the bare conductor of the first wire. The third wire, in turn, is attached to the second wire and so forth.

Chapter 11
Wiring Devices

A *device* — by *National Electrical Code (NEC)* definition — is a unit of an electrical system that is intended to carry, but not utilize, electric energy. This covers a wide assortment of system components that include, but are not limited to, the following:

- Switches
- Relays
- Contactors
- Receptacles
- Conductors

However, for our purpose, we will deal with switching devices and also those items used to connect utilization equipment to electrical circuits; namely, *receptacles*. Both are commonly known as *wiring devices*.

Other devices, such as relays, contactors, and the like, are covered in Chapter 14. Therefore, items covered elsewhere will not be repeated here.

RECEPTACLES

A receptacle is a contact device installed at the outlet for the connection of a single attachment plug. Several types and configurations are available for use with many different attachment plug caps — each designed for a specific application. For example, receptacles are available for two-wire, 120-volt, 15- and 20-ampere circuits; others are designed for use on two- and three-wire, 240-volt, 20-, 30-, 40-, and 50-ampere circuits. There are also many other types, most of which are included in this chapter.

Receptacles are rated according to their voltage and amperage capacity. This rating, in turn, determines the number and configuration of the contacts — both on the receptacle and the receptacle's mating plug. Figure 11-1 (beginning on page 236) shows the most common configurations, along with their applications. This chart was developed by the Wiring Device Section of NEMA and illustrates 75 various configurations, which cover 38 voltage and current ratings. The configurations represent existing devices as well as suggested standards (shown with an asterisk in the chart) for future design. Note that all configurations in Figure 11-1 are for general-purpose nonlocking devices. Locking-type receptacles and caps are covered later in this chapter.

As indicated in the chart, unsafe interchangeability has been eliminated by assigning a unique configuration to each voltage and current rating. All dual ratings have been eliminated, and interchangeability exists only where it does not present an unsafe condition.

Each configuration is designated by a number composed of the chart line number, the amperage, and either "R" for receptacle or "P" for plug cap. For example, a 5-15R is found in line 5 and represents a 15-ampere receptacle.

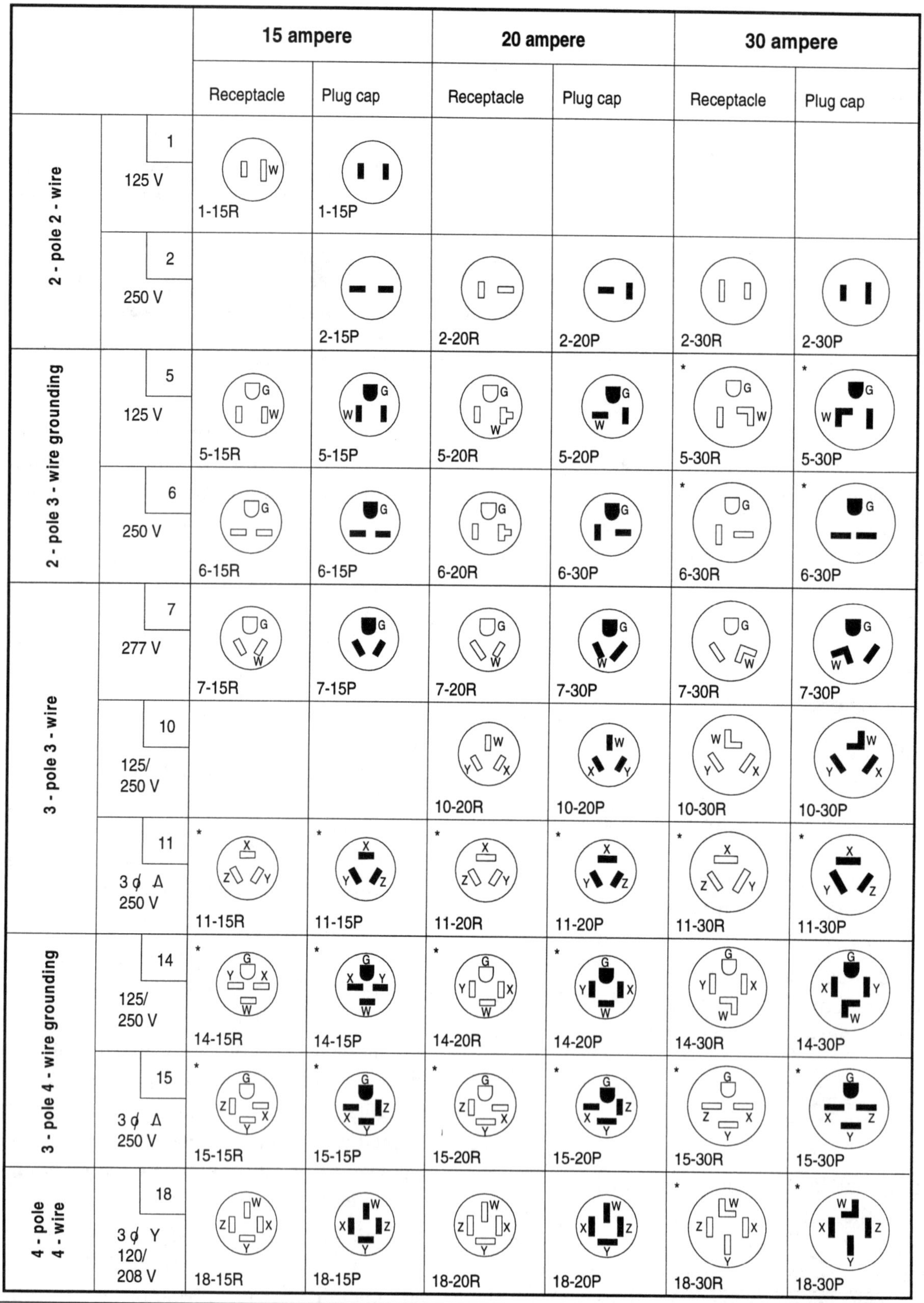

Figure 11-1: NEMA configurations for general-purpose nonlocking receptacles and plug caps

50 ampere		60 ampere	
Receptacle	Plug cap	Receptacle	Plug cap
5-50R	5-50P		
6-50R	6-50P		
7-50R	7-50P		
10-50R	10-50R		
11-50R	11-50P		
14-50R	14-50P	14-60R	14-60P
15-50R	15-50P	15-60R	15-60P
18-50R	18-50P	18-60R	18-60P

Figure 11-1: *Continued*

A clear distinction is made in the configurations between "system grounds" and "equipment grounds." System grounds, referred to as grounded conductors, normally carry current at ground potential, and terminals for such conductors are marked "W" for "White" in the chart. Equipment grounds, referred to as grounding conductors, carry current only during ground-fault conditions, and terminals for such conductors are marked "G" for "grounding" in the chart.

Receptacle Characteristics

Receptacles have various symbols and information inscribed on them that help to determine their proper use and ratings. For example, Figure 11-2 on page 238 shows a standard duplex receptacle and contains the following printed inscriptions:

- The testing laboratory label
- The CSA (Canadian Standards Association) label
- Type of conductor for which the terminals are designed
- Current and voltage ratings, listed by maximum amperage, maximum voltage, and current restrictions

The testing laboratory label is an indication that the device has undergone extensive testing by a nationally recognized testing lab and has met with the minimum safety requirements. The label does not indicate any type of quality rating. The receptacle in Figure 11-2 is marked with the "UL" label which indicates that the device type was tested by Underwriters' Laboratories, Inc. of Northbrook, IL. ETL Testing Laboratories, Inc. of Cortland, NY is another nationally recognized testing laboratory. They provide a labeling, listing and follow-up service for the safety testing of electrical products to nationally recognized safety standards or specifically designated requirements of jurisdictional authorities.

The CSA (Canadian Standards Association) label is an indication that the material or device has

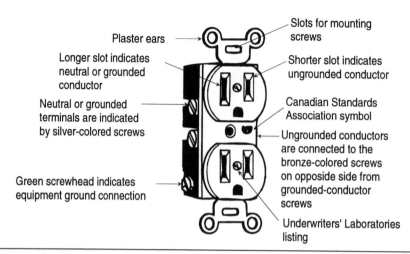

Plaster ears

Longer slot indicates neutral or grounded conductor

Neutral or grounded terminals are indicated by silver-colored screws

Green screwhead indicates equipment ground connection

Slots for mounting screws

Shorter slot indicates ungrounded conductor

Canadian Standards Association symbol

Ungrounded conductors are connected to the bronze-colored screws on opposide side from grounded-conductor screws

Underwriters' Laboratories listing

Figure 11-2: Characteristics of typical duplex receptacles

undergone a similar testing procedure by the Canadian Standards Association and is acceptable for use in Canada.

Current and voltage ratings are listed by maximum amperage, maximum voltage and current restriction. On the device shown in Figure 11-2, the maximum current rating is 15 amperes at 125 volts — the latter of which is the maximum voltage allowed on a device so marked.

Conductor markings are also usually found on duplex receptacles. Receptacles with quick-connect wire clips will be marked "Use #12 or #14 solid wire only." If the inscription "CO/ALR" is marked on the receptacle, either copper, aluminum, or copper-clad aluminum wire may be used. The letters "ALR" stand for "aluminum revised." Receptacles marked with the inscription "CU/AL" should be used for copper only, although they were originally intended for use with aluminum also. However, such devices frequently failed when connected to 15- or 20-ampere circuits. Consequently, devices marked with "CU/AL" are no longer acceptable for use with aluminum conductors.

The remaining markings on duplex receptacles may include the manufacturer's name or logo, "Wire Release" inscribed under the wire-release slots, and the letters "GR" beneath or beside the green grounding screw.

The screw terminals on receptacles are color-coded. For example, the terminal with the green screwhead is the equipment ground connection and is connected to the U-shaped slots on the receptacle. The silver-colored terminal screws are for connecting the grounded or neutral conductors and are associated with the longer of the two vertical slots on the receptacle. The brass-colored terminal screws are for connecting the ungrounded or "hot" conductors and are associated with the shorter vertical slots on the receptacle.

Note: The long vertical slot accepts the grounded or neutral conductor while the shorter vertical slot accepts the ungrounded or "hot" conductor.

Mounting Receptacles

Although no actual *NEC* requirements exist on mounting heights and positioning receptacles, there are certain installation methods that have become "standard" in the electrical industry. Figure 11-3 shows mounting heights of duplex receptacles used on conventinal residential and small commercial installations. However, do not take these dimensions as gospel; they are frequently varied to suit the building structure. For example,

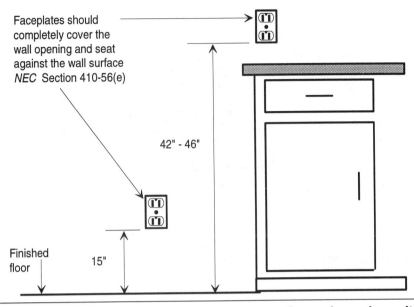

Faceplates should completely cover the wall opening and seat against the wall surface *NEC* Section 410-56(e)

42" - 46"

Finished floor

15"

Figure 11-3: Recommended mounting heights of duplex receptacles under various situations

ceramic tile might be placed above a kitchen or bathroom countertop. If the dimensions in Figure 11-3 puts the receptacle part of the way out of the tile, say, half in and half out, the mounting height should be adjusted to either place the receptacle completely in the tile or completely out of the tile as shown in Figure 11-4.

Refer again to Figure 11-3 and note that the mounting heights are given to the bottom of the outlet box. Many dimensions on electrical draw-ings are given to the center of the outlet box or receptacle. However, during the actual installa-tion, workers installing the outlet boxes can mount them more accurately (and in less time) by lining up the bottom of the box with a chalk mark rather than trying to "eyeball" this mark to the center of the box.

A decade or so ago, most electricians mounted receptacle outlets 12 inches from the finished floor to the center of the outlet box. However, a recent

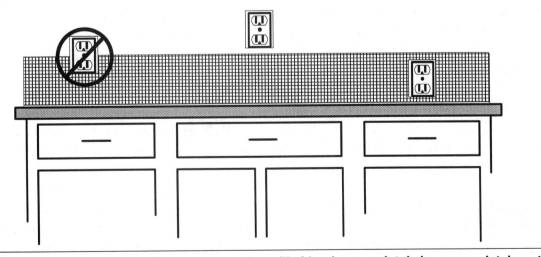

Figure 11-4: Adjust mounting heights so receptacles will either be completely in or completely out of the tile

survey taken of over 500 homeowners shows that they prefer a mounting height of 15 inches from the finished floor to the bottom of the outlet box. It is easier to plug and unplug the cord-and-plug assemblies at this height — especially among senior citizens and those homeowners who are confined to wheelchairs. However, always check the working drawings, written specifications, and details of construction for measurements that may affect the mounting height of a particular receptacle outlet.

There is always the possibility of a metal receptacle cover coming loose and falling downward onto the blades of an attachment plug cap that may be loosely plugged into the receptacle. By the same token, a hairpin, fingernail file, metal fly-swatter handle, or any other metal object may be knocked off a table and fall downward onto the the plug blades. Any of these objects could cause a short-circuit if the falling metal object fell on both the "hot" and grounded neutral blades of the plug at the same time. For these reasons, it is recommended that the equipment grounding slot in receptacles be placed at the top. In this position, any falling metal object would fall onto the grounding blade which would more than likely prevent a short-circuit. The do's and don'ts of mounting methods are shown in Figure 11-5.

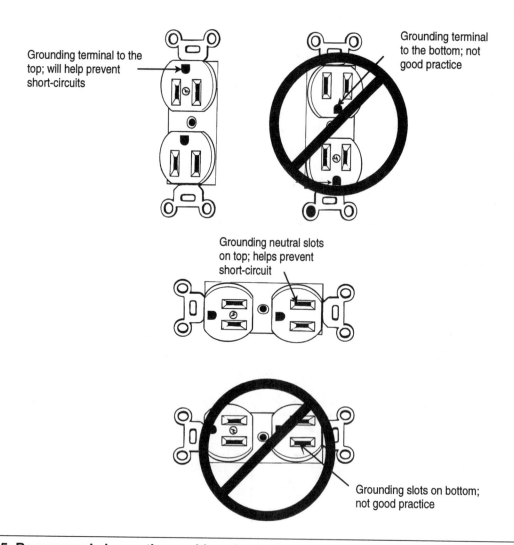

Figure 11-5: Recommended mounting positions for duplex receptacles

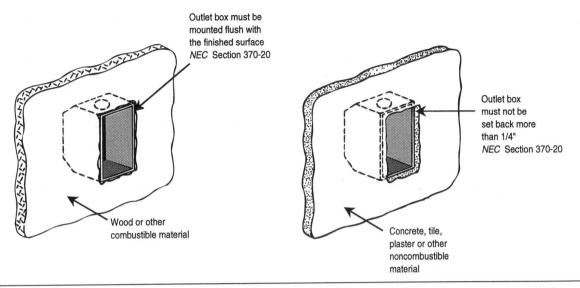

Figure 11-6: *NEC* requirements for mounting outlet boxes in walls or ceilings

When duplex receptacles are mounted in a horizontal position, the grounded neutral slots should be on top for the same reasons as discussed previously. Again, see Figure 11-5.

NEC Section 370-20 requires all outlet boxes installed in walls or ceiling of concrete, tile, or other noncombustible material such as plaster or drywall to be installed in such a matter that the front edge of the box or fitting will not set back of the finished surface more than $\frac{1}{4}$ inch. Where walls and ceilings are constructed of wood or other combustible material, outlet boxes and fittings must be flush with the finished surface of the wall. See Figure 11-6.

Wall surfaces such as drywall, plaster, etc. that contain wide gaps or are broken, jagged, or otherwise damaged must be repaired so there will be no gaps or open spaces greater than $\frac{1}{8}$ inch between the outlet box and wall material. These repairs should be made prior to installing the faceplate. Such repairs are best made with a noncombustible caulking or spackling compound. See Figure 11-7.

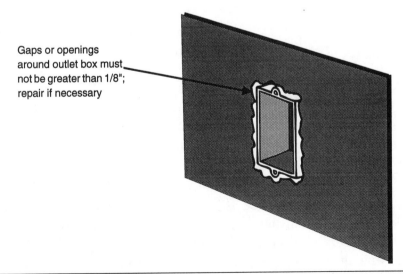

Figure 11-7: Gaps or openings around outlet boxes must be repaired

Types Of Receptacles

There are many types of receptacles. For example, the duplex receptacles that have been discussed are the straight-blade type which accepts a straight blade connector or plug. This is the most common type of receptacle and such receptacles are found on virtually all electrical projects from residential to large industrial installations. Refer again to Figure 11-1 for types of receptacles that fall under this category.

Twist Lock Receptacles: Twist lock receptacles are designed to accept a somewhat "curved blade" connector or plug. The plug/connector and the receptacle will lock together with a slight twist. The locking prevents accidentally unplugging the equipment.

Pin-and-Sleeve Receptacles: Pin-and-sleeve devices have a unique locking feature. These receptacles are made with an extremely heavy-duty plastic housing that makes them highly indestruc-tible. They are manufactured with long brass pins for long life and are color-coded according to voltage for easy identification.

Low-Voltage Receptacles: These receptacles are designed for both ac and dc systems where the maximum potential is 50 volts. Receptacles used for low-voltage systems must have a minimum current-carrying rating of 15 amperes.

440-Volt Receptacles: Portable electrical equipment operating at 440 to 460 volts is common on many industrial installations. Such equipment includes welders, battery chargers, and other types of portable equipment. Special "440-volt" plugs and receptacles are used to connect and disconnect such equipment from a power source. 440-volt receptacles are available in two-wire, single-phase; three-wire, three-phase, and four-wire, three-phase. Equipment grounding is required in all cases, and provisions are provided in each receptacle for such grounding.

WARNING! Make certain that the plug-and-cord assembly is compatible with both the equipment and receptacle before connecting to a 440-volt receptacle. Polarity and equipment-grounding checks on the plug-and-cord assembly should be made on a monthly basis and sooner if subject to hard usage.

LOCATING RECEPTACLES

Several *NEC* sections contain specific requirements for locating receptacles in all types of installations. A summary of these requirements follows.

Residential Occupancies

NEC Section 210-52 should be referred to when laying out outlets for residential and some commerical installations. This section details the general provisions along with small-appliance circuit requirements, laundry requirements, unfinished basements, attached garages, and other areas of the home. Since many of the requirements for residential occupancies parallel those required for commercial installations, a brief review of residential requirements is stated here. For further details, see the book *Residential Electrical Design.* Order forms appear in the back of this book for your convenience.

In general, every dwelling — regardless of its size — must have receptacles located in each habitable area so that no point along the floor line in any wall space (2 feet wide or wider) is more than 6 feet from a receptacle. The purpose of this requirement is to prevent the need for extension cords longer than 6 feet and to minimize the use of cords across doorways, fireplaces, and similar openings.

In addition, a minimum of two 20-ampere small appliance branch circuits are required to serve all receptacle outlets, including refrigeration equipment, in the kitchen, pantry, breakfast room, dining room, or similar area of the dwelling unit. Such circuits, whether two or more are used, must have no other outlets connected to them.

At least one receptacle is required in each laundry area, on the outside of the building at the front and back, in each basement, in each attached and detached garage, in each hallway 10 feet or more in length, and at an accessible location for servicing any HVAC equipment. Figures 11-8 and 11-9 (the latter on page 244) summarize these and other *NEC* requirements regarding the installation of receptacles in dwelling units.

When upgrading existing electrical systems, the *NEC* permits the use of a GFCI receptacle in place of a grounded receptacle. With such an arrangement, additional grounded receptacles may

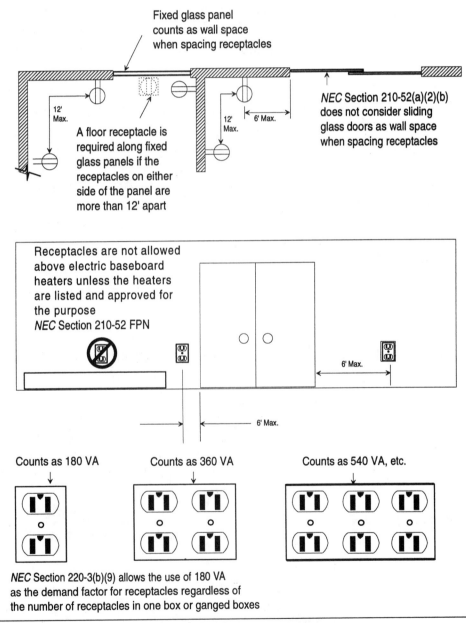

Figure 11-8: Summary of *NEC* requirements for locating receptacles.

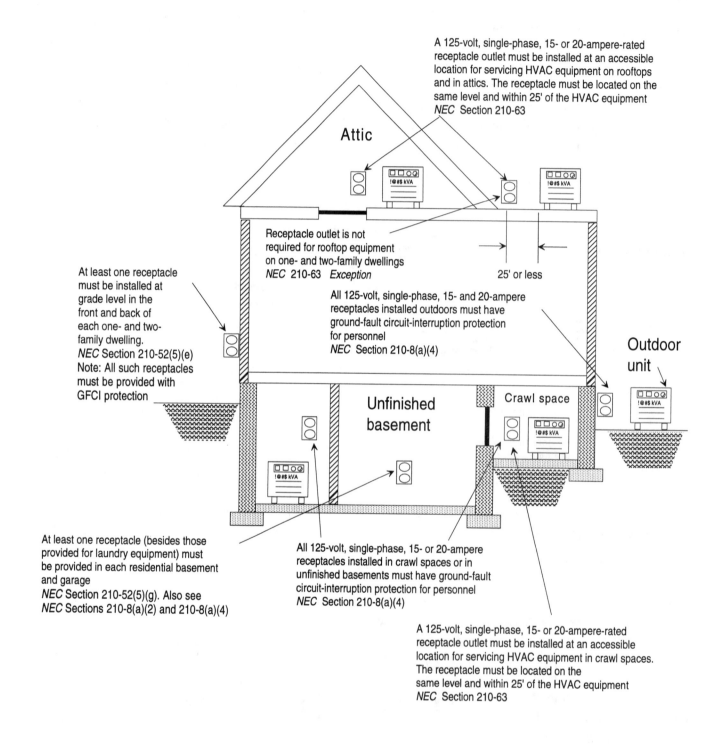

A 125-volt, single-phase, 15- or 20-ampere-rated receptacle outlet must be installed at an accessible location for servicing HVAC equipment on rooftops and in attics. The receptacle must be located on the same level and within 25' of the HVAC equipment *NEC* Section 210-63

Attic

Receptacle outlet is not required for rooftop equipment on one- and two-family dwellings *NEC* 210-63 *Exception*

25' or less

At least one receptacle must be installed at grade level in the front and back of each one- and two-family dwelling. *NEC* Section 210-52(5)(e) Note: All such receptacles must be provided with GFCI protection

All 125-volt, single-phase, 15- and 20-ampere receptacles installed outdoors must have ground-fault circuit-interruption protection for personnel *NEC* Section 210-8(a)(4)

Outdoor unit

Unfinished basement

Crawl space

At least one receptacle (besides those provided for laundry equipment) must be provided in each residential basement and garage *NEC* Section 210-52(5)(g). Also see *NEC* Sections 210-8(a)(2) and 210-8(a)(4)

All 125-volt, single-phase, 15- or 20-ampere receptacles installed in crawl spaces or in unfinished basements must have ground-fault circuit-interruption protection for personnel *NEC* Section 210-8(a)(4)

A 125-volt, single-phase, 15- or 20-ampere-rated receptacle outlet must be installed at an accessible location for servicing HVAC equipment in crawl spaces. The receptacle must be located on the same level and within 25' of the HVAC equipment *NEC* Section 210-63

Figure 11-9: More *NEC* requirements concerning placement of receptacles

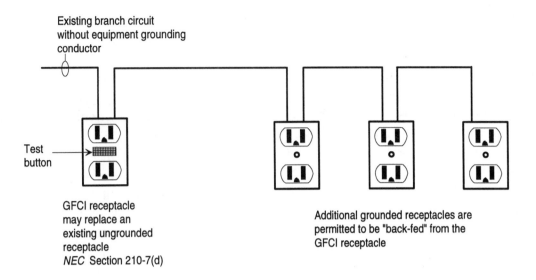

Existing branch circuit
without equipment grounding
conductor

Test
button

GFCI receptacle
may replace an
existing ungrounded
receptacle
NEC Section 210-7(d)

Additional grounded receptacles are
permitted to be "back-fed" from the
GFCI receptacle

Figure 11-10: A GFCI may replace an ungrounded receptacle

be connected on the downstream side of the GFCI as shown in Figure 11-10.

Other receptacles and related circuits are provided as needed according to the load to be served. For example, receptacles are normally provided in residential occupancies for electric ranges, clothes dryers, and similar appliances. Most operate on 120/240-volt branch circuits using 30- to 60-ampere receptacles. See Figure 11-1.

Commercial Applications

Receptacle requirements for commercial installations follow the same general requirements set forth for residential occupancies, with some exceptions. For example, guest rooms in hotels, motels, and similar occupancies must have receptacle outlets installed in accordance with Section 210-52. However, some leaway is given commercial installations. *NEC* Section 210-60 permits receptacle outlets to be located conveniently for permanent furniture layout.

The only other "must" requirement for commercial installations deals with the placement of receptacle outlets in show windows. *NEC* Section 210-62 requires at least one receptacle for each 12

linear feet of show window area measured horizontally at its maximum width. See Figure 11-11. To calculate the number of receptacles required at the top of any show window, measure the total linear feet, and then divide this figure by 12 and any remainder or "fraction thereof" requires an additional receptacle. For example, the show window in Figure 11-11 is 18 feet in length. Consequently, the number of receptacles required may be calculated as follows:

$$\frac{18 \; feet}{12 \; feet} = 1.5 \; receptacles$$

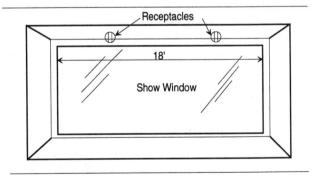

Receptacles

18'

Show Window

Figure 11-11: *NEC* Section 210-62 deals with commercial show windows

Since "1.5" is more than one, to comply with *NEC* Section 210-62, "or major fraction thereof," two receptacles are required in this area. Had the calculation resulted in a figure of, say, 1.01, the local inspection authorities would probably require only one receptacle in the show widow.

Of course, GFCIs are required on all 15- and 20-ampere receptacles installed in commercial bathrooms or toilets, in commercial garages, receptacles installed outdoors where there is direct grade-level access (below 6' 6" inches), crawl spaces, boathouses, and all receptacles installed on roofs.

Other receptacles and related circuits are provided as needed according to the load to be served.

SWITCHES

The purpose of a switch is to make and break an electrical circuit, safely and conveniently. In doing so, a switch may be used to manually control lighting, motors, fans, and other various items connected to an electrical circuit. Switches may also be activated by light, heat, chemicals, motion, and electrical energy for automatic operation. *NEC* Article 380 covers the installation and use of switches.

Although there is some disagreement concerning the actual definitions of the various switches that might fall under the category of *wiring devices*, the most generally accepted ones are as follows:

Bypass Isolation Switch: This is a manually operated device used in conjunction with a transfer switch to provide a means of directly connecting load conductors to a power source, and of disconnecting the transfer switch.

General-Use Switch: A switch intended for use in general distribution and branch circuits. It is rated in amperes, and it is capable of interrupting its rated current at its rated voltage.

General-Use Snap Switch: A form of general-use switch so constructed that it can be installed in flush device boxes or on outlet box covers, or

otherwise used in conjunction with wiring systems recognized by the *NEC*.

Isolating Switch: A switch intended for isolating an electric circuit from the source of power. It has no interrupting rating, and it is intended to be operated only after the circuit has been opened by some other means.

Motor-Circuit Switch: A switch, rated in horsepower, capable of interrupting the maximum operating overload current of a motor of the same horsepower rating as the switch at its rated voltage.

Transfer Switch: A transfer switch is a device for transferring one or more load conductor connections from one power source to another. This type of switch may be either automatic or nonautomatic.

Common Terms

In general, the major terms used to identify the characteristics of switches are:

- Pole or poles
- Throw

The term *pole* refers to the number of conductors that the switch will control in the circuit. For example, a single-pole switch breaks the connection on only one conductor in the circuit. A double-pole switch breaks the connection to two conductors, and so forth.

The term *throw* refers to the number of internal operations that a switch can perform. For example, a single-pole, single-throw switch will "make" one conductor when thrown in one direction — the "ON" direction — and "break" the circuit when thrown in the opposite direction; that is, the "OFF" position. The commonly used ON/OFF toggle switch is an SPST switch (single-pole, single-throw). A two-pole, single-throw switch opens or closes two conductors at the same time. Both conductors are either open or closed; that is, in the ON or OFF position. A two-pole, double-throw switch is used to direct a two-wire circuit through one of two different paths. One application of a

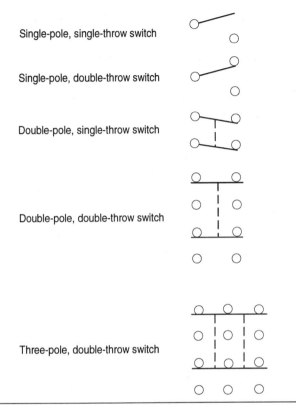

Single-pole, single-throw switch

Single-pole, double-throw switch

Double-pole, single-throw switch

Double-pole, double-throw switch

Three-pole, double-throw switch

Figure 11-12: Common switch configurations

two-pole, double-throw switch is in an electrical transfer switch where certain circuits may be energized from either the main electric service, or from an emergency standby generator. The double-throw switch "makes" the circuit from one or the other and prevents the circuits from being energized from both sources at once. Figure 11-12 shows common switch configurations.

Switch Identification

Switches vary in grade, capacity, and purpose. It is very important that proper types of switches are selected for the given application. For example, most single-pole toggle switches used for the control of lighting are restricted to ac use only. This same switch is not suitable for use on, say, a 32-volt dc emergency lighting circuit. A switch rated for ac only will not extinguish a dc arc quickly enough. Not only is this a dangerous prac-

tice (causing arcing and heating of the device), the switch contacts would probably burn up after only a few operations of the handle, if not the first time.

Figure 11-13 on page 248 shows a typical single-pole toggle switch — the type most often used to control ac lighting in all installations. Note the identifying marks. They are similar to those on the duplex receptacle discussed previously. The main difference is the "T" rating which means that the switch is rated for switching lamps with tungsten filaments (incandescent lamps).

Screw terminals are also color-coded on conventional toggle switches. Switches are typically constructed with a ground screw attached to the metallic strap of the switch. The ground screw is usually a green-colored hex-head screw. This screw is for connecting the equipment-grounding conductor to the switch. On three-way switches, the common or pivot terminal usually has a black or bronze screw head.

The switch shown is the type normally used for residential construction. Heavier-duty switches are usually the type used on commercial wiring — some of which are rated for use on 277-volt circuits with current-carrying ratings up to 30 amperes. Therefore, it is important to check the rating of each switch before it is installed.

The exact type and grade of switch to be used on a specific installation is often dictated by the project drawings or written specifications. Sometimes wall switches are specified by manufacturer and catalog number; other times they are specified by type, grade, voltage, current rating, and the like, leaving the contractor or electrician to select the manufacturer. The naming of a certain brand of switch for a particular project does not necessarily mean that this brand must be used. The naming of a certain brand or make or manufacturer in the specifications is to establish a quality standard for the article desired. The contractor is not restricted to the use of the specific brand of the manufacturer named unless so indicated in the specifications. However, where a substitution is requested, a substitution will be permitted only with the written approval of the engineer. No substitute material or

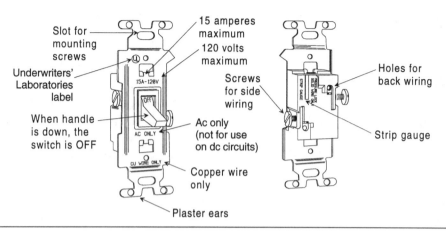

Figure 11-13: Typical identifying marks on a single-pole switch

equipment shall be ordered, fabricated, shipped, or processed in any manner prior to the approval of the architect-engineer. The contractor shall assume all responsibility for additional expenses as required in any way to meet changes from the original material or equipment specified. If notice of substitution is not furnished to the architect-engineer within ten days after the contract is awarded, the equipment and materials named in the specifications are to be used.

Electrical specifications dealing with wall switches are covered in at least two sections of the specifications:

- 16100 Basic Materials and Methods
- 16500 Lighting

NEC REQUIREMENTS FOR SWITCHES

The *NEC* requirements for installing light switches are many and are scattered in various locations in the *NEC* book. For example, wall-switch controlled lighting outlets are required in each habitable room of all residential occupancies. Wall-switch controlled lighting is also required in each bathroom, hallways, stairways, attached garages, and at outdoor entrances. A wall-switch controlled receptacle may be used in place of the lighting outlet in habitable rooms other than the kitchen and bathrooms. Providing a wall switch for

room lighting is intended to prevent an occupant's groping in the dark for table lamps or pull chains. In stairways with six or more steps, the stairway lighting must be controlled at two locations — at the top and also the bottom of the stairway. This is accomplished by using two three-way switches.

Lighting outlets are also required in attics, crawl spaces, utility rooms, and basements when these spaces are used for storage or contain equipment such as HVAC equipment. Again, if the basement or attic stairs have more than six steps, a three-way switch is required at each landing.

At least one wall-switch controlled lighting outlet is required in each guest room in hotels, motels, or similar locations as shown in Figure 11-14. Note that a wall switch-controlled receptacle is permitted in lieu of the lighting outlet.

At least one wall-switch controlled lighting outlet must be installed at or near equipment requiring servicing such as HVAC equipment. The wall switch must be located at the point of entry to the attic or underfloor space.

In many commercial installations, circuit breakers in panelboards are permitted to control main-area lighting where the areas are constantly illuminated during operating hours. Consequently, wall switches are not required in these areas. However, wall switches are normally installed at outdoor entrances, entrances to storerooms, small offices, toilets, and similar locations.

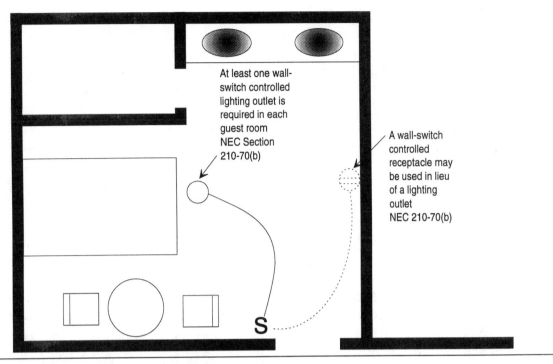

Figure 11-14: *NEC* **switch requirements for guest rooms**

Wiring diagrams of switch circuits — single-pole, three-way, and four-way switches — are thoroughly discussed in Chapter 12 and these diagrams will not be repeated here. However, it is recommended that the reader review these diagrams at this time if deemed necessary.

SAFETY SWITCHES

Enclosed single-throw safety switches are manufactured to meet industrial, commercial, and residential requirements. The two basic types of safety switches are:

- General duty
- Heavy duty

Double-throw switches are also manufactured with enclosures and features similar to the general and heavy-duty, single-throw designs.

The majority of safety switches have visible blades and safety handles. The switch blades are in full view when the enclosure door is open and there is visually no doubt when the switch is OFF.

The only exception is Type 7 and 9 enclosures; these do not have visible blades. Switch handles on all types of enclosures are an integral part of the box, not the cover, so that the handle is in control of the switch blades under normal conditions.

Heavy-Duty Switches

Heavy-duty switches are intended for applications where ease of maintenance, rugged construction, and continued performance are primary concerns. They can be used in atmospheres where general-duty switches would be unsuitable, and are therefore widely used in industrial applications. Heavy-duty switches are rated 30 through 1200 amperes and 240 to 600 volts ac or dc. Switches with horsepower ratings are capable of opening a circuit up to six times the rated current of the switch. When equipped with Class J or Class R fuses for 30 through 600 ampere switches, or Class L fuses in 800 and 1200 ampere switches, many heavy-duty safety switches are UL listed for use on systems with up to 200,000 RMS symmetrical amperes available fault current. This, however, is about the highest short-circuit rating available for any heavy-duty safety switch. Applications include use where the required enclosure is NEMA TYPE 1, 3R, 4, 4X, 5, 7, 9, 12 or 12K.

Operating Mechanism And Cover Latching

Most heavy-duty safety switches have a spring driven quick-make, quick-break mechanism. A quick-breaking action is necessary if the switch is to be safely switched OFF under a heavy load.

The spring action, in addition to making the operation quick-make, quick-break, firmly holds the switch blades in the ON or OFF position. The operating handle is an integral part of the switching mechanism and is in direct control of the switch blades under normal conditions.

A one-piece cross bar, connected to all switch blades, should be provided which adds to the overall stability and integrity of the switching assembly by promoting proper alignment and uniform switch blade operation.

Dual cover interlocks are standard on most heavy-duty switches where the NEMA enclosure permits. However, NEMA Types 7 and 9 have bolted covers and obviously cannot contain dual cover interlocks. The purpose of a dual interlock is to prevent the enclosure door from being opened when the switch handle is in the ON position and prevent the switch from being turned ON while the door is open. A means of bypassing the interlock is provided to allow the switch to be inspected in the ON position by qualified personnel. However, this practice should be avoided if at all possible. Heavy-duty switches can be padlocked in the OFF position with up to three padlocks.

Enclosures

Heavy-duty switches are available in a variety of enclosures which have been designed to conform to specific industry requirements based upon the intended use. Sheet metal enclosures (that is, NEMA Type 1) are constructed from cold-rolled steel which is usually phosphatized and finished with an electrode deposited enamel paint. The Type 3R rainproof and Type 12 and 12K dusttight enclosures are manufactured from galvannealed sheet steel and painted to provide better weather protection. The Type 4, 4X and 5 enclosures are made of corrosion-resistant Type 304 stainless steel and require no painting. Type 7 and 9 enclosures are cast from copper-free aluminum and finished with an enamel paint. Type 1 switches are general purpose and designed for use indoors to protect the enclosed equipment from falling dirt and personnel from live parts. Switches rated through 200 amperes are provided with ample knockouts. Switches rated from 400 through 1200 amperes are provided without knockouts.

The following are the NEMA enclosure types that will be encountered most often. Always make certain that the proper enclosure is chosen for the application.

Type 3R switches are designated "rainproof" and are designed for use outdoors.

Type 3R enclosures for switches rated through 200 amperes have provisions for interchangeable bolt-on hubs at the top endwall. Type 3R switches rated higher than 200 amperes have blank top endwalls. Knockouts are provided (below live parts only) on enclosures for 200 ampere and smaller Type 3R switches. Type 3R switches are available in ratings through 1200 amperes.

Type 4, 4X, and 5 stainless steel switches are designated dusttight, watertight and corrosion resistant and designed for indoor and outdoor use. Common applications include commercial type kitchens, dairies, canneries, and other types of food processing facilities, as well as areas where mildly corrosive liquids are present. All Type 4, 4X, and 5 stainless steel enclosures are provided without knockouts. Use of watertight hubs is required. Available switch ratings are 30 through 600 amperes.

Type 12 and Type 12K switches are designated dusttight (except at knockout locations on Type

12K) and are designed for indoor use. In addition, NEMA Type 12 safety switches are designated as raintight for outdoor use when the supplied drain plug is removed. Common applications include heavy industries where the switch must be protected from such materials as dust, lint, flyings, oil seepage, etc. Type 12K switches have knockouts in the bottom and top endwalls only. Available switch ratings are 30 through 600 amperes in Type 12 and 30 through 200 amperes in Type 12K.

NEMA Type 4, 4X, 5, Type 12 and 12K switch enclosures have positive sealing to provide a dusttight and raintight (watertight with stainless steel) seal. Enclosure doors are supplied with oil resistant gaskets. Switches rated 30 through 200 amperes incorporate spring loaded, quick-release latches. Switches rated at 400 and 600 amperes feature single-stroke sealing by operation of a cover-mounted handle. Those rated at 30, 60, and 100 amperes in these enclosures are provided with factory-installed fuse pullers.

General-Duty Switches

General-duty switches for residential and light commercial applications are used where operation and handling are moderate and where the available fault current is 10,000 RMS symmetrical amperes or less. Some general-duty safety switches, however, exceed this specification in that they are UL listed for application on systems having up to 100,000 RMS symmetrical amperes of available fault current when Class R fuses and Class R fuse kits are used. Class T fusible switches are also available in 400, 600 and 800 ampere ratings. These switches accept 300 VAC Class T fuses only. Some examples of general-duty switch application include residential, farm, and small business services entrances, and light-duty, branch-circuit disconnects.

General-duty switches are rated up to 600 amperes at 240 volts ac in general purpose (Type 1) and rainproof (Type 3R) enclosures. Some general-duty switches are horsepower rated and capable of opening a circuit up to six times the rated current of the switch; others are not. Always check the switch's specifications before using under a horsepower-rated condition.

Principles Of Operation And Latching

Although not required by either the UL or NEMA standards, some general-duty switches have spring-driven quick-make, quick-break operating mechanisms. Operating handles are an integral part of the operating mechanism and are not mounted on the enclosure cover. The handle provides indication of the status of the switch. When the handle is up, the switch is ON. When the handle is down, the switch is OFF. A padlocking bracket is provided which allows the switch handle to be locked in the OFF position. Another bracket is provided which allows the enclosure to be padlocked closed.

Enclosures

General-duty safety switches are available in either Type 1 for general purpose, indoor applications, or Type 3R for rainproof, outdoor applications.

Double-Throw Safety Switches

Double-throw switches are used as manual transfer switches and are not intended for use as motor circuit switches; thus, horsepower ratings are generally unavailable.

Double-throw switches are available as either fused or nonfusible devices and two general types of switch operation are available:

- Quick-make, quick-break
- Slow-make, slow-break

Figure 11-15 on page 252 shows a practical application of a double-throw safety switch used as a transfer switch in conjunction with a stand-by emergency generator system.

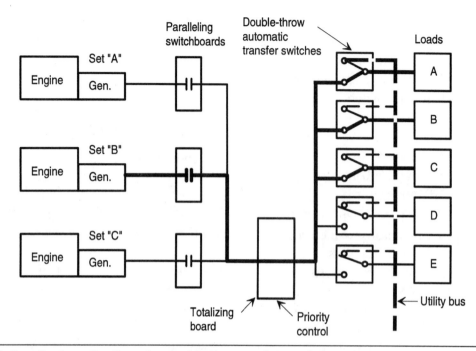

Figure 11-15: Practical application of a double-throw safety switch

NEC Safety-Switch Requirements

Safety switches, in both fusible and nonfusible types, are used as a disconnecting means for services, feeders, and branch circuits. Installation requirements involving safety switches are found in several places throughout the *NEC*, but mainly in the following articles and sections:

- *NEC* Article 373
- *NEC* Article 380
- *NEC* Article 430-H
- *NEC* Article 440-B
- *NEC* Section 450-8(c)

Summary

A *device* — by *NEC* definition — is a unit of an electrical system that is intended to carry, but not utilize, electric energy. This covers a wide assortment of system components that include, but are not limited to, the following: switches, relays, contactors, receptacles, and conductors.

The purpose of a switch is to make and break an electrical circuit, safely and conveniently. *NEC* Article 380 covers most of the installation requirements for switches.

A receptacle is a contact device installed at the outlet for the connection of a single attachment plug. A single receptacle is a single contact device with no other contact device on the same yoke. A multiple receptacle is a single device containing two or more receptacles — the most common being the *duplex receptacle*. *NEC* sections dealing mainly with receptacles are as follows:

- Spacing: 210-52(a)
- Countertops: 210-52(c)
- Bathrooms: 210-52(d)
- Outdoors: 210-52(e)
- Basements: 210-52(g)
- Garages: 210-52(g)
- Hallways: 210-52(h)
- Guest rooms: 210-60
- Show windows: 210-62

252

Chapter 12
Electric Lighting

Electric lighting is one of the greatest conveniences available to society. For example, lighting is used in stores, hotels, and office buildings to improve the efficiency of employees, to aid in the selling of merchandise, and to reduce eye strain.

The exteriors of some commercial buildings are beautifully floodlighted and streets are lighted brightly with electric lamps. The lighting of outdoor sport areas enables us to view football, baseball, and other sports at night. Television would not be possible without electricity and artificial light.

The cases are endless, and almost everyone today realizes the value of better lighting. This field also provides some of the most fascinating and enjoyable work for the electrician.

The electrician will find that fluorescent lighting is used for lighting most commercial installations. However, there is also a need for incandescent lamps where certain accent lighting is required, or for use in restaurants or lounges where the "commercial appearance" of fluorescent lighting is undesirable.

High-intensity discharge lighting systems are gradually finding their way into mass merchandising areas and are replacing both fluorescent and incandescent lamps due to their high efficiency. However, most lighting designers will use combinations of incandescent, fluorescent, and high-intensity discharge equipment to achieve a particular objective. Consequently, the electrician must be familiar with all types of lamps, and know how each is installed and connected. For example, installing lighting in a store window may involve several different types of lighting fixtures as shown in Figure 12-1 on page 254. In such installations, the electrician will normally be provided with a detailed set of electrical drawings to follow. However, sometimes these prints are very sketchy and will require further research. The shop drawings, if available, should be studied prior to the installation, and if there are any questions, either the fixture manufacturer or the lighting designer should be consulted. Many lighting fixtures have installation instructions packed with them; others do not.

Troffers

The recessed or lay-in two- and four-lamp fluorescent fixtures are by far the most popular type of commercial lighting. This type of fixture is used almost exclusively in offices and as general illumination in store buildings and other commercial establishments. Any electrician working on commercial installations will see lots of these fixtures on every job.

In general, rectangular-shaped 2- × 4-foot and square 2- × 2-foot fluorescent lighting fixtures are available in two basic types:

- Flush
- Surface-mounted

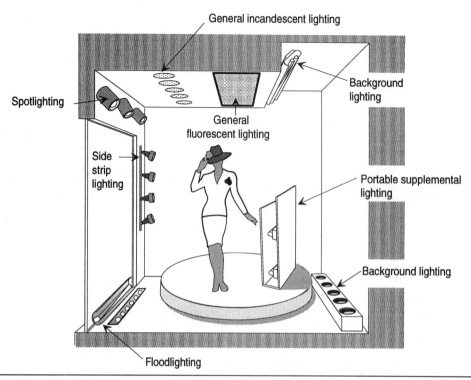

General incandescent lighting

Spotlighting

Background lighting

General fluorescent lighting

Side strip lighting

Portable supplemental lighting

Background lighting

Floodlighting

Figure 12-1: Lighting layout for a store window

The flush-mounted fixtures are further divided into two basic types:

- Those designed for recess mounting in plaster or drywall ceilings.

- Those designed as lay-in fixtures in inverted "T" acoustical ceilings.

This latter type will be encountered the most on commercial electrical installations.

An acoustical, inverted "T" ceiling is shown in Figure 12-2. In applications, electricians will rough-in the electrical system above the ceiling prior to the installation of the ceiling grid. Once

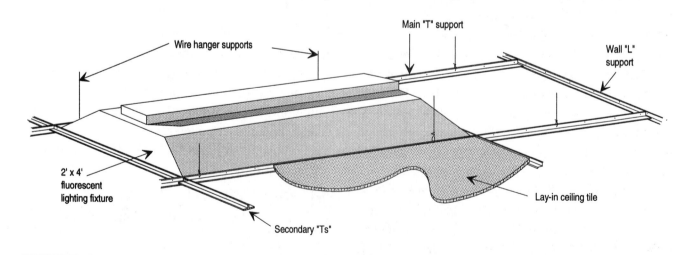

Main "T" support

Wire hanger supports

Wall "L" support

2' x 4' fluorescent lighting fixture

Lay-in ceiling tile

Secondary "Ts"

Figure 12-2: Construction of inverted "T" ceiling with lay-in fluorescent lighting fixture

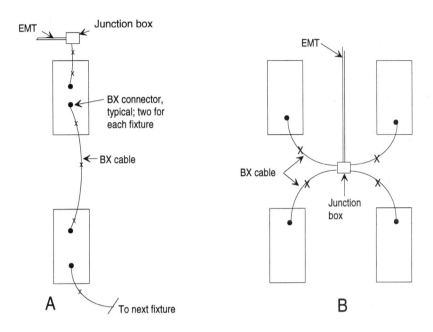

Figure 12-3: Two popular methods of connecting lay-in fluorescent lighting fixtures

the ceiling grid is installed (without the acoustical tiles), electricians install the lay-in fixtures at their appropriate location. In the case of continuous rows of fixtures, the first fixture of each row is fed with one or more branch circuits, and the remaining fixtures are connected from the first by means of the integral wireways in each fixture; that is, building wire, such as THHN, is pulled through the fixture wireways and a connection made to each fixture. No additional conduit or cable is required.

When these fixtures are spaced some distance apart, there are several ways to make the required connections. Two of the most popular appear in Figure 12-3. Figure 12-3A shows EMT terminating into a junction box in the area where the fixtures are to be installed. The connection between fixtures is then made with Type AC ("BX") cable. Figure 12-3B shows EMT terminating into junction boxes in the center of each group of four fixtures. Flexible cable is then used to connect each fixture to the junction box. When this arrangement is used, and all fixtures are spaced the same dis-

tance, the BX cable connections are usually prefabricated "on the ground" with the connections finalized in each fixture. The opposite end of the cable is stripped and an insulating bushing and connector installed. A group of four fixtures is then laid in place in the ceiling grid and all that is required is to connect the fixture leads to the junction box, make the splices, and install the box cover.

In some installations, due to the job specifications or *NEC* requirements, a junction box is installed above each lighting fixture where EMT or rigid conduit connects the junction boxes. Then a short piece of flexible metal conduit or BX cable is used to connect each fixture to the junction box.

Surface-mounted 2- × 4-foot fluorescent fixtures require some type of hanger since they do not depend on the inverted T-bars of the ceiling grid for support. One of the main considerations for mounting lighting fixtures of any type is to select the proper hangers and accessories. Some careful

Figure 12-4: Concrete insert

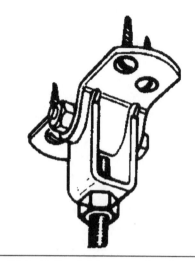

Figure 12-6: Adjustable swinging hanger flange

thought in this area can save sufficient time to make the practice worthwhile.

For example, concrete inserts (Figure 12-4) can be placed in concrete pours for use in hanging rows of fluorescent fixtures at a later stage of the building construction. The type shown is manufactured with a knockout that saves covering or stuffing the opening during the concrete pour. Once the concrete has cured sufficiently, the knockout slots are removed and all-thread rod is screwed into the insert. The fixtures are then hung from the all-thread rod.

The T-bar clip in Figure 12-5 is another time-saving device when installing surface-mounted fixtures to acoustical T-bar ceilings. This clip has an integral boss with a recessed $\frac{1}{4}$-20 stud bolt. The clips are snapped on the T-bar channel quickly and then the fixtures are held in place by a cupped washer and wing-nut. Each clip will safely carry a load of 100 pounds. However, electricians must make certain that the T-bar ceiling is properly secured to hold the weight of lighting fixtures. This also applies to the lay-in fixtures mentioned previously.

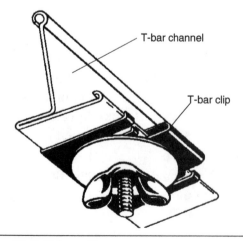

T-bar channel

T-bar clip

Figure 12-5: T-bar clip

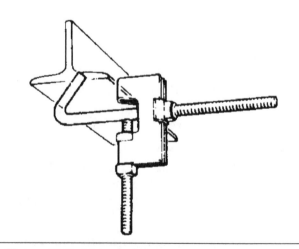

Figure 12-7: I-beam clamp

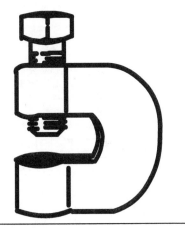

Figure 12-8: Beam clamp

Other hangers include an adjustable swinging hanger flange which accepts ½-inch rod (Figure 12-6); I-beam clamps (Figure 12-7); and regular beam clamps (Figure 12-8). Channel bar is also frequently used to mount lighting fixtures.

The proper mounting of lighting fixtures is very important in any lighting installation. A 50-pound lighting fixture falling from a height of 8 to 16 feet can cause severe bodily injury — if not death — if it should fall on someone. Therefore, it is the electrician's responsibility to make absolutely certain that all fixtures are mounted properly and safely.

NEC Section 410-15 gives specific requirements for lighting fixture support. For example, any lighting fixture that weighs more than 50 pounds must be supported independently of the outlet box. Fixtures weighing less than 50 pounds are permitted to be installed to outlet boxes by means of the box screws. However, always use good judgment when mounting lighting fixtures. Remember, the *NEC* gives the minimum requirements for electrical installations; not necessarily those that make a quality installation. For example, if you are mounting, say, 45-pound lighting fixtures in an auditorium with 16-foot high ceilings, the *NEC* allows the fixtures to be mounted directly to the outlet boxes. However, considering that persons will be seated next to each other and if one of the lighting fixtures should fall while the room

was occupied, one or more persons would surely be injured. Most electricians would provide additional support for the fixtures — just in case!

Practical Application Of Fluorescent Fixtures

Figure 12-9 on page 258 shows the lighting floor plan of a small commercial installation. Let's see how these fixtures might be installed and connected.

In examining the floor plan in Figure 12-9, note that the lighting fixtures in each area are identified by a triangle symbol with a letter inside. These symbols correspond to a lighting-fixture schedule that gives the manufacturer and catalog number of each type of fixture, along with the number and type of lamps in each type. Furthermore, mounting details are also provided; that is, surface, wall, flush, recessed, and the like. The lighting-fixture schedule that corresponds to the fixtures in the floor plan appears in Figure 12-10. From this, we see that Type "A" fixtures are surface-mounted fluorescent fixtures with two 40-watt cool-white lamps.

Fixture Type "B" is a surface-mounted 2- × 4-foot fluorescent fixture with four 40-watt cool-white lamps, while Type "C" fixtures are the same, but are designed to be recessed in the ceiling grid. The remaining fixtures, Type "D" are single-tube, surface-mounted fluorescent fixtures with wrap-around lens.

Beginning with the Type "A" fixtures, these are surface-mounted using T-bar clips as discussed previously. Adjoining end plates from the fixtures are removed so that the fixtures may be mounted in continuous rows — the fixture wireways acting as a raceway for the branch-circuit conductors feeding the fixtures. A junction box is installed above the ceiling at the end of each row of fixtures, and this end fixture is fed from the junction box with BX cable. The homerun from the junction box (also the connections between the junction boxes) consists of EMT and terminates in panel "A." EMT

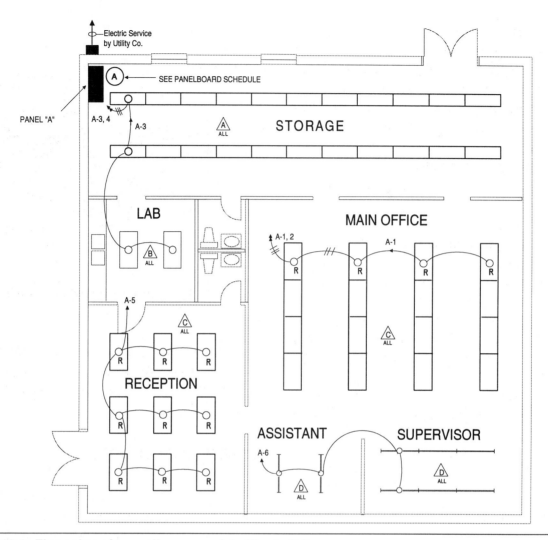

Figure 12-9: Floor plan of a small commercial office building

LIGHTING FIXTURE SCHEDULE

SYMBOL	TYPE	MANUFACTURER AND CATALOG NUMBER	MOUNTING	LAMPS
▭▭▭	A	GOODYLITE 10234	SURFACE	2-40W T-12WWX
▭	B	JETOLIER 10420	SURFACE	4-40W T-12 WWX
▭R	C	JETOLIER RPC-210-6E	RECESSED	4-40W T-12
⊢○⊣	D	ALPINE P 7 S AL 2936	SURFACE	1-40W T-12CW

Figure 12-10: Lighting-fixture schedule for the floor plan in Figure 12-9

is also used from one of these junction boxes to feed the two Type "B" lighting fixtures in the lab area. The drawing shows the connection to one fixture, and then the other fixture is fed from the first. However, in actual practice, a junction box will probably be installed in the center between the two fixtures; a piece of BX cable will then be used to feed each of the two fixtures.

The Type "C" fixtures in the main office are connected similar to the method shown in Figure 12-3A; the same is true for the Type "C" fixtures installed in the reception area. In both the supervisor's and assistant's areas, EMT feeds junction boxes placed in the ceiling above and between the rows of fixtures. Type AC or BX cable is used to make the connection from the junction box to the fixtures. This type of fixture is surface-mounted and since the six fixtures in the supervisor's area are installed in two rows of three, each row can be wired as described for the Type "A" fixtures in the storage area; that is, the adjoining end plates removed from the fixtures and then using the fixtures as a raceways for the fixture wires.

Incandescent Lighting

There are many types of incandescent lighting fixtures used on commercial projects, from tiny step lights to huge chandeliers. However, the electrician will probably see more recessed incandescent fixtures used in modern commercial buildings than any other type.

Recessed "high-hat" or "bullet" type fixtures are used extensively for accent lighting in stores and show windows, as wall-wash, decorative lighting in office buildings, and as general illumination in restaurants and cocktail lounges — the intensity of the light in the latter establishments is usually controlled by dimmers.

The method of installing recessed fixtures will vary, depending upon the type of ceiling. In most cases, however, the housing or "can" is mounted during the rough-in stage and all connections made to the fixture's junction box. After the finished

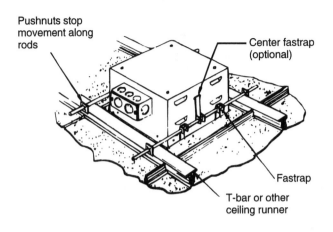

Figure 12-11: Typical recessed incandescent lighting fixture

ceiling is installed, the lampholder, reflector, and trim are installed inside of the housing. One method of installing recessed fixtures is shown in Figure 12-11.

OFFICES AND SCHOOLS

The requirements for office and school lighting are quite similar in that both require a relatively high level of illumination with good visual comfort in order to satisfy the needs of a wide range of seeing tasks over long periods of time.

The benefits of good lighting are:

- It stimulates morale and efficiency
- It increases production
- It improves accuracy
- It helps conserve energy
- Quality of illumination

If an average school classroom was to be lighted with a single 1500-watt incandescent lamp hung in the center of the room, it would certainly produce a sufficient amount of illumination, but the resulting lighting effect would be intolerable. First, the intensely bright filament would repel even a glance. Second, the image of this bright light would be reflected from polished desk tops and other shiny objects. Third, persons sitting at desks and facing away from the center of the room would cast dark shadows on the desks. Fourth, the illumination level in the room would be uneven. The level directly under the lamp would be extremely high, while the light around the perimeter would be relatively weak. This example illustrates what a lighting layout should strive to overcome. Office and school lighting requires that careful consideration be given to the quality of illumination which includes brightness ratios, glare, shadows, and uniformity. Color and color combinations should also be considered for best effects.

Brightness Ratios

Comfortable seeing conditions in offices and classrooms can be obtained only if the brightness of the light source is kept within agreeable limits. The degree of brightness control required is dependent upon the source used, the size of the room, the illumination level, the reflectances, the finish of room surfaces and furniture, and the nature of the seeing task.

A general guide to acceptable brightness limits for fluorescent fixtures in school or office areas is given in the table in Figure 12-12.

If the average brightness of a lighting fixture at each of the angles in Figure 12-12 does not exceed the luminance in any single column, the brightness will meet the generally accepted limits for control of direct glare.

A comfortable balance of perceived brightness in the office requires that the brightness ratios

between areas of appreciable size from normal viewpoints be within the following:

1 to ⅓ — Between task and adjacent surroundings

1 to ¹⁄₁₀ — Between task and more remote darker surfaces

1 to 10 — Between task and more remote lighter surfaces

20 to 1 — Between fixtures and surroundings adjacent to them

40 to 1 — Anywhere within the normal field of view

In school rooms the acceptable ratios are:

1 to ⅓ — Between task and adjacent surroundings and remote darker surfaces

1 to 10 — Between task and remote lighter surfaces at 30 footcandles. This ratio should decrease as the level of illumination increases. At 150 footcandles the ratio is 1 to 3.

Recommended Reflectances

For good brightness ratios in offices and schools, the reflectances in the table in Figure 12-13 are recommended.

Lighting Fixtures

Whether intended for pendant, surface, or recessed mounting, lighting fixtures should be of

Angle	Average Brightness (Footlamberts)								
85°	250	240	230	220	210	200	190	180	165
75°	250	250	250	250	250	250	250	250	250
65°	250	265	280	295	310	325	340	355	375
55°	250	285	315	350	385	415	450	480	535

Figure 12-12: Average brightness for various angles

Surface	Reflectance in Percent	
	Office	Schools
Ceiling	80-90	70-90
Walls	40-60	40-60
Desk tops and other furniture	26-44	35-50
Floors	21-39	30-50

Figure 12-13: Recommended reflectance in offices and schools

low brightness. For this reason, suspended fixtures providing upward light are popular. Surface-mounted fixtures may be provided with luminous side panels to reduce the brightness ratio between fixture and ceiling. Recessed fixtures should have diffusing panels that will give approximately the same brightness as that of the ceiling against which the fixture is viewed.

Glare

Since the eye cannot render clear vision when a bright light source is within field of view, glare is one of the main factors of poor lighting. Glare reduces the sensitivity of the visual sense, and therefore reduces the visibility of a seeing task. In fact, glare is distracting and annoying, often to the extent of causing extreme discomfort and even pain.

Elimination of glare is a matter of proper brightness control. Direct glare is prevented by enclosing the loop with a diffuser or by proper shielding.

Lighting Design

Modern lighting for offices and schools is related to the architectural design and requires careful coordination of the work of the architect and lighting designer in order to obtain a complete and satisfactory lighting system.

Most lighting systems for offices and schools are designed on the basis of being economically justifiable to provide a sufficient level of illumination at the proper environmental brightness. However, aesthetics should also be considered early in the planning stage.

The first point to determine is the proper amount of illumination. Consulting engineers normally perform the necessary calculations using the zonal-cavity method for schools and offices. This subject is too complicated and lengthy to cover thoroughly in one book chapter. However, a brief description of the fundamental approach follows:

ZONAL-CAVITY METHOD

The zonal-cavity method of calculating average illumination levels assumes each room or area to consist of the following three separate cavities:

- Ceiling cavity
- Room cavity
- Floor cavity

Figure 12-14 shows that the *ceiling cavity* extends from the lighting fixture plane upward to the ceiling. The *floor cavity* extends from the work plane downward to the floor, while the *room cavity* is the space between the lighting fixture plane and the working plane.

If the lighting fixtures are recessed or surface-mounted on the ceiling, there will be no ceiling cavity and the ceiling-cavity reflectance will be equal to the actual ceiling reflectances. Similarly, if the work plane is at floor level, there will be no floor cavity and the floor-cavity reflectance will be equal to the actual floor reflectance. The geometric proportions of these spaces become the "cavity ratios."

Cavity Ratio

Rooms are classified according to shape by ten cavity-ratio numbers. The basic equation for ob-

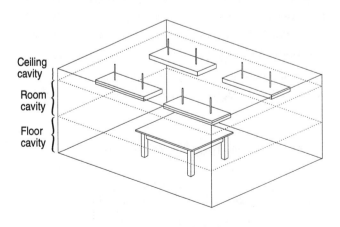

Figure 12-14: Room cavity ratios

taining cavity ratios in rectangular-shaped rooms is:

$$Cavity\ Ratio = \frac{5 \times Height\ (Length + Width)}{Length \times Width}$$

where height is the height of the cavity under consideration — that is, ceiling, floor, or room cavity.

For example, assume the room illustrated in Figure 12-14 is 8 feet wide by 12 feet in length. The lighting fixtures are suspended 1 foot below the ceiling. Find the ceiling cavity ratio. By substituting known values in the previous equation, we have:

$$Ceiling\ Cavity\ Ratio = \frac{5 \times 1\ (12 + 8)}{12 \times 8}$$

$$= 1.04\ or\ 1$$

For rooms composed of more than one rectangle, such as an L-shaped room, the cavity ratio is obtained by the following equation:

$$Cavity\ Ratio = \frac{2.5 \times Wall\ Area}{Floor\ Area}$$

In calculating the ceiling cavity ratio, wall area is determined by multiplying the total linear feet of the walls by the distance between the lighting fixture plane and the ceiling cavity.

For example, an L-shaped room has the physical dimensions as illustrated in Figure 12-15. Notice that the specifications give the ceiling cavity as 2 feet deep. Find the ceiling cavity ratio of this room.

Find the total linear feet of the walls:
15 + 15 + 10 + 10 + 5 + 5 = 60 linear feet

Multiply the total linear feet by the ceiling cavity depth:

60 × 2 = 120 square feet

Find the total floor area by dividing the room into two separate rectangles as shown in Figure 12-16.

A = 5 × 15 = 75 square feet

B = 5 × 10 = 50 square feet

A + B (total floor area) = 75 + 50 = 125 square feet

Substitute these values in the equation.

$$Cavity\ Ratio = \frac{2.5 \times Wall\ Area}{Floor\ Area}$$

$$= \frac{2.5 \times 120}{125} = 2.4$$

Similarly, the floor cavity height in Figure 12-15 is 2.5 feet. Since we already know the perimeter of the room to be 60 feet, we can multiply 2.5 by 60 and obtain a floor-cavity wall area of 150 square feet. The floor area remains the same for all three cavity calculations. Consequently,

$$Floor\ Cavity\ Ratio = \frac{2.5 \times 150}{125} = 3$$

For calculating the room cavity ratio, the wall area is determined by multiplying the total linear feet of the wall (60 feet in this case) by the height of the room cavity. Again refer to Figure 12-15, which shows the height of the room cavity to be 4 feet. The room cavity wall area is then (4 × 60) 240 square feet. The floor area has previously been determined as 125 square feet. Thus,

$$Room\ Cavity\ Ratio = \frac{2.5 \times 240}{125} = 4.8$$

For other than rectangular rooms, the area can be calculated as required. For example, in a circular room, the cavity wall area equals height × $2\pi r$ and the floor area equals $\pi r2$. Thus,

$$Cavity\ Ratio = \frac{2.5 \times height \times 2\pi r}{\pi r^2}\ or\ \frac{5\ height}{r}$$

Effective Reflectance

Before the coefficient of utilization can be selected, the combination of ceiling and wall

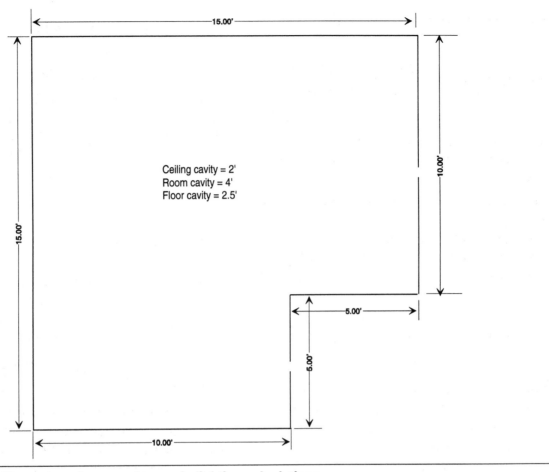

Ceiling cavity = 2'
Room cavity = 4'
Floor cavity = 2.5'

Figure 12-15: Floor plan of an area for use in lighting calculations

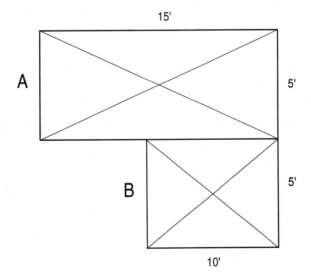

Figure 12-16: Dividing an L-shaped room into two separate rectangles

reflectance as well as floor and wall reflectance must be converted to *effective ceiling* or *floor reflectance*. The effective reflectance of the ceiling and floor cavities takes into account the effect of interreflection of light among the various room surfaces. Charts or tables are provided by lighting-fixture manufacturers for this conversion. A sample table appears in Figure 12-17. To find the effective reflectance, locate the column containing the known percentage of ceiling or floor reflectance and wall reflectance. Then continue down this column and read the effective reflectance opposite the appropriate cavity ratio.

As an example, assume the room illustrated in Figure 12-14 has a ceiling reflectance of 80 percent and a wall reflectance of 50 percent. We have previously calculated the ceiling cavity ratio (1). By following the instructions given in the preceding paragraphs, we find that the effective ceiling reflectance is 66 percent.

Coefficient Of Utilization

Manufacturers usually supply photometric data for their own lighting fixtures. These data contain coefficients of utilization for various surface reflections and room-cavity ratios. An accurate calculation will result in actual applications if the coefficients of utilization of the actual lighting fixture under consideration are used. The table in Figure 12-18 on page 266 is typical of such data. This table of coefficients is based on an effective floor-cavity reflectance of 20 percent. If it is substantially different, an adjustment is made later.

Assume that we want the coefficient of utilization for: effective ceiling-cavity reflectance (%CCR) of 0.78, wall reflectance (%WR) of 0.5, and room-cavity ratio (RCR) of 4 from the table in Figure 12-18. The effective floor-cavity reflectance (%FCR) is 18 percent and is close enough to 20 percent for our purposes. The 78 percent ceiling reflectance is also close enough to the 80 percent reflectance in the table without further adjustment.

For an 80 percent effective ceiling-cavity reflectance, a 50 percent wall reflectance, and a room cavity ratio of 4, the table gives a coefficient of utilization of 0.43.

If the effective floor-cavity reflectance is 18, 19, 21, or 22 percent, direct use of coefficient of utilization tables can be made. However, if the effective floor-cavity reflectance is 17 percent or below or 23 percent and above, an adjustment may be necessary. The table in Figure 12-19 on page 266 lists correction factors for effective floor-cavity reflectances other than 20 percent.

To demonstrate the use of the table in Figure 12-19 and to provide an example of interpolation, assume the effective floor reflectance is 17 percent.

The table shows that, for an 80 percent ceiling, 50 percent wall, and a room-cavity ratio of 4, the correction factor is 1.05. This will be applied to the tentative coefficient of utilization of 0.43 obtained previously. The instructions in the table require interpolation between 1 and 1.05 as follows:

Step 1. Seventeen percent is three-tenths of the way from 20 to 10 percent.
Step 2. A factor of 1 applies to 20 percent and 1.05 applies to 10 percent.
Step 3. Therefore, the correction factor we want is three-tenths of the way from 1 to 1.05.
Step 4. Multiplying three-tenths by (1.05 - 1) gives

$$0.3 \times 0.05 = 0.015$$

Step 5. Adding 0.015 and 1 gives 1.015 as the required correction factor.

The instructions in this table state that the lighting fixture coefficient of utilization is to be divided by this factor. Thus,

$$\frac{0.43}{1.015} = 0.424$$

and the corrected coefficient of utilization for a 17 percent effective floor-cavity reflectance will be 0.42 instead of 0.43.

% Ceiling or Floor Reflectance	90				80				70			50			30				10		
% Wall Reflectance	90	70	50	30	80	70	50	30	70	50	30	70	50	30	65	50	30	10	50	30	10
0	90	90	90	90	80	80	80	80	70	70	70	50	50	50	30	30	30	30	10	10	10
0.1	90	89	88	87	79	79	78	78	69	69	68	59	49	48	30	30	29	29	10	10	10
0.2	89	88	86	85	79	78	77	76	68	67	66	49	48	47	30	29	29	28	10	10	9
0.3	89	87	85	83	78	77	75	74	68	66	64	49	47	46	30	29	28	27	10	10	9
0.4	88	86	83	81	78	76	74	72	67	65	63	48	46	45	30	29	27	26	11	10	9
0.5	88	85	81	78	77	75	73	70	66	64	61	48	46	44	29	28	27	25	11	10	9
0.6	88	84	80	76	77	75	71	68	65	62	59	47	45	43	29	28	26	25	11	10	9
0.7	88	83	78	74	76	74	70	66	65	61	58	47	44	42	29	28	26	24	11	10	8
0.8	87	82	77	73	75	73	69	65	64	60	56	47	43	41	29	27	25	23	11	10	8
0.9	87	81	76	71	75	72	68	63	63	59	55	46	43	40	29	27	25	22	11	9	8
1.0	86	80	74	69	74	71	66	61	63	58	53	46	42	39	29	27	24	22	11	9	8
1.1	86	79	73	67	74	71	65	60	62	57	52	46	41	38	29	26	24	21	11	9	8
1.2	86	78	72	65	73	70	64	58	61	56	50	45	41	37	29	26	23	20	11	9	7
1.3	85	78	70	64	73	69	63	57	61	55	49	45	40	36	29	26	23	20	12	9	7
1.4	85	77	69	62	72	68	62	55	60	54	48	45	40	35	28	26	22	19	12	9	7
1.5	85	76	68	61	72	68	61	54	59	53	47	44	39	34	28	25	22	18	12	9	7
1.6	85	75	66	59	71	67	60	53	59	52	45	44	39	33	28	25	21	18	12	9	7
1.7	84	74	65	58	71	66	59	52	58	51	44	44	38	32	28	25	21	17	12	9	7
1.8	84	73	64	56	70	65	58	50	57	50	43	43	37	32	28	25	21	17	12	9	6
1.9	84	73	63	55	70	65	57	49	57	49	42	43	37	31	28	25	20	16	12	9	6
2.0	83	72	62	53	69	64	56	48	56	48	41	43	37	30	28	24	20	16	12	9	6
2.1	83	71	61	52	69	63	55	47	56	47	40	43	36	29	28	24	20	16	13	9	6
2.2	83	70	60	51	68	63	54	45	55	46	39	42	36	29	28	24	19	15	13	9	6
2.3	83	69	59	50	68	62	53	44	54	46	38	42	35	28	28	24	19	15	13	9	6
2.4	82	68	58	48	67	61	52	43	54	45	37	42	35	27	28	24	19	14	13	9	6
2.5	82	68	57	47	67	61	51	42	53	44	36	41	34	27	27	23	18	14	13	9	6
2.6	82	67	56	46	66	60	50	41	53	43	35	41	34	26	27	23	18	13	13	9	5
2.7	82	66	55	45	66	60	49	40	52	43	34	41	33	26	27	23	18	13	13	9	5
2.8	81	66	54	44	66	59	48	39	52	42	33	41	33	25	27	23	18	13	13	9	5
2.9	81	65	53	43	65	58	48	38	51	41	33	40	33	25	27	23	17	12	13	9	5
3.0	81	64	52	42	65	58	47	38	51	40	32	40	32	24	27	22	17	12	13	8	5
3.1	80	64	51	41	64	57	46	37	50	40	31	40	32	24	27	22	17	12	13	8	5
3.2	80	63	50	40	64	57	45	36	50	39'	30	40	31	23	27	22	16	11	13	8	5
3.3	80	62	49	39	64	56	44	35	49	39	30	39	31	23	27	22	16	11	13	8	5
3.4	80	62	48	38	63	56	44	34	49	38	29	39	31	22	27	22	16	11	13	8	5
3.5	79	61	48	37	63	55	43	33	48	38	29	39	30	22	26	22	16	11	13	8	5
3.6	79	60	47	36	62	54	42	33	48	37	28	39	30	21	26	21	15	100	13	8	5
3.7	79	60	46	35	62	54	42	32	48	37	27	38	30	21	26	21	15	10	13	8	4
3.8	79	59	45	35	62	53	41	31	47	36	27	38	29	21	26	21	15	10	13	8	4
3.9	78	59	45	34	61	53	40	30	47	36	26	38	29	20	26	21	15	10	13	8	4
4.0	78	58	44	33	61	52	40	30	46	35	26	38	29	20	26	21	15	9	13	8	4
4.1	78	57	43	32	60	52	39	29	46	35	25	37	28	20	26	21	14	9	13	8	4
4.2	78	57	43	32	60	51	39	29	46	34	25	37	28	19	26	20	14	9	13	8	4
4.3	78	56	42	31	60	51	38	28	45	34	25	37	28	19	26	20	14	9	13	8	4
4.4	77	56	41	30	59	51	38	28	45	34	24	37	27	19	26	20	14	8	13	8	4
4.5	77	55	41	30	59	50	37	27	45	33	24	37	27	19	25	20	14	8	14	8	4
4.6	77	55	40	29	59	50	37	26	44	33	24	36	27	18	25	20	13	8	14	8	4
4.7	77	54	40	29	58	49	36	26	44	33	23	36	26	18	25	20	13	8	14	8	4
4.8	76	54	39	28	58	49	36	25	44	32	23	36	26	18	25	19	13	7	14	8	4
4.9	76	53	38	28	58	49	35	25	43	32	23	36	26	18	25	19	13	7	14	8	4
5.0	76	53	38	27	57	48	35	25	43	32	22	36	26	17	25	19	13	7	14	8	4

Figure 12-17: Effective reflectance chart as recommended by the *Illuminating Engineering Society*

% CCR	80%		50%	
% FCR	20%		20%	
% WR	70%	50%	50%	30%
RCR 1	.60	.58	.54	.53
2	.56	.52	.49	.47
3	.52	.47	.45	.42
4	.48	.43	.31	.38
5	.45	.39	.37	.34
6	.42	.36	.34	.31
7	.39	.33	.31	.28
8	.36	.30	.29	.25
9	.34	.27	.26	.23
10	.31	.25	.24	.21

Figure 12-18: Typical coefficients of utilization

Maintenance Factor

The maintenance factor takes into account the reduction in light output because of lamp aging and dirt accumulation. The appropriate maintenance factor for any given condition and lighting fixture type may be determined as follows.

Types of lighting fixtures are divided into five categories. The category for each lighting fixture is printed in tables supplied by lighting-fixture

Room Cavity Ratio	Percent Effective Ceiling-Cavity Reflectance											
	80			Percent Wall Reflectance						10		
	50	30	10	50	30	10	50	30	10	50	30	10
1	1.08	1.08	1.07	1.07	1.06	1.06	1.05	1.04	1.04	1.01	1.01	1.01
2	1.07	1.06	1.05	1.06	1.05	1.04	1.04	1.03	1.03	1.01	1.01	1.01
3	1.05	1.04	1.03	1.05	1.04	1.03	1.03	1.03	1.02	1.01	1.01	1.01
4	1.05	1.03	1.02	1.04	1.03	1.02	1.03	1.02	1.02	1.01	1.01	1.00
5	1.04	1.03	1.02	1.03	1.02	1.02	1.02	1.02	1.01	1.01	1.01	1.00
6	1.03	1.02	1.01	1.03	1.02	1.02	1.02	1.02	1.01	1.01	1.01	1.00
7	1.03	1.02	1.01	1.03	1.02	1.01	1.02	1.01	1.01	1.01	1.01	1.00
8	1.03	1.02	1.01	1.02	1.02	1.01	1.02	1.01	1.01	1.01	1.01	1.00
9	1.02	1.01	1.01	1.02	1.01	1.01	1.02	1.01	1.01	1.01	1.01	1.00
10	1.02	1.01	1.01	1.01	1.01	1.01	1.02	1.01	1.01	1.01	1.01	1.00

If the effective floor cavity reflectance is

30%:	Multiply the luminaire CU by the factor shown in this table
23 to 29%:	Multiply the luminaire CU by the result obtained from interpolating between 1.00 and the factor shown in this table
18 to 22%:	Use the luminaire CU directly; do not use this table.
11 to 17%:	Divide the luminaire CU by the result obtained from interpolating between 1.00 and the factor shown in this table
10%:	Divide the luminaire CU by the factor shown in this table

Figure 12-19: Correction factors for effective floor-cavity reflectances

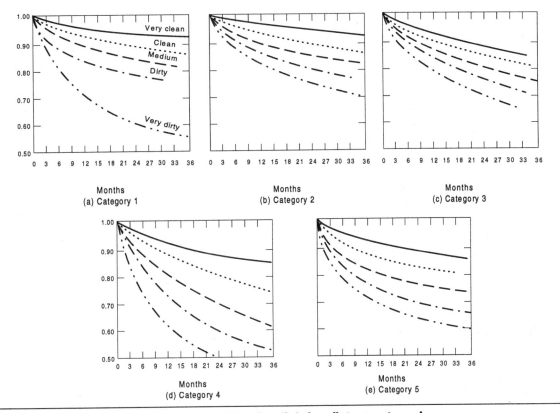

Figure 12-20: Maintenance factor curves of the five lighting-fixture categories

manufacturers. In general, after the category has been determined, the maintenance factor can be read from one of the five curves for each category as shown in Figure 12-20. The point on the curve should be selected on the basis of the estimated number of months between cleaning of the lighting fixtures and lamps. The particular curve selected should be based on the dirt content of the atmosphere under considerataion.

Once all of the preceding data has been collected, calculated, and the results determined, the number of lighting fixtures and lamps can be calculated from the following equation:

$$NF = \frac{FA \times DF}{LPF \times LPL \times CU \times MF}$$

where

NF = number of fixtures

FA = floor area

DF = desired footcandles

LPF = lamps per fixture

LPL = lumens per lamp

CU = coefficient of utilization

MF = maintenance factor

As an example, using our previous room, assume the seeing task to be 150 footcandles. We have previously determined the floor area to be 96 square feet and the coefficient of utilization to be 0.42. The lighting fixtures that will be used will contain four 40-watt fluorescent lamps. Assume a maintenance factor of 0.7. Referring to lamp-data tables supplied by lamp manufacturers, we find that this lamp has approximately 3250 initial lumens. By inserting these values in the equation, we have:

$$NF = \frac{96 \times 150}{4 \times 3250 \times 0.42 \times 0.7}$$

= 3.8 or 4 light fixtures

Four lighting fixtures will then be needed to obtain approximately 150 footcandles of illumination. Since the calculation did not come out even (3.8), four fixtures will give slightly more than 150 footcandles, but this slight difference is insignificant for all practical purposes.

Layout Of Lighting Fixtures

Lighting-fixture locations depend on the general architecture, size of bays, type of lighting fixture under consideration, and so on. To provide even distribution of illumination for an area, the permissible maximum spacing recommendations by the lighting-fixture manufacturers should not be exceeded. These recommended ratios are supplied in terms of maximum spacing to mounting height. In most cases, however, it is desirable to locate fixtures closer together than these maximums to obtain required illumination levels.

In our sample area, the interior dimensions of the room in Figure 12-14 are 12 feet by 8 feet and the floor plan is shown in Figure 12-21.

As mentioned previously, the exact number of lighting fixtures required for this room is 3.8, so either three or four fixtures had to be installed. The amount of actual illumination then can be recomputed by the equation

$$FC = \frac{TL \times LPL \times CU \times MF}{A}$$

where

FC = footcandles

TL = total lamps

LPL = lumens per lamp

CU = coefficient of utilization

MF = maintenance factor

A = area

Substituting all known values in the equation, we have

$$FC = \frac{16 \times 3250 \times 0.42 \times 0.7}{96}$$
$$= 159 \ footcandles$$

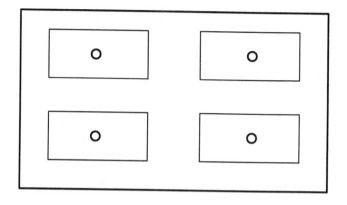

Figure 12-21: Floor plan showing fixture layout

STORE LIGHTING

Light, as any store owner knows, is an important factor in selling. It is needed to attract customers by providing them with pleasant surroundings. By using the many kinds of visual impressions that can be created by lighting effects, the owner may achieve a dramatic presentation of his merchandise. Light also plays an important role in improving the morale and selling efficiency of the employees.

Thus, the lighting of a store interior should combine three major functions:

- Attraction — To take advantage of involuntary attraction, which is the tendency of the eye to turn instinctively toward bright areas.

- Appraisal — To give the customer the proper illumination to make a full appraisal of what is seen when examining the color, texture, and quality of the merchandise.

- Atmosphere—To create atmosphere or feeling in a room that appeals to the customer's emotions and to his sense of beauty.

The principles of store lighting can be applied to lighting systems for restaurants, hotels/motels, banks, and similar establishments.

Illumination Levels

Recommended illumination levels representing modern practice in various store applications are available from lighting-fixture manufacturers and also from the Illuminating Engineering Society. These levels will, however, vary with the atmosphere desired. Too many lighting designs are begun with a selected lighting level which is then computed mathematically to determine the number of fixtures required. A good store lighting design should first begin by using a lighting approach that emphasizes all the basic lighting effects on merchandise, including color brightness control. The question of directional and diffused lighting for general illumination is the second decision.

Choice Of Light Sources

Some stores use fluorescent lamps almost exclusively, while others may use only incandescent lamps. High-intensity discharge lighting systems are gradually finding their way into mass merchandising areas and are replacing both fluorescent and incandescent lamps. Most designers of store lighting, however, use combinations of incandescent, fluorescent, and high-intensity discharge equipment to achieve a particular objective. Basically, this objective should be to provide a lighting system that will, when tied to an overall environmental design, create an overall psychological or emotional response in the customer.

Use Of Fluorescent Lamps

The wide variety of fluorescent lamps available have made this type of lamp the favorite for store lighting designers. These elongated light sources have made possible many applications of valance and vertical-surface lighting, cove lighting, and general illumination.

Fluorescent lamps are provided in several different qualities of white light. In general, the designations warm and cool represent the difference between artificial and natural daylight in the feeling they give to a room. Their deluxe counterparts have a greater content of red light, which is supplied by a second phosphor within the tube. While the additional red light provides a quality fluorescent light which closely approaches that of incandescent light, there is a sacrifice in efficiency.

Many colored fluorescent lamps are also available and are interchangeable with conventional white lamps. However, the relatively large size and low brightness of fluorescent lamps limit the total amount of colored light that can be delivered from a given source, and accurate control of the direction of the light is difficult. Therefore, colored fluorescent lamps are more useful for flooding large areas than for small areas requiring concentrated light.

Projector And Reflector Lamps

Projector and reflector incandescent lamps have proved useful for many special applications in the lighting of stores. All such lamps used in store areas should be carefully aimed to avoid creating annoying glare or reflections from mirror-like surfaces. This type of lighting, when properly used, adds greatly to the sales appeal, decor, and visual environment of store areas.

Reflector color lamps are a convenient and effective source of colored light, especially where control of such light is needed. There are at least six colors available, all of which are designed for maximum effectiveness when used either alone or in combination with each other. For example, pink and blue-white lamps can be used to give an entire display a warm or cool tone with very little color distortion; pink lamps complement complexion tones; blue-white lamps are excellent for lighting displays of silverware, jewelry, or home appliances.

Color Consideration

Color probably presents the biggest design problem in lighting store areas, although the size, shape, and finish of merchandise will also influence the choice of light sources and lighting fixtures in a given store area. For example, a deep-pile rug has a textured finish, and evenly diffused light may illuminate each fiber so uniformly that appraisal of the pattern and fiber depth may be difficult. However, directional and diffused light used together would emphasize the soft texture and deep nap of the rug.

Many kinds of merchandise show up better under a dual system of incandescent/quartz and fluorescent lighting than under one system or the other. Therefore, both types are normally used in the majority of store areas. In the overall lighting of stores, fluorescent lamps are used where the goal is a cool atmosphere or greatest economy. A predominantly incandescent system may be used where a warmer atmosphere is desired, although deluxe warm white fluorescent may be a better choice for this effect.

A guideline for the choice of light sources in stores where color is a determining factor is lighting that shows merchandise as it will appear in actual use. This method will usually bring customer satisfaction and will minimize return of goods because of color change at the point of use.

Lighting For Emphasis Of Merchandise

Light is one of the most effective means of directing attention to merchandise. The cardinal principle in store lighting is to make merchandise brighter than other areas in the field of view. To this end, the light should be directed to the display counters, showcases, wall cases, and feature displays, rather than to circulation areas. The brightness of luminaires, ceilings, columns, and other architectural features should be limited to relatively low values. Higher wall brightnesses are sometimes useful, however, for attracting customers to perimeter areas of the store.

Counter Areas And Showcase Tops

In most stores, counter display areas and showcase tops are satisfactorily lighted with 100 to 200 footcandles, depending on the productivity expected, the general lighting level, and the size of the details that must be readily visible. The display area illumination should be three to ten times the level of the general area lighting. Most stores display high-profit or impulse items at the end of the counter or near the cash register. These preferred locations should have two to five times as much illumination as the rest of the counter. Incandescent spotlights with their accompanying highlights and shadows are very effective.

Showcases

A showcase interior should have more illumination than the top of the case, but not more than about twice as much. Where the illumination on top is in the 30- to 75-footcandle range, T-6 or T-8 slimlines, operated at 200 milliamperes, provide sufficient light inside the case; for higher footcandles on the top, an operating current of 300 milliamperes is sometimes recommended. Too great a differential between the illumination inside the case and that on the top, where the merchandise is normally inspected more closely prior to purchase, is not advisable. Many products examined under an illumination level significantly lower than that under which they were displayed lose some of their attraction. This is not a consideration with feature displays, since they are seldom removed for inspection.

Chapter 13
Electric Motors

The principal means of changing electrical energy into mechanical energy or power is the electric motor — ranging in size from small fractional-horsepower, low-voltage motors to the very large high-voltage synchronous motors.

Electric motors are classified according to the following:

- Size (horsepower)
- Type of application
- Electrical characteristics
- Speed, starting, speed control and torque characteristics
- Mechanical protection
- Method of cooling

In basic terms, electric motors convert electric energy into the productive power of rotary mechanical force. This capability finds many applications in unlimited ways in commercial establishments for powering elevators in public buildings, HVAC and refrigeration fans and compressors, gasoline and water pumps, power tools, and a host of other applications.

All of these and more represent the scope of electric motor participation in powering and controlling machines and equipment used in commercial buildings.

Electric motors are machines that change electrical energy into mechanical energy. They are rated in horsepower. The attraction and repulsion of the magnetic poles produced by sending current through the armature and field windings cause the armature to rotate. The armature rotation produces a twisting power called torque.

Nameplate Information

A typical motor nameplate is shown in Figure 13-1. A nameplate is one of the most important parts of a motor since it gives the motor's electrical and mechanical characteristics; that is, the horsepower, voltage, rpms, etc. Always refer to the motor's nameplate before connecting it to an electric system. The same is true when performing preventative maintenance or troubleshooting motors.

Referring again to the motor nameplate in Figure 13-1, note that the manufacturer's name and

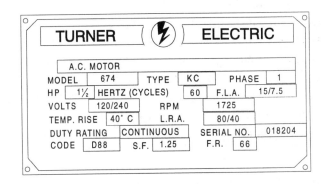

Figure 13-1: Typical motor nameplate

logo is at the top of the plate; these items, of course, will change with each manufacturer. The line directly below the manufacture's name identifies the motor for use on ac systems as opposed to dc or ac-dc systems. The *model number* identifies that particular motor from any other. The *type* or *class* specifies the insulation used to ensure the motor will perform at the rated horsepower and service-factor load. The *phase* indicates whether the motor has been designed for single- or three-phase use.

Horsepower on the nameplate defines the rated output capacity of the motor; hertz (cycles) indicates the alternating current frequency at which the motor is designed to operate. The F.L.A. section gives the amperes of current the motor draws at full load. When two values are shown on the nameplate, the motor usually has a dual voltage rating. Volts and amps are inversely proportional; the higher the voltage, the lower the amperes, and vice versa. The higher amp value corresponds to the lower voltage rating on the nameplate. Two-speed motors will also show two ampere readings.

Volts is the electrical potential "pressure" for which the motor is designed. Sometimes two voltages are listed on the nameplate, such as 120/240. In this case the motor is intended for use on either a 120- or 240-volt circuit. Special instructions are furnished for connecting the motor for each of the different voltages. For example, Figure 13-2 shows connections for a three-phase motor for use on either 208/240 or 480-volt systems.

The *rpm* inscription represents revolutions per minute; that is, the motor speed. The rpm reading on motors is the approximate full-load speed. *Temp. rise* designates the maximum air temperature immediately surrounding the motor. Forty degrees centigrade is the NEMA maximum ambient temperature.

L.R.A. stands for locked rotor amps. It relates to starting current and selection of fuse or circuit breaker size. When two values are shown, the motor usually has a dual-voltage rating. The duty rating designates the duty cycle of a motor. "Continuous" means that the motor is designed for around-the-clock operation.

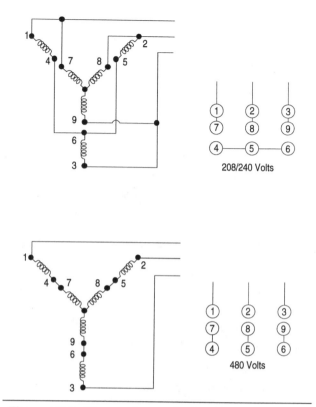

Figure 13-2: Three-phase, dual-voltage motor connections

Each motor is usually given a different *serial number* for identification and tracking purposes. The *code* designation is a serial data code used by the manufacturer. In Figure 13-1, the first letter identifies the month and the last two numbers identify the year of manufacture (D88 is April 88).

A service factor (*S.F.*) is a multiplier which, when applied to the rated horsepower, indicates a permissible horsepower loading which may be carried continuously when the voltage and frequency are maintained at the value specified on the nameplate, although the motor will operate at an increased temperature rise.

The frame (*F.R.*) designation specifies the shaft height and motor mounting dimensions and provides recommendations for standard shaft diameters and usable shaft extension lengths.

Besides the information just mentioned, many larger motors also contain plates with wiring diagrams to facilitate connections, maintenance, and repairs.

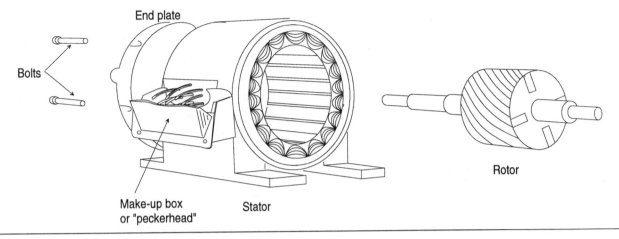

Figure 13-3: Basic parts of a motor

SINGLE-PHASE MOTORS

Single-phase ac motors are usually limited in size to about two or three horsepower. For residential and small commercial applications, these motors will be found in both central and individual room air conditioning units, fans, ventilating units, and refrigeration units such as household refrigerators and the larger units used to cool produce and other foods in market places.

Since there are so many applications of electric motors, there are many types of single-phase motors in use. Some of the more common types are repulsion, universal, and single-phase induction motors. This latter type includes split-phase, capacitor, shaded-pole, and repulsion-induction motors.

Figure 13-3 shows the basic parts of a motor to familiarize you with its makeup.

Split-Phase Motors

Split-phase motors are fractional-horsepower units that use an auxiliary winding on the stator to aid in starting the motor until it reaches its proper rotation speed (see Figure 13-4). This type of motor finds use in small pumps, oil burners, and other applications.

In general, the split-phase motor consists of a housing, a laminated iron-core stator with embedded windings forming the inside of the cylindrical housing, a rotor made up of copper bars set in slots in an iron core and connected to each other by copper rings around both ends of the core, plates that are bolted to the housing and contain the bearings that support the rotor shaft, and a centrifugal switch inside the housing. This type of rotor is often called a *squirrel cage* rotor since the configuration of the copper bars resembles an actual cage. These motors have no windings as such, and a centrifugal switch is provided to open the circuit to the starting winding when the motor reaches running speed.

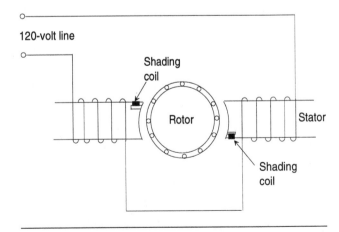

Figure 13-4: Diagram of a split-phase motor

To understand the operation of a split-phase motor, look at the wiring diagram in Figure 13-4. Current is applied to the stator windings, both the main winding and the starting winding, which is in parallel with it through the centrifugal switch. The two windings set up a rotating magnetic field, and this field sets up a voltage in the copper bars of the squirrel-cage rotor. Because these bars are shortened at the ends of the rotor, current flows through the rotor bars. The current-carrying rotor bars then react with the magnetic field to produce motor action. When the rotor is turning at the proper speed, the centrifugal switch cuts out the starting winding since it is no longer needed.

Capacitor Motors

Capacitor motors are single-phase ac motors ranging in size from fractional horsepower (hp) to perhaps as high as 15 hp. This type of motor is widely used in all types of single-phase applications such as powering air compressors, refrigerator compressors, and the like. This type of motor is similar in construction to the split-phase motor,

except a capacitor is wired in series with the starting winding, as shown in Figure 13-5.

The capacitor provides higher starting torque, with lower starting current, than does the split-phase motor, and although the capacitor is sometimes mounted inside the motor housing, it is more often mounted on top of the motor, encased in a metal compartment.

In general, two types of capacitor motors are in use: the capacitor-start motor and the capacitor start-and-run motor. As the name implies, the former utilizes the capacitor only for starting; it is disconnected from the circuit once the motor reaches running speed, or at about 75 percent of the motor's full speed. Then the centrifugal switch opens to cut the capacitor out of the circuit.

The capacitor start-and-run motor keeps the capacitor and starting winding in parallel with the running winding, providing a quiet and smooth operation at all times.

Capacitor split-phase motors require the least maintenance of all single-phase motors, but they have a very low starting torque, making them unsuitable for many applications. Their high maximum torque,

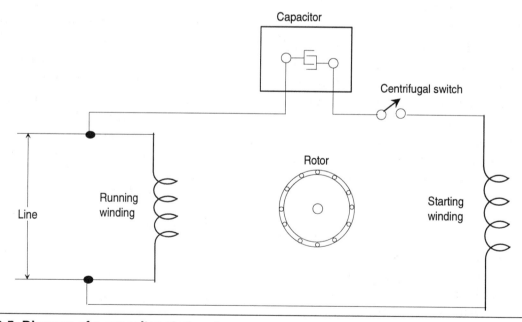

Figure 13-5: Diagram of a capacitor-start motor

however, makes them especially useful in HVAC systems to power slow-speed, direct-connected fans.

Repulsion-Type Motors

Repulsion-type motors are divided into several groups, including (1) repulsion-start, induction-run motors, (2) repulsion motors, and (3) repulsion-induction motors. The repulsion-start, induction-run motor is of the single-phase type, ranging in size from about $\frac{1}{10}$ hp to as high as 20 hp. It has high starting torque and a constant-speed characteristic, which makes it suitable for such applications as commercial refrigerators, compressors, pumps, and similar applications requiring high starting torque.

The repulsion motor is distinguished from the repulsion-start, induction-run motor by the fact that it is made exclusively as a brush-riding type and does not have any centrifugal mechanism. Therefore, this motor both starts and runs on the repulsion principle. This type of motor has high starting torque and a variable-speed characteristic. It is reversed by shifting the brush holder to either side of the neutral position. Its speed can be decreased by moving the brush holder farther away from the neutral position.

The repulsion-induction motor combines the high starting torque of the repulsion-type and the good speed regulation of the induction motor. The stator of this motor is provided with a regular single-phase winding, while the rotor winding is similar to that used on a dc motor. When starting, the changing single-phase stator flux cuts across the rotor windings and induces currents in them; thus, when flowing through the commutator, a continuous repulsive action on the stator poles is present.

This motor starts as a straight repulsion-type and accelerates to about 75 percent of normal full speed when a centrifugally operated device connects all the commutator bars together and converts the winding to an equivalent squirrel-cage type. The same mechanism usually raises the brushes to reduce noise and wear. Note that, when

the machine is operating as a repulsion-type, the rotor and stator poles reverse at the same instant, and that the current in the commutator and brushes is ac.

This type of motor will develop four to five times normal full-load torque and will draw about three times normal full-load current when starting with full-line voltage applied. The speed variation from no load to full load will not exceed 5 percent of normal full-load speed.

The repulsion-induction motor is used to power air compressors, refrigeration (compressor and fans), pumps, stokers, and the like. In general, this type of motor is suitable for any load that requires a high starting torque and constant-speed operation. Most motors of this type are less than 5 hp.

Universal Motors

This type of motor is a special adaptation of the series-connected dc motor, and it gets its name *"universal"* from the fact that it can be connected on either ac or dc and operates the same. All are single-phase motors for use on 120 or 240 volts.

In general, the universal motor contains field windings on the stator within the frame, an armature with the ends of its windings brought out to a commutator at one end, and carbon brushes that are held in place by the motor's end plate, allowing them to have a proper contact with the commutator.

When current is applied to a universal motor, either ac or dc, the current flows through the field coils and the armature windings in series. The magnetic field set up by the field coils in the stator react with the current-carrying wires on the armature to produce rotation.

Universal motors are frequently used on small fans.

Shaded-Pole Motors

A shaded-pole motor is a single-phase induction motor provided with an uninsulated and permanently short-circuited auxiliary winding displaced in magnetic position from the main winding. The auxiliary winding is known as the shading coil and usually surrounds from one-third to one-half of the pole. The main winding surrounds the entire pole and may consist of one or more coils per pole.

Applications for this motor include small fans, timing devices, relays, instrument dials, or any constant-speed load not requiring high starting torque.

POLYPHASE MOTORS

Three-phase motors offer extremely efficient and economical application and are usually the preferred type for commercial and industrial applications when three-phase service is available. In fact, the great bulk of motors sold are standard ac three-phase motors. These motors are available in ratings from fractional horsepower up to thousands of horsepower in practically every standard voltage and frequency. In fact, there are few applications for which the three-phase motor cannot be put to use.

Three-phase motors are noted for their relatively constant speed characteristic and are available in designs giving a variety of torque characteristics; that is, some have a high starting torque and others a low starting torque. Some are designed to draw a normal starting current, others a high starting current.

The three main parts of a three-phase motor are the stator, rotor, and end plates. It is very similar in construction to conventional split-phase motors except that the three-phase motor has no centrifugal switch.

The stator consists of a steel frame and a laminated iron core with windings formed of individual coils placed in slots. The rotor may be a squirrel-cage or wound-rotor type. Both types contain a laminated core pressed onto a shaft. The squirrel-cage rotor is similar to a split-phase motor. The wound rotor has a winding on the core that is connected to three slip rings mounted on the shaft.

The end plates or brackets are bolted to each side of the stator frame and contain the bearings in which the shaft revolves. Either ball bearings or sleeve bearings are used.

Induction Motors

Induction motors, both single-phase and polyphase, get their name from the fact that they utilize the principle of electromagnetic induction. An induction motor has a stationary part, or stator, with windings connected to the ac supply, and a rotation part, or rotor, which contains coils or bars. There is no electrical connection between the stator and rotor. The magnetic field produced in the stator windings induces a voltage in the rotor coils or bars.

Since the stator windings act in the same way as the primary winding of a transformer, the stator of an induction motor is sometimes called the *primary*. Similarly, the rotor is called the *secondary* because it carries the induced voltage in the same way as the secondary of a transformer.

The magnetic field necessary for induction to take place is produced by the stator windings. Therefore, the induction-motor stator is often called the *field* and its windings are called *field windings*.

The terms primary and secondary relate to the electrical characteristics and the terms stator and rotor to the mechanical features of induction motors.

The rotor transfers the rotating motion to its shaft, and the revolving shaft drives a mechanical load or a machine, such as a pump, spindle, or clock.

Commutator segments, which are essential parts of dc motors, are not needed on induction motors. This simplifies greatly the design and the maintenance of induction motors as compared to dc motors.

The turning of the rotor in an induction motor is due to induction. The rotor, or secondary, is not connected to any source of voltage. If the magnetic field of the stator, or primary, revolves, it will induce a voltage in the rotor, or secondary. The magnetic field produced by the induced voltage acts in such a way that it makes the secondary follow the movement of the primary field.

The stator, or primary, of the induction motor does not move physically. The movement of the primary magnetic field must thus be achieved electrically. A rotating magnetic field is made possible by a combination of two or more ac voltages that are out of phase with each other and applied to the stator coils. Direct current will not produce a rotating magnetic field. In three-phase induction motors, the rotating magnetic field is obtained by applying a three-phase system to the stator windings.

The direction of rotation of the rotor in an ac motor is the same as that of its rotating magnetic field. In a three-phase motor the direction can be reversed by interchanging the connections of any two supply leads. This interchange will reverse the sequence of phases in the stator, the direction of the field rotation, and therefore the direction of rotor rotation.

Synchronous Motors

A synchronous polyphase motor has a stator constructed in the same way as the stator of a conventional induction motor. The iron core has slots into which coils are wound, which are also arranged and connected in the same way as the stator coils of the induction motor. These are, in turn, grouped to form a three-phase connection and the three free leads are connected to a three-phase source. Frames are equipped with air ducts, which aid in cooling of the windings, and coil guards protect the windings from damage.

The rotor of a synchronous motor carries poles that project toward the armature; they are called *salient poles.* The coils are wound on laminated pole bodies and connected to slip rings on the shaft. A squirrel-cage winding for starting the motor is embedded in the pole faces.

The pole coils are energized by direct current, which is usually supplied by a small dc generator called the *exciter.* This exciter may be mounted directly on the shaft to generate dc voltage, which is applied through brushes to slip rings. On low-speed synchronous motors, the exciter is normally belted or of a separate high-speed, motor-driven type.

The dimensions and construction of synchronous motors vary greatly, depending on the rating of the motors. However, synchronous motors for industrial power applications are rarely built for less than 25 horsepower (hp) or so. In fact, many are 100 hp or more. All are polyphase motors when built in this size. Vertical and horizontal shafts with various bearing arrangements and various enclosures cause wide variations in the appearance of the synchronous motor.

Synchronous motors are used in electrical systems where there is need for improvement in power factor or where low power factor is not desirable. This type of motor is especially adapted to heavy loads that operate for long periods of time without stopping, such as for air compressors, pumps, ship propulsion, and the like.

The construction of the synchronous motor is well adapted for high voltages, as it permits good insulation. Synchronous motors are frequently used at 2300 volts or more. Its efficient slow-running speed is another advantage.

MOTOR ENCLOSURES

Electric motors differ in construction and appearance, depending on the type of service for which they are to be used. Open and closed frames are quite common. In the former enclosure, the motor's parts are covered for protection, but the air can freely enter the enclosure. Further designations for this type of enclosure include drip-proof, weather-protected, and splash-proof.

Totally enclosed motors have an airtight enclosure. They may be fan cooled or self-ventilated. An enclosed motor equipped with a fan has the fan as an integral part of the machine, but external to the enclosed parts. In the self-ventilated enclosure, no external means of cooling is provided.

The type of enclosure to use will depend on the ambient and surrounding conditions. In a drip-proof machine, for example, all ventilating openings are so constructed that drops of liquid or solid particles falling on the machine at an angle of not greater than 15 degrees from the vertical cannot enter the machine, even directly or by striking and running along a horizontal or inclined surface of the machine. The application of this machine would lend itself to areas where liquids are processed.

An open motor having all air openings that give direct access to live or rotating parts, other than the shaft, limited in size by the design of the parts or by screen to prevent accidental contact with such parts is classified as a drip-proof, fully guarded machine. In such enclosures, openings shall not permit the passage of a cylindrical rod $\frac{1}{2}$ inch in diameter, except where the distance from the guard to the live rotating parts is more than 4 inches, in which case the openings shall not permit the passage of a cylindrical rod $\frac{3}{4}$ inch in diameter.

There are other types of drip-proof machines for special applications such as externally ventilated and pipe ventilated, which as the names imply are either ventilated by a separate motor-driven blower or cooled by ventilating air from inlet ducts or pipes.

An enclosed motor whose enclosure is designed and constructed to withstand an explosion of a specified gas or vapor that may occur within the motor and to prevent the ignition of this gas or vapor surrounding the machine is designated "*explosionproof*" (XP) motors.

Hazardous atmospheres (requiring XP enclosures) of both a gaseous and dusty nature are classified by the *National Electrical Code (NEC)* as follows:

- Class I, Group A: atmospheres containing acetylene.

- Class I, Group B: atmospheres containing hydrogen gases or vapors of equivalent hazards such as manufactured gas.

- Class I, Group C: atmospheres containing ethyl ether vapor.

- Class I, Group D: atmospheres containing gasoline, petroleum, naphtha, alcohols, acetone, lacquer-solvent vapors, and natural gas.

- Class II, Group E: atmospheres containing metal dust.

- Class II, Group F: atmospheres containing carbon-black, coal, or coke dust.

- Class II, Group G: atmospheres containing grain dust.

The proper motor enclosure must be selected to fit the particular atmosphere. However, explosion-proof equipment is not generally available for Class I, Groups A and B, and it is therefore necessary to isolate motors from the hazardous area.

MOTOR TYPE

The type of motor will determine the electrical characteristics of the design. NEMA-designated designs for polyphase motors are given in the table in Figure 13-6.

An "A" motor is a three-phase, squirrel-cage motor designed to withstand full-voltage starting with locked rotor current higher than the values for a B motor and having a slip at rated load of less than 5 percent.

A "B" motor is a three-phase, squirrel-cage motor designed to withstand full-voltage starting and developing locked rotor and breakdown torques adequate for general application, and having a slip at rated load of less than 5 percent.

A "C" motor is a three-phase, squirrel-cage motor designed to withstand full-voltage starting, developing locked rotor torque for special high-

NEMA Design	Starting Torque	Starting Current	Breakdown Torque	Full-Load Slip
A	Normal	Normal	High	Low
B	Normal	Low	Medium	Low
C	High	Low	Normal	Low
D	Very High	Low	—	High

Figure 13-6: NEMA-designated motor designs

torque applications, and having a slip at rated load of less than 5 percent.

Design "D" is also a three-phase, squirrel-cage motor designed to withstand full-voltage starting, developing 275 percent locked rotor torque, and having a slip at rated load of 5 percent or more.

SELECTION OF ELECTRIC MOTORS

Each type of motor has its particular field of usefulness. Because of its simplicity, economy, and durability, the induction motor is more widely used for industrial purposes than any other type of ac motor, especially if a high-speed drive is desired.

If ac power is available, all drives requiring constant speed should use squirrel-cage induction or synchronous motors on account of their ruggedness and lower cost. Drives requiring varying speed, such as fans, blowers, or pumps may be driven by wound-rotor induction motors. However, if there are applications requiring adjustable speed or a wide range of speed control, it will probably be desirable to install dc motors on such equipment and supply them from the ac system by motor-generator sets of electronic rectifiers.

Practically all constant-speed machines may be driven by ac squirrel-cage motors because they are made with a variety of speed and torque characteristics. When large motors are required or when power supply is limited, the wound-rotor is used even for driving constant-speed machines. A wound-rotor motor, with its controller and resis-

tance, can develop full-load torque at starting with not more than full-load amperes at starting, depending on the type of motor and the starter used.

For varying-speed service, wound-rotor motors with resistance control are used for fans, blowers, and other apparatus for continuous duty, and for other intermittent duty applications. The controller and resistors must be properly chosen for the particular application.

Cost is an important consideration where more than one type of ac motor is applicable. The squirrel-cage motor is the least expensive ac motor of the three types considered and requires very little control equipment. The wound-rotor is more expensive and requires additional secondary control.

NEC REQUIREMENTS

NEC Article 430 covers application and installation of motor circuits and motor-control connections, including conductors, short-circuit and ground-fault protection, starters, disconnects, and overload protection.

NEC Article 440 contains provisions for motor-driven equipment and for branch circuits and controllers for HVAC equipment.

All motors must be installed in a location that allows adequate ventilation to cool the motors. Furthermore, the motors should be located so that maintenance, troubleshooting and repairs can be readily performed. Such work could consist of lubricating the motor's bearings, or perhaps replacing worn brushes. Testing the motor for open circuits and ground faults is also necessary from time-to-time.

When motors must be installed in locations where combustible material, dust, or similar material may be present, special precautions must be taken in selecting and installing motors.

Exposed live parts of motors operating at 50 volts or more between terminals must be guarded; that is, they must be installed in a room, enclosure, or location so as to allow access by only qualified persons (electrical maintenance personnel). If such a room, enclosure, or location is not feasible, an alternative is to elevate the motors not less than 8 feet above the floor. In all cases, adequate space must be provided around motors with exposed live parts — even when properly grounded — to allow for maintenance, troubleshooting, and repairs.

The chart in Figure 13-7 summarizes installation rules from the 1999 *NEC*.

MOTOR INSTALLATION

The best motors on the market will not operate properly if they are installed incorrectly. Therefore, all personnel involved with the installation of electric motors should thoroughly understand the proper procedures for installing the various types of motors that will be used. Furthermore, proper maintenance of each motor is essential to keep it functioning properly once it is installed.

When an electric motor is received at the job site, always refer to the manufacturer's instructions and follow them to the letter. Failure to do so could result in serious injury or fatality. In general, disconnect all power before servicing. Install and ground according to *NEC* requirements and good practices. Consult qualified personnel with any questions or services required.

Uncrating: Once the motor has been carefully uncrated, check to see if any damage has occurred during handling. Be sure that the motor shaft and armature turn freely. This time is also a good time to check to determine if the motor has been exposed to dirt, grease, grit, or excessive moisture in either shipment or storage before installation. Motors in storage should have shafts turned over once each month to redistribute grease in the bearings.

The measure of insulation resistance is a good dampness test. Clean the motor of any dirt or grit.

Lifting: Eyebolts or lifting lugs on motors are intended only for lifting the motor and factory motor-mounted standard accessories. These lifting provisions should never be used when lifting or handling the motor when the motor is attached to other equipment as a single unit.

The eyebolt lifting-capacity rating is based on a lifting alignment coincident with the eyebolt centerline. The eyebolt capacity reduces as deviation from this alignment increases.

Guards: Rotating parts such as pulleys, couplings, external fans, and unusual shaft extensions should be permanently guarded against accidental contact with clothing or body extremities.

Requirements: All motors should be installed, protected, and fused in accordance with the latest *NEC*, NEMA Standard Publication No. MG-2, and any and all local requirements.

Frames and accessories of motors should be grounded in accordance with *NEC* Article 430. For general information on grounding, refer to *NEC* Article 250.

Thermal Protector Information: A space on the motor's nameplate may or may not be stamped to indicate the following:

- The motor is thermally protected
- The motor is not thermally protected
- The motor has an overheat-protective device

Summary

Always select the appropriate motor for the application. In selecting a location for the motor, the first consideration should be given to ventilation. It should be far enough from walls or other objects to permit a free passage of air.

Applications	*NEC* Regulation	*NEC* Section
Location	Motors must be installed in areas with adequate ventilation. The must also be arranged so that sufficient work space is provided for replacement and maintenance.	430-14(a)
	Open motors must be located and protected so that sparks cannot reach combustible materials.	430-16(b)
	In locations where dust or flying material will collect on or in motors in such quantities as to seriously interfere with the ventilation or cooling of motors and thereby cause dangerous temperatures, suitable types of enclosed motors that will not overheat under the prevailing conditions must be used.	430-16
Disconnecting means	A motor disconnecting means must be within sight from the controller location (with exceptions) and disconnect both the motor and controller. The disconnect must be readily accessible and clearly indicate the *Off/On* positions (open/closed).	Article 430-102(a)
	Motor-control circuits require a disconnecting means to disconnect them from all supply sources.	430-103
	The disconnecting means must be a motor-circuit safety switch rated in horsepower or a circuit breaker.	430-109
Wiring methods	Flexible connections such as Type AC cable, "Greenfield," flexible metallic tubing, etc. are standard for motor connections.	Articles 333 and 349
Motor-control circuits	All conductors or a remote motor control circuit outside of the control device must be installed in a raceway or otherwise protected. The circuit must be wired so that an accidental ground in the control device will not start the motor.	430-73
Guards	Exposed live parts of motors and controllers operating at 50 volts or more must be guarded by installation in a room, enclosure, or location so as to allow access by only qualified persons, or elevated 8 feet or more above the floor.	Article 430-132
Motors operating over 600 volts	Special installation rules apply to motors operating at over 600 volts.	Article 430(k)
Controller grounding	Motor controllers must have their enclosures grounded.	430-144

Figure 13-7: Summary of *NEC* requirements for motor installations

The motor should never be placed in an area with a hazardous process or where flammable gasses or combustible material may be present unless it is specifically designed for this type of service.

- Drip-proof motors are intended for use where the atmosphere is relatively clean, dry, and noncorrosive. If the atmosphere is not like the preceding, then request approval of the motor for the use intended.

- Totally enclosed motors may be installed where dirt, moisture, and corrosion are present or in outdoor locations.

- Explosionproof motors are built for use in hazardous locations as indicated by the Underwriters' label on the motor.

The ambient temperature of the air surrounding the motor should not exceed 40°C or 104°F unless the motor has been especially designed for high-ambient-temperature applications. The free flow of air around the motor should not be obstructed.

After a location has been decided upon, the mounting follows: For floor mounting, motors should be provided with a firm, rigid foundation, with the plane of four mounting stud pads flat within .010 inch for a 56 to 210 frame and .015 inch for a 250 to 680 frame. This may be accomplished by shims under the motor feet.

Chapter 14
Motor Controllers

Electric motors provide one of the principal sources for driving all types of equipment and machinery. Every motor in use, however, must be controlled, if only to start and stop it, before it becomes of any value.

Motor controllers cover a wide range of types and sizes, from a simple toggle switch to a complex system with such components as relays, timers, and switches. The common function, however, is the same in any case, that is, to control some operation of an electric motor. A motor controller will include some or all of the following functions:

- Starting and stopping
- Overload protection
- Overcurrent protection
- Reversing
- Changing speed
- Jogging
- Plugging
- Sequence control
- Pilot light indication

The controller can also provide the control for auxiliary equipment such as brakes, clutches, solenoids, heaters, and signals, and may be used to control a single motor or a group of motors.

The term *motor starter* is often used and means practically the same thing as a *controller*. Strictly, a motor starter is the simplest form of controller and is capable of starting and stopping the motor and providing it with overload protection.

TYPES OF MOTOR CONTROLLERS

A large variety of motor controllers are available that will handle almost every conceivable application. However, all of them can be grouped in the following categories.

Plug-and-Receptacle

The *National Electrical Code (NEC)* defines a controller as any switch or device normally used to start and stop a motor by making and breaking the motor circuit current. The simplest form of controller allowed by the *NEC* is an attachment plug and receptacle. See Figure 14-1 on page 284. However, such an arrangement is limited to portable motors rated at $1/3$ horsepower (hp) or less.

Referring again to Figure 14-1, note that drawing (A) is a pictorial view of a portable drill motor with a cord-and-plug assembly attached. If this motor is portable and less than $1/3$ horsepower (hp), then the plug and receptacle may act as the motor's controller as permitted in *NEC* Section 430-81(c).

Drawing (B) in Figure 14-1 is the same circuit, but this time depicted in the form of a wiring diagram. Note that symbols have been used to represent the various circuit items rather than

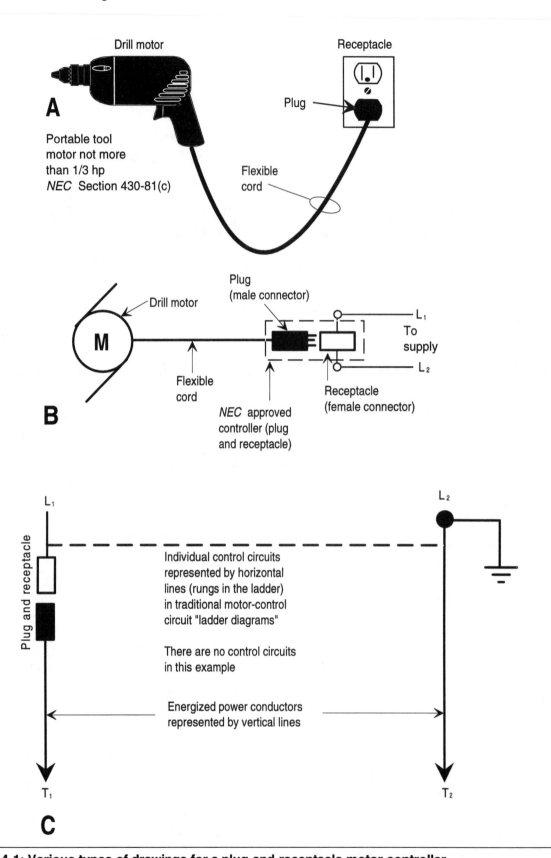

Figure 14-1: Various types of drawings for a plug-and-receptacle motor controller

actually drawing the items in life form; yet they are arranged on the basis of their physical relationship to each other. This simplifies the drawing — both from a drafter's point of view and also those who must interpret the drawing.

Another form of drawing for this same circuit is shown in (C) of Figure 14-1. This type of drawing has become known as *ladder diagram*; it's a schematic representation of the electrical circuit in question — the same as the drawing in (B). However, ladder diagrams are drawn in an "H" format, with the energized power conductors represented by vertical lines and the individual circuits represented by horizontal lines. Rather than physically representing the circuit items as in drawing (B), a ladder diagram arranges the conductors and electrical components according to their electrical function in the circuits; that is, schematically. Therefore, ladder diagrams merely represent the current paths (shown as the rungs of a ladder) to each of the controlled or energized output devices.

Where stationary motors rated at $\frac{1}{8}$ hp or less and which are normally left running (clock motors, fly fans, and the like), and are so constructed that they cannot be damaged by overload or failure to start, the branch-circuit protective device may serve as the the controller. Consequently, the branch-circuit breaker or fusible disconnect serves as both branch-circuit overcurrent protection and motor controller. Such a circuit appears in Figure 14-2.

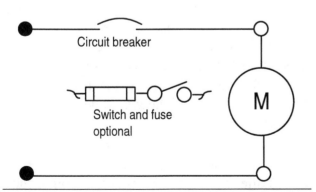

Figure 14-2: Branch-circuit protective device serving as the motor controller

Manual Starters

A manual starter is a motor controller whose contact mechanism is operated by a mechanical linkage from a toggle handle or pushbutton, which is in turn operated by hand. A thermal unit and direct-acting overload mechanism provide motor running overload protection. Basically, a manual starter is an ON-OFF switch with overload relays.

Manual starters are used mostly on small machine tools, fans and blowers, pumps, compressors, and conveyors. They have the lowest cost of all motor starters, have a simple mechanism, and provide quiet operation with no ac magnet hum. The contacts, however, remain closed and the lever stays in the ON position in the event of a power failure, causing the motor to automatically restart when the power returns. Therefore, low-voltage protection and low-voltage release are not possible with these manually operated starters. However, this action is an advantage when the starter is applied to motors that run continuously.

Fractional Horsepower Manual Starters

Fractional-horsepower manual starters are designed to control and provide overload protection for motors of 1 hp or less on 120- or 240-volt single-phase circuits. They are available in single- and two-pole versions and are operated by a toggle handle on the front. When a serious overload occurs, the thermal unit trips to open the starter contacts, disconnecting the motor from the line. The contacts cannot be reclosed until the overload relay has been reset by moving the handle to the full OFF position, after allowing about 2 minutes for the thermal unit to cool. The open-type starter will fit into a standard outlet box and can be used with a standard flush plate. The compact construction of this type of device makes it possible to mount it directly on the driven machinery and in various other places where the available space is small. Figure 14-3 on page 286 shows fractional horsepower (FHP) manual motor-starter diagrams for both 120- and 240-volt single-phase motors.

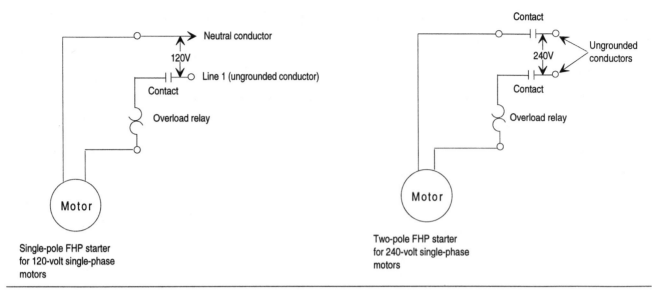

Single-pole FHP starter for 120-volt single-phase motors

Two-pole FHP starter for 240-volt single-phase motors

Figure 14-3: Wiring diagram for fractional-horsepower motor starters

Note that the single-pole FHP starter has only one contact to trip and disconnect the motor from the line; the grounded or neutral conductor is not opened when the handle is in the OFF position. This single-pole starter also has one overload relay connected in series with the ungrounded conductor.

The two-pole FHP starter has two contacts to open both phases when connected to a 240-volt circuit. When the toggle handle is in the off position, no current flows to the motor. However, only one overload relay is needed, since one will shut down the motor if the relay detects an overload and opens.

Manual Motor Starting Switches

Manual motor starting switches provide ON-OFF control of single- or three-phase ac motors where overload protection is not required or is separately provided. Two- or three-pole switches are available with ratings up to 10 hp, 600 volts, three phase. The continuous current rating is 30 amperes at 250 volts maximum and 20 amperes at 600 volts maximum. The toggle operation of the manual switch is similar to the fractional-horse-power starter, and typical applications of the switch include pumps, fans, conveyors, and other electrical machinery that have separate motor protection. They are particularly suited to switch non-motor loads, such as resistance heaters.

Integral Horsepower Manual Starters

The integral horsepower manual starter is available in two- and three-pole versions to control single-phase motors up to 5 hp and polyphase motors up to 10 hp, respectively.

Two-pole starters have one overload relay and three-pole starters usually have three overload relays. When an overload relay trips, the starter mechanism unlatches, opening the contacts to stop the motor. The contacts cannot be reclosed until the starter mechanism has been reset by pressing the STOP button or moving the handle to the RESET position, after allowing time for the thermal unit to cool.

Integral horsepower manual starters with low-voltage protection prevent automatic start-up of motors after a power loss. This is accomplished with a continuous-duty solenoid, which is energized whenever the line-side voltage is present. If

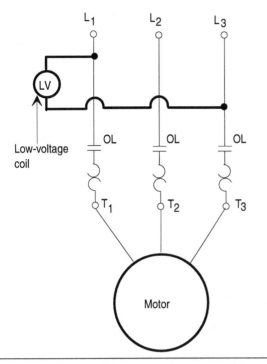

Figure 14-4: Integral hp manual starter with low-voltage protection

the line voltage is lost or disconnected, the solenoid de-energizes, opening the starter contacts. The contacts will not automatically close when the voltage is restored to the line. To close the contacts, the device must be manually reset. This manual starter will not function unless the line terminals are energized. This is a safety feature that can protect personnel or equipment from damage and is used on such equipment as conveyors, grinders, metal-working machines, mixers, woodworking, etc. Figure 14-4 shows a wiring diagram of an integral hp manual starter with low-voltage protection.

MAGNETIC CONTROLLERS

Magnetic motor controllers use electromagnetic energy for closing switches. The electromagnet consists of a coil of wire placed on an iron core. When current flows through the coil, the iron of the magnet becomes magnetized and attracts the iron bar, called the *armature*. An interruption of the current flow through the coil of wire causes the

armature to drop out due to the presence of an air gap in the magnetic circuit.

Line-voltage magnetic motor starters are electromechanical devices that provide a safe, convenient, and economic means for starting and stopping motors, and they have the advantage of being controlled remotely. The great bulk of motor controllers are of this type. Therefore, the operating principles and applications of magnet motor controllers should be fully understood.

In the construction of a magnetic controller, the armature is mechanically connected to a set of contacts so that, when the armature moves to its closed position, the contacts also close. When the coil has been energized and the armature has moved to the closed position, the controller is said to be *picked up* and the armature is seated or sealed in. Some of the magnet and armature assemblies in current use are as follows:

- *Clapper type:* In this type, the armature is hinged. As it pivots to seal in, the movable contacts close against the stationary contacts.

- *Vertical action:* The action is a straight line motion with the armature and contacts being guided so that they move in a vertical plane.

- *Horizontal action:* Both armature and contacts move in a straight line through a horizontal plane.

- *Bell crank:* A bell crank lever transforms the vertical action of the armature into a horizontal contact motion. The shock of armature pickup is not transmitted to the contacts, resulting in minimum contact bounce and longer contact life.

These four types of assemblies are shown in Figure 14-5 on page 288.

The magnetic circuit of a controller consists of the magnet assembly, the coil, and the armature. It is so named from a comparison with an electrical circuit. The coil and the current flowing in it causes

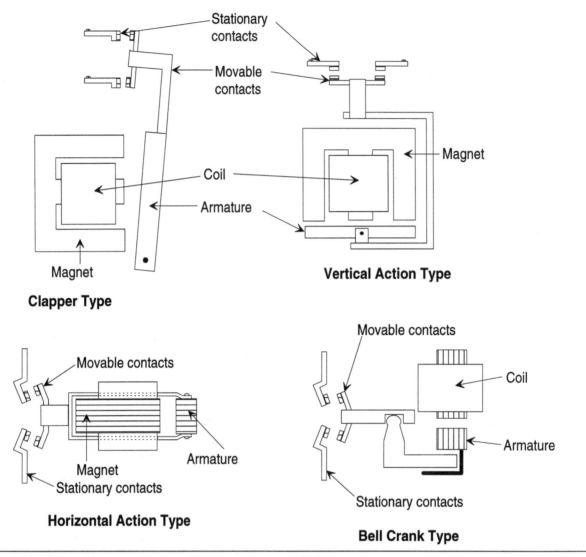

Figure 14-5: Several types of magnetic-armature assemblies

magnetic flux to be set up through the iron in a similar manner to a voltage causing current to flow through a system of conductors. The changing magnetic flux produced by alternating currents results in a temperature rise in the magnetic circuit. The heating effect is reduced by laminating the magnet assembly and armature by placing a coil of many turns of wire around a soft iron core. The magnetic flux set up by the energized coil tends to be concentrated; therefore, the magnetic field effect is strengthened. Since the iron core is the path of least resistance to the flow of the magnetic lines

of force, magnetic attraction will concentrate according to the shape of the magnet.

The magnetic assembly is the stationary part of the magnetic circuit. The coil is supported by and surrounds part of the magnet assembly in order to induce magnetic flux into the magnetic circuit.

The armature is the moving part of the magnetic circuit. When it has been attracted into its sealed-in position, it completes the magnetic circuit. To provide maximum pull and to help ensure quietness, the faces of the armature and the magnetic assembly are ground to a very close tolerance.

When a controller's armature has sealed in, it is held closely against the magnet assembly. However, a small gap is always deliberately left in the iron circuit. When the coil becomes de-energized, some magnetic flux (residual magnetism) always remains, and if it were not for the gap in the iron circuit, the residual magnetism might be sufficient to hold the armature in the sealed-in position. See Figure 14-6.

The shaded-pole principle is used to provide a time delay in the decay of flux in dc coils, but it is used more frequently to prevent a chatter and wear in the moving parts of ac magnets. A shading coil is a single turn of conducting material mounted in the face of the magnet assembly or armature. The alternating main magnetic flux induces currents in the shading coil, and these currents set up auxiliary magnetic flux that is out of phase from the pull due to the main flux, and this keeps the armature sealed-in when the main flux falls to zero (which occurs 120 times per second with 60-cycle ac). Without the shading coil, the armature would tend to open each time the main flux goes through zero. Excessive noise, wear on magnet faces, and heat would result. A magnet assembly and armature showing shading coils is shown in Figure 14-7.

Magnetic Coils

The magnetic coil used in motor controllers has many turns of insulated copper wire wound on a spool. Most coils are protected by an epoxy mold-

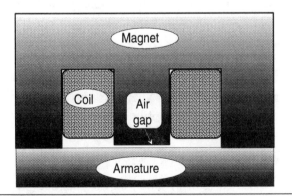

Figure 14-6: A small air gap is always deliberately left in the iron-core circuit

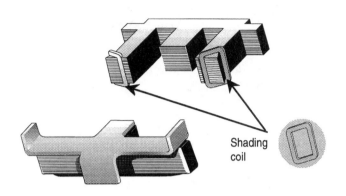

Figure 14-7: Magnet assembly and armature along with shading coils

ing which makes them very resistant to mechanical damage.

When the controller is in the open position there is a large air gap (not to be confused with the built-in gap discussed previously) in the magnet circuit; this is when the armature is at its furthest distance from the magnet. The impedance of the coil is relatively low, due to the air gap, so that when the coil is energized, it draws a fairly high current. As the armature moves closer to the magnet assembly, the air gap is progressively reduced, and with it, the coil current, until the armature has sealed in. The final current is referred to as the sealed current. The inrush current is approximately 6 to 10 times the sealed current. The ratio varies with individual designs. After the controller has been energized for some time, the coil will become hot. This will cause the coil current to fall to approximately 80 percent of its value when cold.

AC magnetic coils should never be connected in series. If one device were to seal in ahead of the other, the increased circuit impedance will reduce the coil current so that the "slow" device will not pick up or, having picked up, will not seal. Consequently, ac coils are always connected in parallel.

Magnet coil data is usually given in volt-amperes (VA). For example, given a magnetic starter whose coils are rated at 600 VA inrush and 60 VA sealed, the inrush current of a 120-volt coil is 600/120 or 5 amperes. The same starter with a 480-volt coil will only draw 600/480 or 1.25 amperes inrush and 60/480 or .125 amperes sealed.

Pick-up Voltage: The minimum voltage which will cause the armature to start to move is called the pick-up voltage.

Sealed-in Voltage: The seal-in voltage is the minimum control voltage required to cause the armature to seat against the pole faces of the magnet. On devices using a vertical action magnet and armature, the seal-in voltage is higher than the pick-up voltage to provide additional magnetic pull to insure good contact pressure.

Control devices using the bell-crank armature and magnet arrangement are unique in that they have different force characteristics. Devices using this operating principle are designed to have a lower seal-in voltage than pick-up voltage. Contact life is extended, and contact damage under abnormal voltage conditions is reduced, for if the voltage is sufficient to pick-up, it is also high enough to seat the armature.

If the control voltage is reduced sufficiently, the controller will open. The voltage at which this happens is called the *drop-out* voltage. It is somewhat lower than the seal-in voltage.

Voltage Variation

NEMA standards require that the magnetic device operate properly at varying control voltages from a high of 110 percent to a low of 85 percent of rated coil voltage. This range, established by coil design, insures that the coil will withstand given temperature rises at voltages up to 10 percent over rated voltage, and that the armature will pick up and seal in, even though the voltage may drop to 15 percent under the nominal rating.

If the voltage applied to the coil is too high, the coil will draw more than its designed current. Excessive heat will be produced and will cause early failure of the coil insulation. The magnetic

pull will be too high, which will cause the armature to slam home with excessive force. The magnet faces will wear rapidly, leading to a shortened life for the controller. In addition, contact bounce may be excessive, resulting in reduced contact life.

Low control voltage produces low coil currents and reduced magnetic pull. On devices with vertical action assemblies, if the voltage is greater than pick-up voltage, but less than seal-in voltage, the controller may pick up but will not seal. With this condition, the coil current will not fall to the sealed value. As the coil is not designed to carry continuously a current greater than its sealed current, it will quickly get very hot and burn out. The armature will also chatter. In addition to the noise, wear on the magnet faces result.

In both vertical action and bell-crank construction, if the armature does not seal, the contacts will not close with adequate pressure. Excessive heat, with arcing and possible welding of the contacts, will occur as the controller attempts to carry current with insufficient contact pressure.

AC Hum

All ac devices which incorporate a magnetic effect produce a characteristic hum. This hum or noise is due mainly to the changing magnetic pull (as the flux changes) inducing mechanical vibrations. Contactors, starters and relays could become excessively noisy as a result of some of the following operating conditions:

- Broken shading coil.
- Operating voltage too low.
- Misalignment between the armature and magnet assembly — the armature is then unable to seat properly.
- Wrong coil.
- Dirt, rust, filings, etc. on the magnet faces — the armature is unable to seal in completely.

- Jamming or binding of moving parts so that full travel of the armature is prevented.

- Incorrect mounting of the controller, as on a thin piece of plywood fastened to a wall; such mounting may cause a "sounding board" effect.

POWER CIRCUITS IN MOTOR STARTERS

The power circuit of a starter includes the stationary and movable contacts, and the thermal unit or heater portion of the overload relay assembly. The number of contacts (or "poles") is determined by the electrical service. In a three-phase, three-wire system, for example, a three-pole starter is required. See Figure 14-8.

To be suitable for a given motor application, the magnetic starter selected should equal or exceed the motor horsepower and full-load current rat-

ings. For example, let's assume that we want to select a motor starter for a 50-hp motor to be supplied by a 240-volt, three-phase service, and the full-load current of the motor is 125 amperes. Refer to manufacturers' tables, available from your local electrical supplier. It can be seen that a NEMA Size 4 starter would be required for normal motor duty. If the motor were to be used for jogging or plugging duty, a NEMA Size 5 starter should be chosen.

For three-phase motors having locked-rotor kVA per horsepower in excess of that for the motor code letters in the table in Figure 14-9 on page 292, do not apply the controller at its maximum rating without consulting the manufacturer. In most cases, the next higher horsepower rated controller should be used.

Power circuit contacts handle the motor load. The ability of the contacts to carry the full-load current without exceeding a rated temperature rise, and their isolation from adjacent contacts, corre-

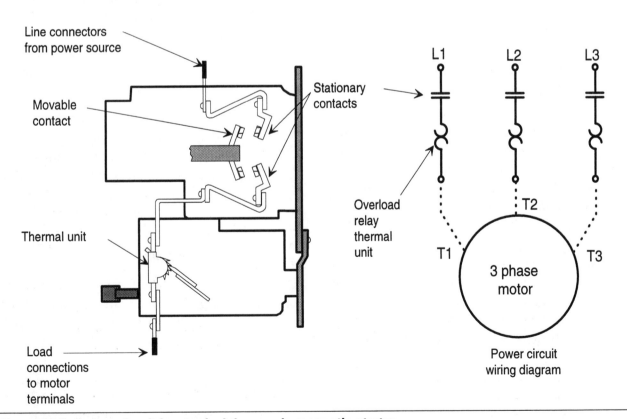

Figure 14-8: Power circuit in a typical three-pole magnetic starter

Controller HP Rating	Maximum Allowable Motor Code Letter
1½	L
3 - 5	K
7½ and above	H

Figure 14-9: Using motor code letters to size motor controllers

sponds to NEMA Standards established to categorize the NEMA Size of the starter. The starter must also be capable of interrupting the motor circuit under locked rotor current conditions.

OVERLOAD PROTECTION

Overload protection for an electric motor is necessary to prevent burnout and to ensure maximum operating life. Electric motors will, if permitted, operate at an output of more than rated capacity. Conditions of motor overload may be caused by an overload on driven machinery, by a low line voltage, or by an open line in a polyphase system, which results in single-phase operation. Under any condition of overload, a motor draws excessive current that causes overheating. Since motor winding insulation deteriorates when subjected to overheating, there are established limits on motor operating temperatures. To protect a motor from overheating, overload relays are employed on a motor control to limit the amount of current drawn. This is overload protection, or running protection.

The ideal overload protection for a motor is an element with current-sensing properties very similar to the heating curve of the motor, which would act to open the motor circuit when full-load current is exceeded. The operation of the protective device should be such that the motor is allowed to carry harmless overloads, but is quickly removed from the line when an overload has persisted too long.

Fuses are not designed to provide overload protection. Their basic function is to protect against short circuits (overcurrent protection). Motors draw a high inrush current when starting and con-

ventional single-element fuses have no way of distinguishing between this temporary and harmless inrush current and a damaging overload. Such fuses, chosen on the basis of motor full-load current, would blow every time the motor is started. On the other hand, if a fuse were chosen large enough to pass the starting or inrush current, it would not protect the motor against small, harmful overloads that might occur later.

Dual-element or time-delay fuses can provide motor overload protection, but suffer the disadvantage of being nonrenewable so they must be replaced.

The overload relay is the heart of motor protection. It has inverse trip-time characteristics, permitting it to hold in during the accelerating period (when inrush current is drawn), yet providing protection on small overloads above the full-load current when the motor is running. Unlike dual-element fuses, overload relays are renewable and can withstand repeated trip and reset cycles without need of replacement. They cannot, however, take the place of overcurrent protective equipment.

The overload relay consists of a current-sensing unit connected in the line to the motor, plus a mechanism, actuated by the sensing unit, that serves to directly or indirectly break the circuit. In a manual starter, an overload trips a mechanical latch and causes the starter contacts to open and disconnect the motor from the line. In magnetic starters, an overload opens a set of contacts within the overload relay itself. These contacts are wired in series with the starter coil in the control circuit of the magnetic starter. Breaking the coil circuit causes the starter contacts to open, disconnecting the motor from the line.

Overload relays can be classified as being either thermal or magnetic. Magnetic overload relays react only to current excesses and are not affected by temperature. As the name implies, thermal overload relays rely on the rising temperatures caused by the overload current to trip the overload mechanism. Thermal overload relays can be further subdivided into two types, melting alloy and bimetallic.

Melting Alloy Thermal Overload Relays

The melting alloy assembly of the heater element overload relay and solder pot is shown in Figure 14-10. Excessive overload motor current passes through the heater element, thereby melting an eutectic alloy solder pot. The ratchet wheel will then be allowed to turn in the molten pool, and a tripping action of the starter control circuit results, stopping the motor. A cooling off period is required to allow the solder pot to "freeze" before the overload relay assembly may be reset and motor service restored.

Melting alloy thermal units are interchangeable and of a one-piece construction, which ensures a constant relationship between the heater element and solder pot and allows factory calibration, mak-

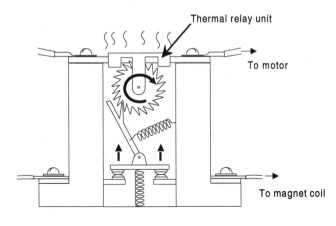

Figure 14-10: Operating characteristics of a melting-alloy overload relay

ing them virtually tamper-proof in the field. These important features are not possible with any other type of overload relay construction. A wide selection of these interchangeable thermal units is available to give exact overload protection of any full-load current to a motor.

Bimetallic Thermal Overload Relays

Bimetallic overload relays are designed specifically for two general types of application: the automatic reset feature is of decided advantage when devices are mounted in locations not easily accessible for manual operation and, second, these relays can easily be adjusted to trip within a range of 85 to 115 percent of the nominal trip rating of the heater unit. This feature is useful when the recommended heater size might result in unnecessary tripping, while the next larger size would not give adequate protection. Ambient temperatures affect overload relays operating on the principle of heat.

SELECTING OVERLOAD RELAYS

When selecting thermal overload relays, the following must be considered:

- Motor full-load current
- Type of motor
- Difference in ambient temperature between motor and controller

Motors of the same horsepower and speed do not all have the same full-load current, and the motor nameplate must always be checked to obtain the full-load amperes for a particular motor. Do not use a published table. Thermal unit selection tables are published on the basis of continuous-duty motors, with 1.15 service factor, operating under normal conditions. The tables are shown in the catalogs of manufacturers and also appear on the inside of the door or cover of the motor controller. These selections will properly protect the motor and allow the motor to develop its full horsepower, allowing for the service factor, if the ambient temperature is the same at the motor as at the controller. If the temperatures are not the same, or if the motor service factor is less than 1.15, a special procedure is required to select the proper thermal unit. Standard overload relay contacts are closed under normal conditions and open when the relay trips. An alarm signal is sometimes required to indicate when a motor has stopped due to an overload trip. Also, with some machines, particularly those associated with continuous processing, it may be required to signal an overload condition, rather than have the motor and process stop automatically. This is done by fitting the overload relay with a set of contacts that close when the relay trips, thus completing the alarm circuit. These contacts are appropriately called alarm contacts.

A magnetic overload relay has a movable magnetic core inside a coil that carries the motor current. The flux set up inside the coil pulls the core upward. When the core rises far enough, it trips a set of contacts on the top of the relay. The movement of the core is slowed by a piston working in an oil-filled dashpot mounted below the coil. This produces an inverse-time characteristic. The effective tripping current is adjusted by moving the core on a threaded rod. The tripping time is varied by uncovering oil bypass holes in the piston. Because of the time and current adjustments, the magnetic overload relay is sometimes used to protect motors having long accelerating times or unusual duty cycles.

Summary

The first of the motor-control arrangements is a plug and receptacle; next comes a fusible disconnect or circuit breaker, and then the manual and fractional-horsepower starters. The magnetic-contactor controller, however, is the type most used in electrical installations. This latter type of controller opens or closes circuits automatically when their control coil is energized. The contactors may be normally open or normally closed.

Protective devices such as overload relays, low-voltage protection devices and low-voltage release devices are an important part of a motor controller. An overload relay will open the contactors in motor circuits when current is too high; a low-voltage protective device will prevent the motor from starting as long as the full-rated voltage is not available, and manual restarting of the motor is necessary after the low-voltage protective device has operated; a low-voltage release device will disconnect the motor during a voltage dip, but the motor will start automatically when the normal voltage returns.

Controllers also contain braking arrangements, accelerators, and reversing switches which reverse the rotation of the motor.

The intent of this chapter is to familiarize electrical workers with terms and concepts which are fundamental to an understanding of motor-control equipment and its applications. However, the subject of motor controllers could fill volumes and we have but touched upon the subject here.

Chapter 15
Special Occupancies

National Electrical Code (NEC) Articles 500 through 504 cover the requirements of electrical equipment and wiring for all voltages in locations where fire or explosion hazards may exist due to flammable gases or vapor, flammable liquids, combustible dust, or ignitable fibers or flyings. Locations are classified depending on the properties of the flammable vapors, liquids, gases or combustible dusts or fibers that may be present, as well as the likelihood that a flammable or combustible concentration or quantity is present.

Any area in which the atmosphere or a material in the area is such that the arcing of operating electrical contacts, components, and equipment may cause an explosion or fire is considered as a hazardous location. In all such cases, explosionproof equipment, raceways, and fittings are used to provide an explosionproof wiring system.

Hazardous locations have been classified in the *NEC* into certain class locations. Various atmospheric groups have been established on the basis of the explosive character of the atmosphere for the testing and approval of equipment for use in the various groups. However, it must be understood that considerable skill and judgment must be applied when deciding to what degree an area contains hazardous concentrations of vapors, combustible dusts or easily ignitable fibers and flyings. Furthermore, many factors — such as temperature, barometric pressure, quantity of release, humidity, ventilation, distance from the vapor source,

and the like — must be considered. When information on all factors concerned is properly evaluated, a consistent classification for the selection and location of electrical equipment can be developed.

Class I Locations

Class I atmospheric hazards are divided into two Divisions (1 and 2) and also into four groups (A, B, C, and D). The Divisions are summarized in the paragraphs to follow while the groups are summarized in the table in Figure 15-1.

Those locations in which flammable gases or vapors may be present in the air in quantities sufficient to produce explosive or ignitable mixtures are classified as Class I locations. If these gases or vapors are present under normal operations, under frequent repair or maintenance operations, or where breakdown or faulty operation of process equipment might also cause simultaneous failure of electrical equipment, the area is designated as Class I, Division 1. Examples of such locations are interiors of paint spray booths where volatile, flammable solvents are used, inadequately ventilated pump rooms where flammable gas is pumped, anesthetizing locations of hospitals (to a height of 5 feet above floor level), and drying rooms for the evaporation of flammable solvents.

Class I, Division 2 covers locations where flammable gases, vapors or volatile flammable gases, vapors or volatile liquids are handled either in a

Class	Division	Group	Typical Atmosphere/Ignition Temps.	Devices Covered	Temperature Measured	Limiting Value
I	1	A	Acetylene (305C, 581F)	All electrical devices and wiring	Maximum external temperature in 40C ambient	See Sect. 500-3 of NEC
Gases, vapors	Normally hazardous	B	1,3-Butadiene (420C, 788F)			
			Ethylene Oxide (429C, 804F)			
			Hydrogen (520C, 968F)			
			Manufactured Gas (containing more than 30% hydrogen by volume)			
			Propylene Oxide (449C, 840F)			
		C	Acetaldehyde (175C, 347F)			
			Diethyl Ether (160C, 320F)			
			Ethylene (450C, 842F)			
			Unsymmetrical Dimethyl Hydrazine (UDMH) (249C, 480F)			
		D	Acetone (465C, 869F)			
			Acrylonitrile (481C, 898F)			
			Ammonia (498C, 928F)			
			Benzene (498C, 928F)			
			Butane (288C, 550F)			
			1-Butanol (343C, 650F)			
			2-Butanol (405C, 761F)			
			n-Butyl Acetate (421C, 790F)			
			Cyclopropane (503C, 938F)			
			Ethane (472C, 882F)			
			Ethanol (363C, 685F)			
			Ethyl Acetate (427C, 800F)			
			Ethylene Dichloride (413C, 775F)			
			Gasoline (280-471C, 536-880F)			
			Heptane (204C, 399F)			
			Hexane (225C, 437F)			
			Isoamyl Alcohol (350C, 662F)			
			Isoprene (220C, 428F)			
			Methane (630C, 999F)			
			Methanol (385C, 725F)			
			Methyl Ethyl Ketone (404C, 759F)			
			Methyl Isobutyl Ketone (449C, 840F)			
			2-Methyl-1-Propanol (416C, 780F)			
			2-Methyl-2-Propanol (478C, 892F)			
			Naphtha (petroleum) (288C, 550F)			
			Octane (206C, 403F)			
			Pentane (243C, 470F)			
			1-Pentanol (300C, 572F)			

Figure 15-1: Summary of hazardous atmospheres

Class	Division	Group	Typical Atmosphere/Ignition Temps.	Devices Covered	Temperature Measured	Limiting Value
1	1	D	Propane (450C, 842F) 1-Propanol (413C, 775F) 2-Propanol (399C, 750F) Propylene (455C, 851F) Styrene (490C, 914F) Toluene (480C, 896F) Vinyl Acetate (402C, 756F) Vinyl Chloride (472C, 882F) Xylenes (464-529C, 867-984F)			
I	2	A	Same as Division 1	Lamps, resistors, coils, etc., other than arcing devices. (see Div. 1)	Max. internal or external temp. not to exceed the ignition temperature in degrees Celsius (°C) of the gas or vapor involved	See Sect. 500-3
Gases, vapors	Not normally hazardous	B	Same as Division 1			
		C	Same as Division 1			
		D	Same as Division 1			
II Combustible dusts	1 Normally hazardous	E	Atmospheres containing combustible metal dusts regardless of resistivity, or other combustible dusts of similarly hazardous characteristics having resistivity of less than 10^2 ohm-centimeter	Devices not subject to overloads (switches, meters)	Max. external temp. in 40C ambient with a dust blanket	Shall be less than ignition temperature of dust but not more than: No overload: E—200C (392F) F—200C (392F) G—165C (329F) Possible overload in operation: Normal E—200C (392F) F—150C (302F) G—120C (248F) Abnormal E—200C (392F) F—200C (392F) G—165C (329F)
		F	Atmospheres containing carbonaceous dusts having resistivity between 10^2 and 10^8 ohm-centimeter			
		G	Atmospheres containing combustible dusts having resistivity of 10^8 ohm-centimeter or greater			
	2 Not normally hazardous	F	Atmospheres containing carbonaceous dusts having resistivity of 10^5 ohm-centimeter or greater	Lighting fixtures	Max. external temp under conditions of use	Same as Division 1
		G	Same as Division 1			
III Easily ignitible fibers and flyings	1&2			Lighting fixtures	Max. external temp. under conditions of use	165C (329F)

Figure 15-1: Summary of hazardous atmospheres (*cont.*)

297

closed system, or confined within suitable enclosures, or where hazardous concentrations are normally prevented by positive mechanical ventilation. Areas adjacent to Division 1 locations, into which gases might occasionally flow, would also belong in Division 2.

Class II Locations

Class II locations are those that are hazardous because of the presence of combustible dust. Class II, Division 1 locations are areas where combustible dust, under normal operating conditions, may be present in the air in quantities sufficient to produce explosive or ignitable mixtures; examples are working areas of grain-handling and storage plants and rooms containing grinders or pulverizers. Class II, Division 2 locations are areas where dangerous concentrations of suspended dust are not likely, but where dust accumulations might form.

Besides the two Divisions (1 and 2), Class II atmospheric hazards cover three groups of combustible dusts. The groupings are based on the resistivity of the dust. Group E is always Division 1. Group F, depending on the resistivity, and Group G may be either Division 1 or 2. Since the *NEC* is considered the definitive classification tool and contains explanatory data about hazardous atmospheres, refer to *NEC* Section 500-8 for exact definitions of Class II, Divisions 1 and 2. Again, these groups are summarized in the table in Figure 15-1.

Class III Locations

These locations are those areas that are hazardous because of the presence of easily ignitable fibers or flyings, but such fibers and flyings are not likely to be in suspension in the air in these locations in quantities sufficient to produce ignitable mixtures. Such locations usually include some parts of rayon, cotton, and textile mills, clothing manufacturing plants, and woodworking plants.

In Class I and Class II locations the hazardous materials are further divided into groups; that is, Groups A, B, C, D in Class I and Groups E, F, and G in Class II. For a more complete listing of flammable liquids, gases and solids, see *Classification of Gases, Vapors and Dusts for Electrical Equipment in Hazardous (Classified) Locations*, NFPA Publication No. 497M.

Once the class of an area is determined, the conditions under which the hazardous material may be present determines the division. In Class I and Class II, Division 1 locations, the hazardous gas or dust may be present in the air under normal operating conditions in dangerous concentrations. In Division 2 locations, the hazardous material is not normally in the air, but it might be released if there is an accident or if there is faulty operation of equipment.

PREVENTION OF EXTERNAL IGNITION/EXPLOSION

The main purpose of using explosionproof fittings and wiring methods in hazardous areas is to prevent ignition of flammable liquids or gases and to prevent an explosion.

Sources Of Ignition

In certain atmospheric conditions when flammable gases or combustible dusts are mixed in the proper proportion with air, any source of energy is all that is needed to touch off an explosion.

One prime source of energy is electricity. Equipment such as switches, circuit breakers, motor starters, pushbutton stations, or plugs and receptacles, can produce arcs or sparks in normal operation when contacts are opened and closed. This could easily cause ignition.

Other hazards are devices that produce heat, such as lighting fixtures and motors. Here surface temperatures may exceed the safe limits of many flammable atmospheres.

Finally, many parts of the electrical system can become potential sources of ignition in the event of insulation failure. This group would include wiring (particularly splices in the wiring), transformers, impedance coils, solenoids, and other low-temperature devices without make-or-break contacts.

Non-electrical hazards such as sparking metal can also easily cause ignition. A hammer, file or other tool that is dropped on masonry or on a ferrous surface can cause a hazard unless the tool is made of non-sparking material. For this reason, portable electrical equipment is usually made from aluminum or other material that will not produce sparks if the equipment is dropped.

Electrical safety, therefore, is of crucial importance. The electrical installation must prevent accidental ignition of flammable liquids, vapors and dusts released to the atmosphere. In addition, since much of this equipment is used outdoors or in corrosive atmospheres, the material and finish must be such that maintenance costs and shutdowns are minimized.

Combustion Principles

Three basic conditions must be satisfied for a fire or explosion to occur:

- A flammable liquid, vapor or combustible dust must be present in sufficient quantity.

- The flammable liquid, vapor or combustible dust must be mixed with air or oxygen in the proportions required to produce an explosive mixture.

- A source of energy must be applied to the explosive mixture.

In applying these principles, the quantity of the flammable liquid or vapor that may be liberated and its physical characteristics must be recognized.

Vapors from flammable liquids also have a natural tendency to disperse into the atmosphere, and rapidly become diluted to concentrations below the lower explosion limit particularly when there is natural or mechanical ventilation.

CAUTION: *The possibility that the gas concentration may be above the upper explosion limit does not afford any degree of safety, as the concentration must first pass through the explosive range to reach the upper explosion limit.*

EXPLOSIONPROOF EQUIPMENT

Each area that contains gases or dusts that are considered hazardous must be carefully evaluated to make certain the correct electrical equipment is selected. Many hazardous atmospheres are Class I, Group D, or Class II, Group G. However, certain areas may involve other groups, particularly Class I, Groups B and C. Conformity with the *NEC* requires the use of fittings and enclosures approved for the specific hazardous gas or dust involved.

The wide assortment of explosionproof equipment now available makes it possible to provide adequate electrical installations under any of the various hazardous conditions. However, the electrician must be thoroughly familiar with all *NEC* requirements and know what fittings are available, how to install them properly, and where and when to use the various fittings.

For example, some workers are under the false belief that a fitting rated for Class I, Division 1 can be used under any hazardous conditions. However, remember the groups! A fitting rated for, say, Class I, Division 1, Group C cannot be used in areas classified as Groups A or B. On the other hand, fittings rated for use in Group A may be used for any group beneath A; fittings rated for use in Class I, Division 1, Group B can be used in areas rated as Group B areas or below, but not vice versa.

Explosionproof fittings (Figure 15-2) are rated for both classification and groups. All parts of these fittings, including covers, are rated accordingly. Therefore, if a Class I, Division 1, Group A fitting is required, a Group B (or below) fitting cover must not be used. The cover itself must be rated for Group A locations. Consequently, when working on electrical systems in hazardous locations, always make certain that fittings and their related components match the condition at hand.

Intrinsically Safe Equipment

Intrinsically safe equipment is equipment and wiring that are incapable of releasing sufficient electrical energy under normal or abnormal conditions to cause ignition of a specific hazardous atmospheric mixture in its most easily ignited concentration.

The use of intrinsically safe equipment is primarily limited to process control instrumentation, since these electrical systems lend themselves to the low energy requirements.

Installation rules for intrinsically safe equipment are covered in *NEC* Article 504. In general, intrinsically safe equipment and its associated wiring must be installed so they are positively separated from the non-intrinsically safe circuits because induced voltages could defeat the concept of intrinsically safe circuits. Underwriters' Laboratories Inc. and Factory Mutual list several devices in this category.

EXPLOSIONPROOF CONDUIT AND FITTINGS

In hazardous locations where threaded metal conduit is required, the conduit must be threaded with a standard conduit cutting die that provides 3/4-inch taper per foot. The conduit should be made up wrench-tight in order to minimize sparking in the event fault current flows through the raceway system (*NEC* Section 500-3(d)). Where it is imprac-

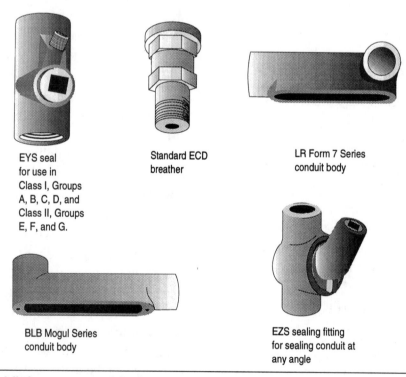

EYS seal
for use in
Class I, Groups
A, B, C, D, and
Class II, Groups
E, F, and G.

Standard ECD
breather

LR Form 7 Series
conduit body

BLB Mogul Series
conduit body

EZS sealing fitting
for sealing conduit at
any angle

Figure 15-2: Typical fittings approved for hazardous areas

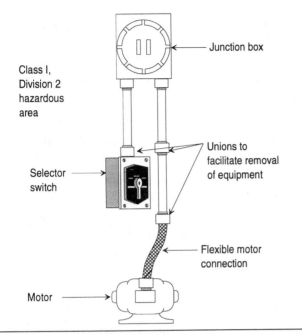

Figure 15-3: Explosionproof flexible connectors are frequently used for motor terminations

tical to make a threaded joint tight, a bonding jumper shall be used. All boxes, fittings, and joints shall be threaded for connection to the conduit system and shall be an approved, explosionproof type (Figure 15-2). Threaded joints must be made up with at least five threads fully engaged. Where it becomes necessary to employ flexible connectors at motor or fixture terminals (Figure 15-3), flexible fittings approved for the particular class location shall be used.

Seals And Drains

Seal-off fittings (Figure 15-4) are required in conduit systems to prevent the passage of gases, vapors, or flames from one portion of the electrical installation to another at atmospheric pressure and normal ambient temperatures. Furthermore, seal-offs (seals) limit explosions to the sealed-off enclosure and prevent precompression of "pressure

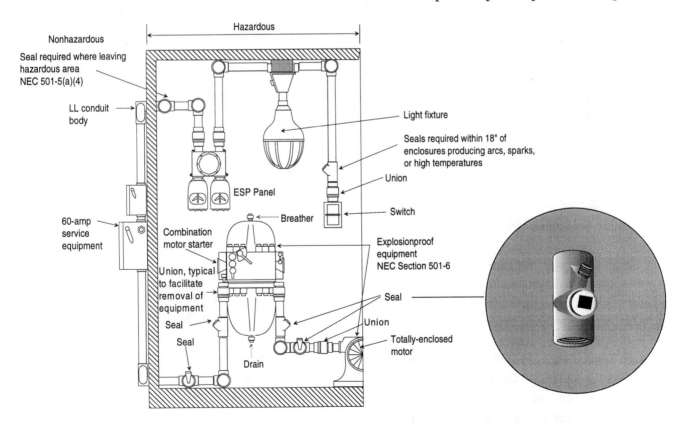

Figure 15-4: Seals must be installed at various locations in Class I, Division 1 locations

301

piling" in conduit systems. For Class I, Division 1 locations, the *NEC* (Section 501-5(a)(1)) states:

- In each conduit run entering an enclosure for switches, circuit breakers, fuses, relays, resistors, or other apparatus which may produce arcs, sparks, or high temperatures, seals shall be installed within 18 inches from such enclosures. Explosionproof unions, couplings, reducers, elbows, capped elbows and conduit bodies similar to "L," "T," and "Cross" types shall be the only enclosures or fittings permitted between the sealing fitting and the enclosure. The conduit bodies shall not be larger than the largest trade size of the conduits.

There is, however, one exception to this rule: Conduits $1\frac{1}{2}$ inches and smaller are not required to be sealed if the current-interrupting contacts are either enclosed within a chamber hermetically sealed against the entrance of gases or vapors, or immersed in oil in accordance with *NEC* Section 501-6.

Seals are also required in Class II locations under the following condition (*NEC* Section 502-5):

- Where a raceway provides communication between an enclosure that is required to be dust-ignitionproof and one that is not. . .

A permanent and effective seal is one method of preventing the entrance of dust into the dust-ignitionproof enclosure through the raceway. A horizontal raceway, not less than 10 feet long, is another approved method, as is a vertical raceway not less than 5 feet long and extending downward from the dust-ignitionproof enclosure.

Where a raceway provides communication between an enclosure that is required to be dust-ignitionproof and an enclosure in an unclassified location, seals are not required.

Where sealing fittings are used, all must be accessible.

While not an *NEC* requirement, many electrical designers and workers consider it good practice to sectionalize long conduit runs by inserting seals not more than 50 to 100 feet apart, depending on the conduit size, to minimize the effects of "pressure piling."

In general, seals are installed at the same time as the conduit system. However, the conductors are installed after the raceway system is complete and *prior* to packing and sealing the seal-offs.

Drains

In humid atmospheres or in wet locations, where it is likely that water can gain entrance to the interiors of enclosures or raceways, the raceways should be inclined so that water will not collect in enclosures or on seals but will be led to low points where it may pass out through integral drains.

Frequently, the arrangement of raceway runs makes this method impractical — if not impossible. In such instances, special drain/seal fittings should be used, such as Crouse-Hinds Type EZDs as shown in Figure 15-5. These fittings prevent harmful accumulations of water above the seal and meets the requirements of *NEC* Section 501-5(f). See Figure 15-6 for a typical hazardous area.

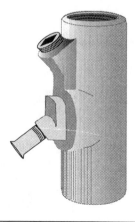

Figure 15-5: Typical drain seal

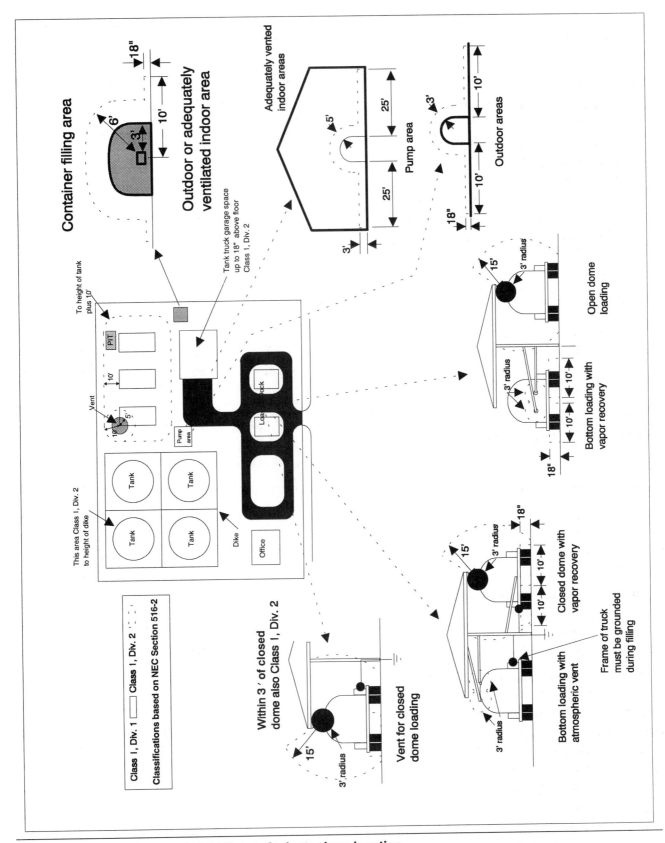

Figure 15-6: Floor plan and elevations of a hazardous location

In locations which usually are considered dry, surprising amounts of water frequently collect in conduit systems. No conduit system is airtight; therefore, it may "breathe." Alternate increases and decreases in temperature and/or barometric pressure due to weather changes or due to the nature of the process carried on in the location where the conduit is installed will cause "breathing."

Outside air is drawn into the conduit system when it "breathes in." If this air carries sufficient moisture, it will be condensed within the system when the temperature decreases and chills this air. The internal conditions being unfavorable to evaporation, the resultant water accumulation will remain and be added to by repetitions of the breathing cycle.

In view of this likelihood, it is good practice to insure against such water accumulations and probable subsequent insulation failures by installing drain/seal fittings with drain covers or fittings with inspection covers even though conditions prevailing at the time of planning or installing do not indicate their need.

Selection Of Seals And Drains

The primary considerations for selecting the proper sealing fittings are as follows:

- Select the proper sealing fitting for the hazardous vapor involved; that is, Class I, Groups A, B, C, or D.

- Select a sealing fitting for the proper use in respect to mounting position. This is particularly critical when the conduit runs between hazardous and nonhazardous areas. Improper positioning of a seal may permit hazardous gases or vapors to enter the system beyond the seal, and permit them to escape into another portion of the hazardous area, or to enter a nonhazardous area. Some seals are designed to be mounted in any position;

others are restricted to horizontal or vertical mounting.

- Install the seals on the proper side of the partition or wall as recommended by the manufacturer.

- Installation of seals should be made *only* by trained personnel in strict compliance with the instruction sheets furnished with the seals and sealing compound.

- It should be noted that *NEC* Section 501-5(c)(4) prohibits splices or taps in sealing fittings.

- Sealing fittings are listed by UL for use in Class I hazardous locations with an approved compound only. This compound, when properly mixed and poured, hardens into a dense, strong mass which is insoluble in water, is not attacked by chemicals, and is not softened by heat. It will withstand, with ample safety factor, pressure of the exploding trapped gases or vapor.

- Conductors sealed in the compound may be approved thermoplastic or rubber insulated type. Both may or may not be lead covered.

Types Of Seals And Fittings

Certain seals, such as Crouse-Hinds EYS seals, are designed for use in vertical or nearly vertical conduit in sizes for $\frac{1}{2}$ through 1 inch. Other styles are available in sizes $\frac{1}{2}$ inch through 6 inches for use in vertical or horizontal conduit. In horizontal runs, these are limited to face-up openings. This, and other types of seals are shown in Figure 15-7.

Seals ranging in sizes from $1\frac{1}{4}$ inch through 6 inches have extra large work openings, and separate filling holes, so that fiber dams are easy to make. However, the overall diameter of these fittings is scarcely greater than that of unions of

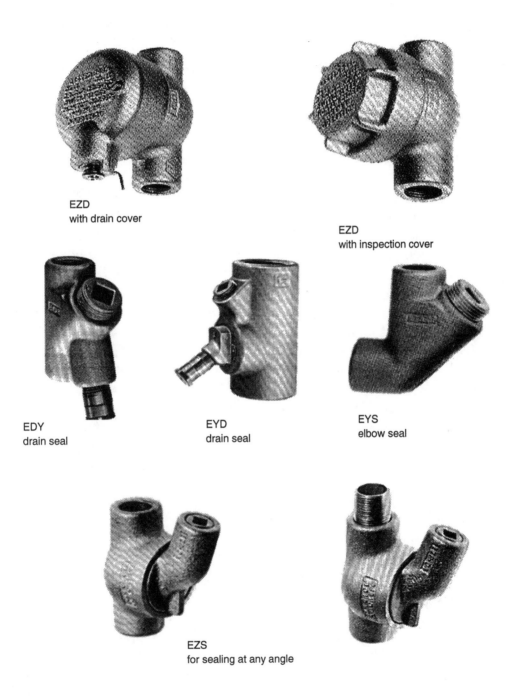

EZD
with drain cover

EZD
with inspection cover

EDY
drain seal

EYD
drain seal

EYS
elbow seal

EZS
for sealing at any angle

Figure 15-7: Various types of seals and related components

corresponding sizes, permitting close conduit spacing.

Crouse-Hinds EZS seals are for use with conduit running at any angle, from vertical through horizontal.

EYD drain seals provide continuous draining and thereby prevent water accumulation. EYD seals are for vertical conduit runs and range in size from $\frac{1}{2}$ inch to 4 inches inclusive. They are provided with one opening for draining and filling, a rubber tube to form drain passage and a drain fitting.

EZD drain seals provide continuous draining and thereby prevent water accumulation. The covers should be positioned so that the drain will be at the bottom. A set screw is provided for locking the cover in this position.

EZD fittings are suitable for sealing vertical conduit runs between hazardous and nonhazardous areas, but must be installed in the hazardous area when it is above the nonhazardous area. They must be installed in the nonhazardous area when it is above the hazardous area.

EZD drain seals are designed so that the covers can be removed readily, permitting inspection during installation or at any time thereafter. After the fittings have been installed in the conduit run and conductors are in place, the cover and barrier are removed. After the dam has been made in the lower hub opening with packing fiber, the barrier must be replaced so that the sealing compound can be poured into the sealing chamber.

EZD inspection seals are identical to EZD drain seals to provide all inspection, maintenance and installation advantages except that the cover is not provided with an automatic drain. Water accumulations can be drained periodically by removing the cover (when no hazards exist). The cover must be replaced immediately.

Sealing Compounds And Dams

Poured seals should be made only by trained personnel in strict compliance with the specific instruction sheets provided with each sealing fitting. Improperly poured seals are worthless.

Sealing compound shall be approved for the purpose; it shall not be affected by the surrounding atmosphere or liquids; and it shall not have a melting point of less than 200 degrees F. (93 degrees C.). The sealing compound and dams must also be approved for the type and manufacturer of fitting. For example, Crouse-Hinds CHICO® sealing compound is the only sealing compound approved for use with Crouse-Hinds ECM sealing fittings.

To pack the seal-off, remove the threaded plug or plugs from the fitting and insert the fiber supplied with the packing kit. Tamp the fiber between the wires and the hub before pouring the sealing compound into the fitting. Then pour in the sealing cement and reset the threaded plug tightly. The fiber packing prevents the sealing compound (in the liquid state) from entering the conduit lines.

Most sealing-compound kits contain a powder in a polyethylene bag within an outer container. To mix, remove the bag of powder, fill the outside container, and pour in the powder and mix.

CAUTION!   *Always make certain that the sealing compound is compatible for use with the packing material, brand and type of fitting, and also with the type of conductors used in the system.*

In practical applications, there may be dozens of seals required for a particular installation. Consequently, after the conductors are pulled, each seal in the system is first packed. To prevent the possibility of overlooking a seal, one color of paint is normally sprayed on the seal hub at this time. This indicates that the seal has been packed. When the sealing compound is poured, a different color paint is once again sprayed on the seal hub to indicate a finished job. This method permits the job supervisor to visually inspect the conduit run, and if a seal is not painted the appropriate color, he or she knows that proper installation on this seal

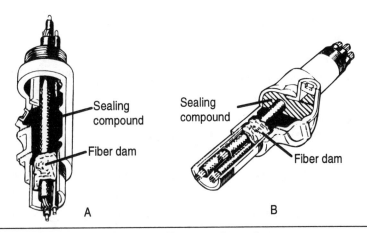

Figure 15-8: Seals made with fiber dams and sealing compound

was not done; therefore, action can be taken to correct the situation immediately.

The seal-off fittings in Figure 15-8 are typical of those used. The type in Figure 15-8A is for vertical mounting and is provided with a threaded, plugged opening into which the sealing cement is poured. The seal-off in Figure 15-8B has an additional plugged opening in the lower hub to facilitate packing fiber around the conductors to form a dam for the sealing cement.

The following procedures are to be observed when preparing sealing compound:

- Use a clean mix vessel for every batch. Particles of previous batches or dirt will spoil the seal.

- Recommended proportions are by volume — usually two parts powder to one part clean water. Slight deviations in these proportions will not affect the result.

- Do not mix more than can be poured in 15 minutes after water is added. Use cold water. Warm water increases setting speed. Stir immediately and thoroughly.

- If batch starts to set do not attempt to thin it by adding water or by stirring. Such a procedure will spoil the seal. Discard partially set material and make

a fresh batch. After pouring, close the opening immediately.

- Do not pour compound in sub-freezing temperatures, or when these temperatures will occur during curing.

- See that compound level is in accordance with the instruction sheet for that specific fitting.

Most other explosionproof fittings are provided with threaded hubs for securing the conduit as described previously. Typical fittings include switch and junction boxes, conduit bodies, union and connectors, flexible couplings, explosionproof lighting fixtures, receptacles, and panelboard and motor starter enclosures.

GARAGES AND SIMILAR LOCATIONS

Garages and similar locations where volatile or flammable liquids are handled or used as fuel in self-propelled vehicles (including automobiles, buses, trucks, and tractors) are not usually considered critically hazardous locations. However, the entire area up to a level 18 inches above the floor is considered a Class I, Division 2 location, and certain precautionary measures are required by the *NEC*. Likewise, any pit or depression below floor level shall be considered a Class I, Division 2 location, and the pit or depression may be judged as Class I, Division 1 location if it is unvented.

Normal raceway (conduit) and wiring may be used for the wiring method above this hazardous level, except where conditions indicate that the area concerned is more hazardous than usual. In this case, the applicable type of explosionproof wiring may be required.

Approved seal-off fittings should be used on all conduit passing from hazardous areas to nonhazardous areas. The requirements set forth in *NEC* Sections 501-5 and 501-5(b)(2) shall apply to horizontal as well as vertical boundaries of the defined hazardous areas. Raceways embedded in a masonry floor or buried beneath a floor are considered to be within the hazardous area above the floor if any connections or extensions lead into or through such an area. However, conduit systems terminating to an open raceway, in an outdoor unclassified area, shall not be required to be sealed between the point at which the conduit leaves the classified location and enters the open raceway.

Figure 15-9 shows a typical automotive service station with applicable *NEC* requirements. Note that space in the immediate vicinity of the gasoline-dispensing island is denoted as Class I, Division 1, to a height of 4 feet above grade. The surrounding area, within a radius of 20 feet of the island, falls under Class I, Division 2, to a height of 18 inches above grade. Bulk storage plants for gasoline are subject to comparable restrictions.

A summary of *NEC* rules governing the installation of electrical wiring at and about gasoline dispensing pumps is shown in Figure 15-10.

THEATERS

The *NEC* recognizes that hazards to life and property due to fire and panic exist in theaters, cinemas, and the like. The *NEC* therefore requires certain precautions in these areas in addition to those for commercial installations. These requirements include the following:

- Proper wiring of motion picture projection rooms (Article 540).

- Heat resistant, insulated conductors for certain lighting equipment (Section 520-42).

- Adequate guarding and protection of the stage switchboard and proper control and overcurrent protection of circuits (Section 520-22).

- Proper type and wiring of lighting dimmers (Sections 520-52(e) and 520-25).

- Use of proper types of receptacles and flexible cables for stage lighting equipment (Section 520-45).

- Proper smoke ventilator control (Section 520-49).

- Proper dressing-room wiring and control (Sections 520-71, 72, and 73).

- Fireproof projection rooms with automatic projector port closures, ventilating equipment, emergency lighting, guarded work lights, and proper location of related equipment (Article 540).

Outdoor or drive-in motion picture theaters do not present the inherent hazards of enclosed auditoriums. However, the projection rooms must be properly ventilated and wired for the protection of the operating personnel.

HOSPITALS

Hospitals and other health-care facilities fall under Article 517 of the *NEC*. Part B of Article 517 covers the general wiring of health-care facilities. Part C covers essential electrical systems for hospitals. Part D gives the performance criteria and wiring methods to minimize shock hazards to patients in electrically susceptible patient areas. Part E covers the requirements for electrical wiring and

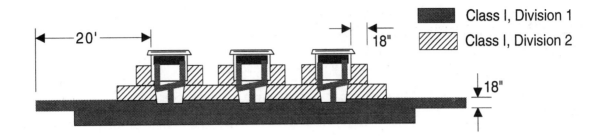

Gasoline dispensing units

Class I, Division 1
Class I, Division 2

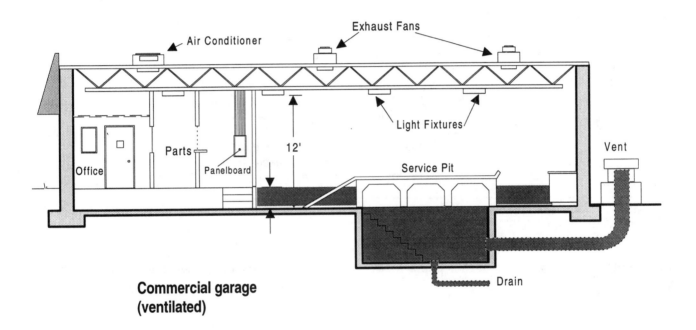

**Commercial garage
(ventilated)**

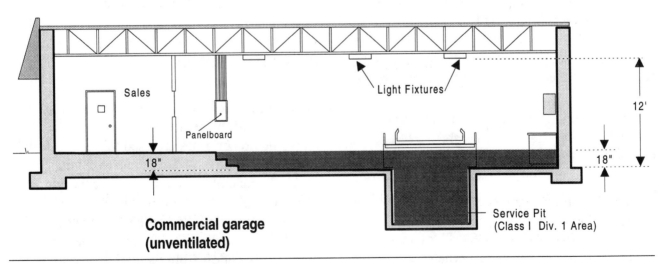

**Commercial garage
(unventilated)**

Figure 15-9: Commercial garage classifications

Application	NEC Regulation	NEC Section
Equipment in hazardous locations	All wiring and components must conform to the rules for Class I locations	514-3
Equipment above hazardous locations	All wiring must conform to the rules for such equipment in commercial garages	514-4
Gasoline dispenser	A disconnecting means must be provided for each circuit leading to or through a dispensing pump to disconnect all conductors including the grounded neutral. An approved seal (seal-off) is required in each conduit entering or leaving a dispenser	514-5(a)
Grounding	Metal portions of all noncurrent-carrying parts of dispensers must be effectively grounded	515-16
Underground wiring	Underground wiring must be installed within 2 feet of ground level — in rigid metal or IMC. If underground wiring is buried 2 feet or more, rigid nonmetallic conduit may be used along with the types mentioned above; Type MI cable may also be used in some cases	514-8

Figure 15-10: *NEC* application rules for service stations

equipment used in inhalation anesthetizing locations.

With the widespread use of x-ray equipment of varying types in health-care facilities, electricians are often required to wire and connect equipment such as discussed in Article 660 of the *NEC*. Conventional wiring methods are used, but provisions should be made for 50- and 60-ampere receptacles for medical x-ray equipment (Section 660-4b).

Anesthetizing locations of hospitals are deemed to be Class I, Division 1, to a height of 5 feet above floor level. Gas storage rooms are designated as Class I, Division 1, throughout. Most of the wiring in these areas, however can be limited to lighting fixtures only — locating all switches and other devices outside of the hazardous area.

The *NEC* recommends that wherever possible electrical equipment for hazardous locations be located in less hazardous areas. It also suggests that by adequate, positive-pressure ventilation from a clean source of outside air the hazards may be reduced or hazardous locations limited or eliminated. In many cases the installation of dust-collecting systems can greatly reduce the hazards in a Class II area.

AIRPORT HANGARS

Buildings used for storing or servicing aircraft in which gasoline, jet fuels, or other volatile flammable liquids or gases are used fall under Article 513 of the *NEC* and are summarized in Figure 15-11. In general, any depression below the level of the hangar floor is considered to be a Class I, Division 1 location. The entire area of the hangar including any adjacent and communicating area not suitably cut off from the hangar is considered to be a Class I, Division 2 location up to a level of 18 inches above the floor. The area within 5 feet

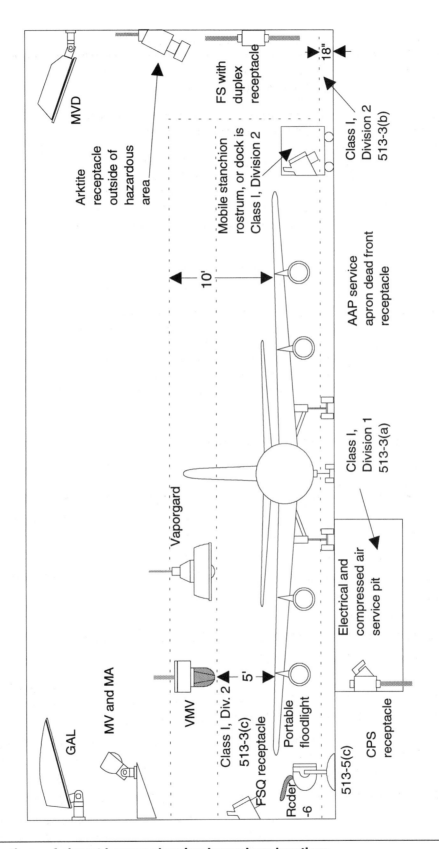

Figure 15-11: Sections of airport hangar showing hazardous locations

horizontally from aircraft power plants, fuel tanks, or structures containing fuel is considered to be a Class I, Division 2 hazardous location; this area extends upward from the floor to a level 5 feet above the upper surface of wings and of engine enclosures.

Adjacent areas in which hazardous vapors are not likely to be released, such as stock rooms and electrical control rooms, should not be classed as hazardous when they are adequately ventilated and effectively cut off from the hangar itself by walls or partitions. All fixed wiring in a hangar not within a hazardous area as defined in Section 513-5(a) must be installed in metallic raceways or shall be Type MI or Type ALS cable; the only exception is wiring in nonhazardous locations as defined in Section 513-3(d), which may be of any type recognized in Chapter 3 (Wiring Methods and Materials) in the *NEC*.

MANUFACTURERS' DATA

Manufacturers of explosionproof equipment and fittings expend a lot of time, energy, and expense in developing guidelines and brochures to ensure that their products are used correctly and in accordance with the latest *NEC* requirements. The many helpful charts, tables, and application guidelines are invaluable to anyone working on projects involving hazardous locations. Therefore, it is recommended that the electrician obtain as much of this data as possible. Once obtained, study this data thoroughly. Doing so will enhance your qualifications for working in hazardous locations of any type.

Manufacturers' data is usually available to qualified personnel (electrical workers) at little or no cost and can be obtained from local distributors of electrical supplies, or directly from the manufacturer.

Summary

Any area in which the atmosphere or a material in the area is such that the arcing of operating electrical contacts, components, and equipment may cause an explosion or fire is considered as a hazardous location. In all such cases, explosionproof equipment, raceways, and fittings are used to provide an explosionproof wiring system.

The wide assortment of explosionproof equipment now available makes it possible to provide adequate electrical installations under any of the various hazardous conditions. However, the electrician must be thoroughly familiar with all *NEC* requirements and know what fittings are available, how to install them properly, and where and when to use the various fittings.

Many factors — such as temperature, barometric pressure, quantity of release, humidity, ventilation, distance from the vapor source, and the like — must be considered. When information on all factors concerned is properly evaluated, a consistent classification for the selection and location of electrical equipment can be developed.

Index

Practical References for Builders

Basic Engineering for Builders

If you've ever been stumped by an engineering problem on the job, yet wanted to avoid the expense of hiring a qualified engineer, you should have this book. Here you'll find engineering principles explained in non-technical language and practical methods for applying them on the job. With the help of this book you'll be able to understand engineering functions in the plans and how to meet the requirements, how to get permits issued without the help of an engineer, and anticipate requirements for concrete, steel, wood and masonry. See why you sometimes have to hire an engineer and what you can undertake yourself: surveying, concrete, lumber loads and stresses, steel, masonry, plumbing, and HVAC systems. This book is designed to help the builder save money by understanding engineering principles that you can incorporate into the jobs you bid.
400 pages, 8¹/₂ x 11, $34.00

Residential Wiring to the 1999 *NEC*

Shows how to install rough and finish wiring in new construction, alterations, and additions. Complete instructions on troubleshooting and repairs. Every subject is referenced to the most recent *National Electrical Code*, and there's 22 pages of the most-needed *NEC* tables to help make your wiring pass inspection — the first time. **352 pages, 5¹/₂ x 8¹/₂, $27.00**

Construction Forms & Contracts

125 forms you can copy and use — or load into your computer (from the FREE disk enclosed). Then you can customize the forms to fit your company, fill them out, and print. Loads into *Word* for *Windows*™, *Lotus 1-2-3*, *WordPerfect*, *Works*, or *Excel* programs. You'll find forms covering accounting, estimating, fieldwork, contracts, and general office. Each form comes with complete instructions on when to use it and how to fill it out. These forms were designed, tested and used by contractors, and will help keep your business organized, profitable and out of legal, accounting and collection troubles. Includes a CD-ROM for *Windows*™ and Mac. **400 pages, 8¹/₂ x 11, $41.75**

Estimating with Microsoft *Excel*

Most builders estimate with *Excel* because it's easy to learn, quick to use, and can be customized to your style of estimating. Here you'll find step-by-step how to create your own customized automated spreadsheet estimating program for use with *Excel*. You'll learn how to use the magic of *Excel* in creating detail sheets, cost breakdown summaries, and linking. You can even create your own macros. Includes a CD-ROM that illustrates examples in the book and provides you with templates you can use to set up your own estimating system. **148 pages, 8¹/₂ x 11, $49.95**

Electrician's Exam Preparation Guide

Need help in passing the apprentice, journeyman, or master electrician's exam? This is a book of questions and answers based on actual electrician's exams over the last few years. Almost a thousand multiple-choice questions — exactly the type you'll find on the exam — cover every area of electrical installation: electrical drawings, services and systems, transformers, capacitors, distribution equipment, branch circuits, feeders, calculations, measuring and testing, and more. It gives you the correct answer, an explanation, and where to find it in the latest *NEC*. Also tells how to apply for the test, how best to study, and what to expect on examination day.
352 pages, 8¹/₂ x 11, $32.00

Estimating Electrical Construction

Like taking a class in how to estimate materials and labor for residential and commercial electrical construction. Written by an A.S.P.E. National Estimator of the Year, it teaches you how to use labor units, the plan take-off, and the bid summary to make an accurate estimate, how to deal with suppliers, use pricing sheets, and modify labor units. Provides extensive labor unit tables and blank forms for your next electrical job.
272 pages, 8¹/₂ x 11, $19.00

Electrical Blueprint Reading Revised

Shows how to read and interpret electrical drawings, wiring diagrams, and specifications for constructing electrical systems. Shows how a typical lighting and power layout would appear on a plan, and explains what to do to execute the plan. Describes how to use a panelboard or heating schedule, and includes typical electrical specifications.
208 pages, 8¹/₂ x 11, $18.00

Construction Estimating Reference Data

Provides the 300 most useful manhour tables for practically every item of construction. Labor requirements are listed for sitework, concrete work, masonry, steel, carpentry, thermal and moisture protection, doors and windows, finishes, mechanical and electrical. Each section details the work being estimated and gives appropriate crew size and equipment needed. Includes a CD-ROM with an electronic version of the book with *National Estimator*, a stand-alone *Windows*™ estimating program, plus an interactive multimedia video that shows how to use the disk to compile construction cost estimates. **432 pages, 11 x 8¹/₂, $39.50**

CD Estimator

If your computer has *Windows*™ and a CD-ROM drive, *CD Estimator* puts at your fingertips 85,000 construction costs for new construction, remodeling, renovation & insurance repair, electrical, plumbing, HVAC and painting. You'll also have the *National Estimator* program — a stand-alone estimating program for *Windows*™ that *Remodeling* magazine called a "computer wiz." Quarterly cost updates are available at no charge on the Internet. To help you create professional-looking estimates, the disk includes over 40 construction estimating and bidding forms in a format that's perfect for nearly any word processing or spreadsheet program for *Windows*™. And to top it off, a 70-minute interactive video teaches you how to use this CD-ROM to estimate construction costs. **CD Estimator is $68.50**

Rough Framing Carpentry

If you'd like to make good money working outdoors as a framer, this is the book for you. Here you'll find shortcuts to laying out studs; speed cutting blocks, trimmers and plates by eye; quickly building and blocking rake walls; installing ceiling backing, ceiling joists, and truss joists; cutting and assembling hip trusses and California fills; arches and drop ceilings — all with production line procedures that save you time and help you make more money. Over 100 on-the-job photos of how to do it right and what can go wrong. **304 pages, 8¹/₂ x 11, $26.50**

How to Succeed With Your Own Construction Business

Everything you need to start your own construction business: setting up the paperwork, finding the work, advertising, using contracts, dealing with lenders, estimating, scheduling, finding and keeping good employees, keeping the books, and coping with success. If you're considering starting your own construction business, all the knowledge, tips, and blank forms you need are here. **336 pages, 8¹/₂ x 11, $28.50**

Markup & Profit: A Contractor's Guide

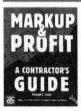

In order to succeed in a construction business, you have to be able to price your jobs to cover all labor, material and overhead expenses, and make a decent profit. The problem is knowing what markup to use. You don't want to lose jobs because you charge too much, and you don't want to work for free because you've charged too little. If you know how to calculate markup, you can apply it to your job costs to find the right sales price for your work. This book gives you tried and tested formulas, with step-by-step instructions and easy-to-follow examples, so you can easily figure the markup that's right for your business. Includes a CD-ROM with forms and checklists for your use. **320 pages, 8¹/₂ x 11, $32.50**

The Contractor's Legal Kit

Stop "eating" the costs of bad designs, hidden conditions, and job surprises. Set ground rules that assign those costs to the rightful party ahead of time. And it's all in plain English, not "legalese." For less than the cost of an hour with a lawyer you'll learn the exclusions to put in your agreements, why your insurance company may pay for your legal defense, how to avoid liability for injuries to your sub and his employees or damages they cause, how to collect on lawsuits you win, and much more. It also includes a FREE computer disk with contracts and forms you can customize for your own use. **352 pages, 8¹/₂ x 11, $59.95**

National Electrical Estimator

This year's prices for installation of all common electrical work: conduit, wire, boxes, fixtures, switches, outlets, loadcenters, panelboards, raceway, duct, signal systems, and more. Provides material costs, manhours per unit, and total installed cost. Explains what you should know to estimate each part of an electrical system. Includes a CD-ROM with an electronic version of the book with *National Estimator*, a stand-alone *Windows*™ estimating program, plus an interactive multimedia video that shows how to use the disk to compile construction cost estimates.
544 pages, 8¹/₂ x 11, $47.75. Revised annually

Contractor's Guide to the Building Code Revised

This new edition was written in collaboration with the International Conference of Building Officials, writers of the code. It explains in plain English exactly what the latest edition of the *Uniform Building Code* requires. Based on the 1997 code, it explains the changes and what they mean for the builder. Also covers the *Uniform Mechanical Code* and the *Uniform Plumbing Code*. Shows how to design and construct residential and light commercial buildings that'll pass inspection the first time. Suggests how to work with an inspector to minimize construction costs, what common building shortcuts are likely to be cited, and where exceptions may be granted. **320 pages, 8¹/₂ x 11, $39.00**

Craftsman's Illustrated Dictionary of Construction Terms

Almost everything you could possibly want to know about any word or technique in construction. Hundreds of up-to-date construction terms, materials, drawings and pictures with detailed, illustrated articles describing equipment and methods. Terms and techniques are explained or illustrated in vivid detail. Use this valuable reference to check spelling, find clear, concise definitions of construction terms used on plans and construction documents, or learn about little-known tools, equipment, tests and methods used in the building industry. It's all here.
416 pages, 8¹/₂ x 11, $36.00

1996 *National Electrical Code* Interpretive Diagrams

Based on the 1996 *NEC*, this new edition visually and graphically explains and interprets the key provisions of the *National Electrical Code* as seen through the eyes of an electrical inspector. All drawings have been carefully prepared and reviewed by inspectors and the Codes and Standards Committee. Here you'll find approved installations of just about every type of electrical part. **270 pages, 5 x 8, $34.95**

Illustrated Guide to the 1999 *National Electrical Code*

This fully-illustrated guide offers a quick and easy visual reference for installing electrical systems. Whether you're installing a new system or repairing an old one, you'll appreciate the simple explanations written by a code expert, and the detailed, intricately-drawn and labeled diagrams. A real time-saver when it comes to deciphering the current *NEC*.
360 pages, 8¹/₂ x 11, $38.75

Vest Pocket Guide to HVAC Electricity

This handy guide will be a constant source of useful information for anyone working with electrical systems for heating, ventilating, refrigeration, and air conditioning. Includes essential tables and diagrams for calculating and installing electrical systems for powering and controlling motors, fans, heating elements, compressors, transformers and every electrical part of an HVAC system. **304 pages, 3¹/₂x 5¹/₂, $18.00**

Residential Electrical Estimating

A fast, accurate pricing system proven on over 1000 residential jobs. Using the manhours provided, combined with material prices from your wholesaler, you quickly work up estimates based on degree of difficulty. These manhours come from a working electrical contractor's records — not some pricing agency. You'll find prices for every type of electrical job you're likely to estimate — from service entrances to ceiling fans.
320 pages, 8¹/₂ x 11, $29.00

National Construction Estimator

Current building costs for residential, commercial, and industrial construction. Estimated prices for every common building material. Provides manhours, recommended crew, and gives the labor cost for installation. Includes a CD-ROM with an electronic version of the book with *National Estimator*, a stand-alone *Windows*™ estimating program, plus an interactive multimedia video that shows how to use the disk to compile construction cost estimates. **616 pages, 8¹/₂ x 11, $47.50. Revised annually**

CD Estimator Heavy

CD Estimator Heavy has a complete 780-page heavy construction cost estimating volume for each of the 50 states. Select the cost database for the state where the work will be done. Includes thousands of cost estimates you won't find anywhere else, and in-depth coverage of demolition, hazardous materials remediation, tunneling, site utilities, precast concrete, structural framing, heavy timber construction, membrane waterproofing, industrial windows and doors, specialty finishes, built-in commercial and industrial equipment, and HVAC and electrical systems for commercial and industrial buildings. **CD Estimator Heavy is $69.00**

Contractor's Guide to QuickBooks Pro 2000

This user-friendly manual walks you through QuickBooks Pro's detailed setup procedure and explains step-by-step how to create a first-rate accounting system. You'll learn in days, rather than weeks, how to use QuickBooks Pro to get your contracting business organized, with simple, fast accounting procedures. On the CD included with the book you'll find a QuickBooks Pro file preconfigured for a construction company (you drag it over onto your computer and plug in your own company's data). You'll also get a complete estimating program, including a database, and a job costing program that lets you export your estimates to QuickBooks Pro. It even includes many useful construction forms to use in your business. **304 pages, 8¹/₂ x 11, $44.50**

Planning Drain, Waste & Vent Systems

How to design plumbing systems in residential, commercial, and industrial buildings. Covers designing systems that meet code requirements for homes, commercial buildings, private sewage disposal systems, and even mobile home parks. Includes relevant code sections and many illustrations to guide you through what the code requires in designing drainage, waste, and vent systems. **192 pages, 8¹/₂ x 11, $19.25**

Contractor's Survival Manual

How to survive hard times and succeed during the up cycles. Shows what to do when the bills can't be paid, finding money and buying time, transferring debt, and all the alternatives to bankruptcy. Explains how to build profits, avoid problems in zoning and permits, taxes, time-keeping, and payroll. Unconventional advice on how to invest in inflation, get high appraisals, trade and postpone income, and stay hip-deep in profitable work. **160 pages, 8¹/₂ x 11, $22.25**

Steel-Frame House Construction

Framing with steel has obvious advantages over wood, yet building with steel requires new skills that can present challenges to the wood builder. This new book explains the secrets of steel framing techniques for building homes, whether pre-engineered or built stick by stick. It shows you the techniques, the tools, the materials, and how you can make it happen. Includes hundreds of photos and illustrations, plus a CD-ROM with steel framing details. **304 pages, 8¹/₂ x 11, $39.75**

Contracting in All 50 States

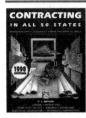

Every state has its own licensing requirements that you must meet to do business there. These are usually written exams, financial requirements, and letters of reference. This book shows how to get a building, mechanical or specialty contractor's license, qualify for DOT work, and register as an out-of-state corporation, for every state in the U.S. It lists addresses, phone numbers, application fees, requirements, where an exam is required, what's covered on the exam and how much weight each area of construction is given on the exam. You'll find just about everything you need to know in order to apply for your out-of-state license. **416 pages, 8¹/₂ x 11, $36.00**

Building Contractor's Exam Preparation Guide

Passing today's contractor's exams can be a major task. This book shows you how to study, how questions are likely to be worded, and the kinds of choices usually given for answers. Includes sample questions from actual state, county, and city examinations, plus a sample exam to practice on. This book isn't a substitute for the study material that your testing board recommends, but it will help prepare you for the types of questions — and their correct answers — that are likely to appear on the actual exam. Knowing how to answer these questions, as well as what to expect from the exam, can greatly increase your chances of passing. **320 pages, 8¹/₂ x 11, $35.00**

Getting Financing & Developing Land

Developing land is a major leap for most builders - yet that's where the big money is made. This book gives you the practical knowledge you need to make that leap. Learn how to prepare a market study, select a building site, obtain financing, guide your plans through approval, then control your building costs so you can ensure yourself a good profit. Includes a CD-ROM with forms, checklists, and a sample business plan you can customize and use to help you sell your idea to lenders and investors. **232 pages, 8¹/₂ x 11, $39.00**

Commercial Metal Stud Framing

Framing commercial jobs can be more lucrative than residential work. But most commercial jobs require some form of metal stud framing. This book teaches step-by-step, with hundreds of job site photos, high-speed metal stud framing in commercial construction. It describes the special tools you'll need and how to use them effectively, and the material and equipment you'll be working with. You'll find the shortcuts, tips and tricks-of-the-trade that take most steel framers years on the job to discover. Shows how to set up a crew to maintain a rhythm that will speed progress faster than any wood framing job. If you've framed with wood, this book will teach you how to be one of the few top-notch metal stud framers. **208 pages, 8¹/₂ x 11, $45.00**

Craftsman Book Company
6058 Corte del Cedro
P.O. Box 6500
Carlsbad, CA 92018

☎ **24 hour order line**
1-800-829-8123
Fax (760) 438-0398

In A Hurry?
We accept phone orders charged to your
○ Visa, ○ MasterCard, ○ Discover or ○ American Express

Card#_____

Exp. date_____Initials_____

Tax Deductible: Treasury regulations make these references tax deductible when used in your work. Save the canceled check or charge card statement as your receipt.

Name_____

Company_____

Address_____

City/State/Zip_____
○ This is a residence

Total enclosed_____(In California add 7.25% tax)
We pay shipping when your check covers your order in full.

Order online http://www.craftsman-book.com
Free on the Internet! Download any of Craftsman's estimating costbooks for a 30-day free trial! http://costbook.com

10-Day Money Back Guarantee

- ○ 34.00 Basic Engineering for Builders
- ○ 35.00 Building Contractor's Exam Preparation Guide
- ○ 68.50 CD Estimator
- ○ 69.00 CD Estimator Heavy
- ○ 45.00 Commercial Metal Stud Framing
- ○ 39.50 Construction Estimating Reference Data with FREE *National Estimator* on a CD-ROM.
- ○ 41.75 Construction Forms & Contracts with a CD-ROM for *Windows*™ and Macintosh.
- ○ 36.00 Contracting in All 50 States
- ○ 44.50 Contractor's Guide to QuickBooks Pro 2000
- ○ 39.00 Contractor's Guide to the Building Code Revised

- ○ 59.95 Contractor's Legal Kit
- ○ 22.25 Contractor's Survival Manual
- ○ 36.00 Craftsman's Illustrated Dictionary of Construction Terms
- ○ 18.00 Electrical Blueprint Reading Revised
- ○ 32.00 Electrician's Exam Preparation Guide
- ○ 19.00 Estimating Electrical Construction
- ○ 49.95 Estimating with Microsoft *Excel*
- ○ 39.00 Getting Financing & Developing Land
- ○ 28.50 How to Succeed w/Your Own Construction Business
- ○ 38.75 Illustrated Guide to the 1999 *NEC*
- ○ 32.50 Markup & Profit: A Contractor's Guide

- ○ 47.50 National Construction Estimator with FREE *National Estimator* on a CD-ROM.
- ○ 34.95 1996 *NEC* Interpretive Diagrams
- ○ 47.75 National Electrical Estimator with FREE *National Estimator* on a CD-ROM.
- ○ 19.25 Planning Drain, Waste & Vent Systems
- ○ 29.00 Residential Electrical Estimating
- ○ 27.00 Residential Wiring to the 1999 *NEC*
- ○ 26.50 Rough Framing Carpentry
- ○ 39.75 Steel-Frame House Construction
- ○ 18.00 Vest Pocket Guide to HVAC Electricity
- ○ 36.50 Commercial Electrical Wiring
- ○ FREE Full Color Catalog

Prices subject to change without notice

Craftsman Book Company
6058 Corte del Cedro
P.O. Box 6500
Carlsbad, CA 92018

☎ **24 hour order line**
1-800-829-8123
Fax (760) 438-0398

In A Hurry?
We accept phone orders charged to your
○ Visa, ○ MasterCard, ○ Discover or ○ American Express

Card#_____

Exp. date_____Initials_____

Tax Deductible: Treasury regulations make these references tax deductible when used in your work. Save the canceled check or charge card statement as your receipt.

Name_____

Company_____

Address_____

City/State/Zip_____
○ This is a residence

Total enclosed_____(In California add 7.25% tax)
We pay shipping when your check covers your order in full.

Order online http://www.craftsman-book.com
Free on the Internet! Download any of Craftsman's estimating costbooks for a 30-day free trial! http://costbook.com

10-Day Money Back Guarantee

- ○ 34.00 Basic Engineering for Builders
- ○ 35.00 Building Contractor's Exam Preparation Guide
- ○ 68.50 CD Estimator
- ○ 69.00 CD Estimator Heavy
- ○ 45.00 Commercial Metal Stud Framing
- ○ 39.50 Construction Estimating Reference Data with FREE *National Estimator* on a CD-ROM.
- ○ 41.75 Construction Forms & Contracts with a CD-ROM for *Windows*™ and Macintosh.
- ○ 36.00 Contracting in All 50 States
- ○ 44.50 Contractor's Guide to QuickBooks Pro 2000
- ○ 39.00 Contractor's Guide to the Building Code Revised

- ○ 59.95 Contractor's Legal Kit
- ○ 22.25 Contractor's Survival Manual
- ○ 36.00 Craftsman's Illustrated Dictionary of Construction Terms
- ○ 18.00 Electrical Blueprint Reading Revised
- ○ 32.00 Electrician's Exam Preparation Guide
- ○ 19.00 Estimating Electrical Construction
- ○ 49.95 Estimating with Microsoft *Excel*
- ○ 39.00 Getting Financing & Developing Land
- ○ 28.50 How to Succeed w/Your Own Construction Business
- ○ 38.75 Illustrated Guide to the 1999 *NEC*
- ○ 32.50 Markup & Profit: A Contractor's Guide

- ○ 47.50 National Construction Estimator with FREE *National Estimator* on a CD-ROM.
- ○ 34.95 1996 *NEC* Interpretive Diagrams
- ○ 47.75 National Electrical Estimator with FREE *National Estimator* on a CD-ROM.
- ○ 19.25 Planning Drain, Waste & Vent Systems
- ○ 29.00 Residential Electrical Estimating
- ○ 27.00 Residential Wiring to the 1999 *NEC*
- ○ 26.50 Rough Framing Carpentry
- ○ 39.75 Steel-Frame House Construction
- ○ 18.00 Vest Pocket Guide to HVAC Electricity
- ○ 36.50 Commercial Electrical Wiring
- ○ FREE Full Color Catalog

Prices subject to change without notice

Mail This Card Today
For a Free Full Color Catalog

Over 100 books, annual cost guides and estimating software packages at your fingertips with information that can save you time and money. Here you'll find information on carpentry, contracting, estimating, remodeling, electrical work, and plumbing.

All items come with an unconditional 10-day money-back guarantee. If they don't save you money, mail them back for a full refund.

Name_____

Company_____

Address_____

City/State/Zip_____

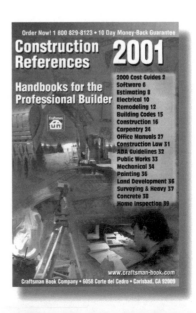

Craftsman Book Company / 6058 Corte del Cedro / P.O. Box 6500 / Carlsbad, CA 92018